Advances in the Theory of Benzenoid Hydrocarbons

AKADEMIE-VERLAG BERLIN

Topics in Current Chemistry · Vol. 153

Advances in the Theory of Benzenoid Hydrocarbons

Advances in the Theory of Benzenoid Hydrocarbons

With 127 Figures and 3 Tables

AKADEMIE-VERLAG BERLIN

Die Originalausgabe erscheint im Springer-Verlag
Berlin · Heidelberg · New York · London · Paris · Tokyo · Hongkong
in der Schriftenreihe "Topics in Current Chemistry", Volume 153

Vertriebsrechte für die sozialistischen Länder:
Akademie-Verlag Berlin

Vertriebsrechte für alle Staaten mit Ausnahme der sozialistischen Länder
Springer-Verlag Berlin · Heidelberg · New York · London · Paris · Tokyo · Hongkong

ISBN 3-540-51505-4 Springer-Verlag Berlin · Heidelberg · New York
ISBN 0-387-51505-4 Springer-Verlag New York · Berlin · Heidelberg

ISBN 3-05-500801-4 Akademie-Verlag Berlin

Erschienen im Akademie-Verlag Berlin, Leipziger Straße 3–4, Berlin, DDR-1086
Lizenznummer: 202 · 100/536/90
Printed in the German Democratic Republic
Gesamtherstellung: VEB Druckerei „Thomas Müntzer", 5820 Bad Langensalza
LSV 1275
Bestellnummer: 764 109 1 (3072/153)
17800

Guest Editors

Editorial Board

Preface

The editors of this volume of Topics in Current Chemistry have recently completed a book devoted to the theory of benzenoid molecules (Gutman I, Cyvin SJ (1989) Introduction to the theory of benzenoid hydrocarbons, Springer, Berlin Heidelberg New York). Due to its introductory nature the book could not embrace a number of relevant topics in which vigorous research activity is nowadays taking place. The aim of the present issue is to fill this gap. Outstanding and currently active researchers were invited to report on their contributions to the theoretical chemistry of benzenoid compounds.

Theoretical investigations of benzenoid molecules have a long history and are usually considered as a traditional, but somewhat obsolete, area of theoretical organic chemistry. This volume of Topics in Current Chemistry should document that in this field there still exists a variety of unsolved and partially solved problems, that there is still room for new ideas and that research activity shows no signs of slowing down.

Nashville and Trondheim, Spring 1989

Ivan Gutman
Sven J. Cyvin

Table of Contents

Benzenoid Hydrocarbons in Space: The Evidence and Implications

Louis J. Allamandola

NASA Ames Research Center 245-6, Moffett Field, California 94035, USA

Table of Contents

Many different celestial objects emit an infrared sprectrum which has been attributed to infrared fluorescence from a family of highly vibrationally excited benzenoid hydrocarbons referred to as polycyclic aromatic hydrocarbons (PAHs). The most intense emitters contain between 20 to 50 carbon atoms, although larger species also contribute to the emission. This assignment is based on a rough resemblance of the interstellar emission spectra to the vibrational spectra of PAHs and related materials such as chars and soots which contain PAH mixtures. The spectroscopic assignments of the features between 3200 and 700 cm^{-1} are discussed in detail.

Much laboratory work on PAHs which are larger than those previously studied, isolated, ionized and dehydrogenated is called for to fully exploit this model. As PAHs are thought to be ubiquitous throughout the interstellar medium and more abundant than all other known polyatomic, interstellar molecules, they possess great potential as important probes of conditions in many different kinds of astronomical objects. Conversely, astronomical observations are extending our knowledge of these exotic materials by probing regions in which some conditions may be impossible to duplicate in the laboratory.

1 Introduction

An intriguing and exciting chapter of modern astrophysics was opened by Gillett, Forrest and Merrill in 1973 [1] with their discovery that some astronomical objects emitted a broad band which peaks near 3050 cm^{-1}. In the ensuing years, astronomers around the world found that this was part of a family of infrared emission features which were emitted by a large number of very different types of astronomical objects. Reviews of the ground breaking observations are in Ref [2] and [3]. The other prominent, well-known bands, or features as they are often called, peak near 1610, "1310", 1160 and 890 cm^{-1}. They are broad, with a FWHH on the order of 30 to 50 cm^{-1} or more. The peak frequencies do not seem to vary by more than a few wavenumbers from object to object except for the case of the "1310" cm^{-1} feature in which it can differ by as much as 50 cm^{-1}.

As the list of astronomical objects which emit these features grew, it became clear that the emission came from regions where ultraviolet radiation was impinging on vast areas of space in which dust was known to be present. Surprisingly, the age and history of the dust seemed to be unimportant. Illumination from hot, UV-rich stars which had formed in – and emerged from – dense, dark dust clouds (age $\sim 10^6 - 10^7$ years) could excite the emission from the surface of the cloud. Examples of this type of object include H–II regions and some reflection nebulae. Similar spectra are emitted from some much younger, isolated objects known as planetary nebulae. These are stars near the end of their normal life which have produced copious amounts of dust during the past $10^4 - 10^5$ years. This dust has a very different history and is much younger than the dust in dense clouds. During the late 1970s and early 1980s, the known extent of the phenomenon was extended to other galaxies as well. The intensities of the bands detected however, were far greater than those which would be expected from the individual objects which comprised the galaxies such as planetary nebulae, reflection nebulae, H–II regions and so on. The implication was that a substantial fraction of this extragalactic emission came from regions which were not associated with individual stars, but from the dark matter in the interstellar medium in these galaxies [4]. Subsequent observations of our galaxy, the Milky Way, by the Infrared Astronomical Satellite (IRAS) have shown that there are strips of interstellar IR emitting clouds, known as the IR Cirrus, that lie slightly above and below the galactic plane which seem to emit these bands as well [5]. Obviously, understanding the source of this completely unexpected, yet widespread, phenomenon became an important problem in astrophysics.

As the number and variety of objects found to emit the features increased, identifying the carrier and determining the emission mechanism became the key challenges. During the fifteen years since their discovery, a number of models had been proposed to account for both aspects of the problem. Many of the models proposed up to 1984 are reviewed in Ref [6]. The comparable intensity of the 3050 cm^{-1} feature to those at lower frequencies, and the close association with ultraviolet radiation led us to propose that the emission was due to the infrared fluorescence from molecule sized emitters excited by the absorption of ultraviolet and visible photons [7, 8]. Although the carrier we intially proposed (molecules, frozen on 10K dust grains) was incorrect, this excitation-emission mechanism is now generally accepted because of the important and analysis observations of reflection nebulae made by Sellgren which showed that the ultraviolet and visible photon fluxes present were unable to sufficiently excite larger species to emit in the 3000 cm^{-1} region [9]. The idea, now gaining wide acceptance, that aromatic material may be the carrier of the features began with the suggestion made by Duley and Williams in 1981 that they arise from vibrations of chemical groups attached to the aromatic moieties which make up small (< 0.01 micron radius) amorphous carbon particles [10]. Subsequently, Léger and Puget [11] and Allamandola, Tielens, and Barker [12] proposed that individual polycyclic aromatic hydrocarbon molecules, referred to as PAHs in the astrophysical literature, were the band carriers.

The pricipal reason for this assignment was the suggestive, but not perfect, resemblance of the interstellar emission spectra with the infrared absorption spectra of benzenoid hydrocarbons in KBr pellets. This assigment was supported by the better match of the interstellar features with the vibrational spectra of chars and soots which are comprised of mixtures of polycyclic aromatic hydrocarbons [13]. The need to invoke emission from free molecules rather than from the aromatic building blocks of amorphous carbon particles was driven by the fluorescent nature of the emission. Energy deposited in a particular molecular structural unit of a larger particle or cluster is thought to become thermalized on timescales on the order of 10^{-11} to 10^{-13} seconds, many orders of magnitude shorter than the 10^{-3} to 10^{-2} second timescale appropriate for the emission of IR photons. Emission from molecular units in a particle implies extremely weak intermolecular coupling [14] and has yet to be demonstrated. This question is presently the matter of some debate.

The imperfect match between the IR spectra of benzenoid hydrocarbons with the interstellar emission bands has been taken to indicate that rather than pure benzenoid compounds being responsible, benzenoid-like (or PAH-like) species are present in the interstellar medium. In many regions of the interstellar medium PAHs are expected to be ionized [12] and those containing less than about 20 carbon atoms are expected to be dehydrogenated [10, 11, 12].

In addition to the spectroscopic evidence, related observations support an aromatic hydrocarbon carrier as well. In planetary nebula, the fraction of the IR emitted in the "1310" cm^{-1} feature, which is by far the most intense of the bands, is strongly correlated with the amount of carbon available [15]. As the carriers must be produced under harsh conditions in planetary nebulae, they must be carbon-rich compounds which are extremely stable. Finally, although there is some variation among the relative band intensities, the bands are correlated, implying that a single class of chemical species is responsible [15].

Apart from the "curiosity" value, why is the discovery of a previously unrecognized and surprisingly ubiquitous component of the interstellar medium important astrophysically? There are several reasons. Current estimates are that between 1 and 10% of all of the carbon in the galaxy is in PAHs comprised of roughly 15 to, say, 500 carbon atoms [16]. The PAHs responsible for the IR features are thought to be more abundant than all of the other known interstellar, gaseous, polyatomic molecules combined [11, 12]. Their proposed ubiquity and high abundance has serious ramifications for other spectral regions. They may well contribute to the strong ultraviolet interstellar extinction, measured between 100 to 200 nm, the weak, diffuse, visible absorption bands [17, 18, 19], and the far-IR Cirrius [12, 20]. Similary their influence on many physical processes may be profound [21]. They may comprise much of the carbon in meteorites [22] and be responsible for maintaining interstellar cloud temperature [23], moderating interstellar cloud chemistry [24], and contributing to the deuterium enrichment found in interplanetary dust particles and meteorites [25]. Comprehensive treatments of various aspects of the polycyclic aromatic hydrocarbon model can be found in Ref [26] and [27].

In view of the previous paragraph, it is important to realize that precious little information is available on important properties of individual PAHs. For example, little is known about the spectroscopic properties of PAH ions and radicals from the UV through the IR. Other important, but poorly understood properties include photo-ionization cross sections, electron capture cross sections, cluster geometry, n-mer binding energies and so on. Recently, important information of fragmentation patterns of doubly ionized PAHs has become available [28, 29]. The carbon chemistry in the stellar atmospheres in which they are produced is just beginning to be modelled and poses significant chemical kinetic challenges [30, 31]. There is much to be done both experimentally and theoretically before the PAH model can be reliably extended from a means to account for certain observations to the point where it can be exploited as a probe of interstellar and circumstellar processes and conditions.

It is becoming increasingly apparent that the interstellar emission includes contributions from both free PAHs and carbonaceous particles as well [13, 32]. As amorphous carbon is primarily made up of randomly oriented clusters of PAHs, cross-linked and interconnected by saturated and unsaturated hydrocarbon chains, the infrared spectra of amorphous carbons and hydrogenated amorphous carbons should resemble spectra of PAH mixtures with some of the individual bands blurred out due to solid-state intermolecular interactions. If the H content is high, in addition to the aromatic features, aliphatic CH bands between 3000 to 2900 cm^{-1} and 1500 to 1400 cm^{-1} should become evident in the spectra of carbonaceous materials. The aromatic signature of small amorphous carbon particles and PAH clusters will be largely determined by the properties of the PAHs of which they are made. Individual bands overlap as the particles get larger and broad features are produced which may retain some substructure indicative of the individual PAHs. For still larger particles, bulk properties dominate and broad components appear as substructure on a strong continuum. As with PAH molecules, the spectroscopic and physical properties of amorphous carbon must be studied in detail.

In this article the infrared spectroscopic evidence for interstellar PAHs will be reviewed. The spectroscopic properties of PAHs studied in salt pellets rather than amorphous carbons will be primarily used since a wealth of very detailed information is available (thanks to the sustained, dedicated effort of Cyvin and his coleagues over many years) and molecule-sized emitters can account for many details of the interstellar spectra. Infrared spectra of amorphous carbon particles and carbonaceous films, synthesized to study the connections with interstellar carbonaceous material, are just now becoming available. The work of Bussoletti and coworkers ([33] and references therein) and Sakata and colleagues ([34] and references therein) is particularly noteworthy in this regard.

2 The Interstellar Emission Spectrum

Infrared spectra of celestial objects are usually measured in narrow frequency ranges because the Earth's atmosphere is opaque in certain infrared regions and the most sensitive instrumentation is region specific. Consequently, although nearly complete mid-IR spectra have been published for only a few objects, detailed spectroscopic information is available in certain frequency ranges for many objects. Spectra have been measured in regions which are obscured by telluric H_2O using NASA's Kuiper Airborne Observatory which can operate a telescope at 12 to 14 km altitude. Observations in the 2000 to 1000 cm^{-1} region made in this aircraft have played a very important role in unraveling the mystery of the interstellar infrared emission bands. Reliable observations in the 2500 to 2100 cm^{-1} and 700 cm^{-1} to lower frequency regions must await spaceborne instrumentation. The former region is blocked by atmospheric CO_2 and the latter by a combination of CO_2 and the very rich H_2O rotation spectrum. Fortunately, the Infrared Space Observatory (ISO), an infrared satellite under construction by the European Space Agency (ESA), is planned to be launched in the early 1990s, and the Space Infrared Telescope Facility (SIRTF), a complementary infrared satellite at an earlier stage of development by the National Aeronautics and Space Administration (NASA) in the United States, is planned to be launched in the latter part of the 1990s.

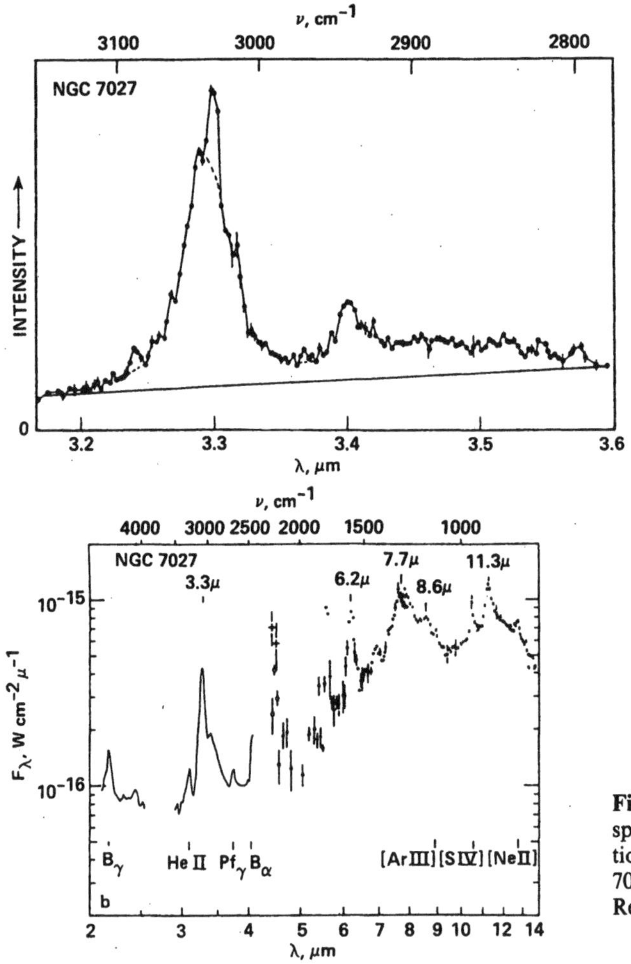

Fig. 1. The infrared emission spectrum from the high excitation planetary nebula NGC 7027 (**a** from Ref. [61], **b** from Ref. [60].)

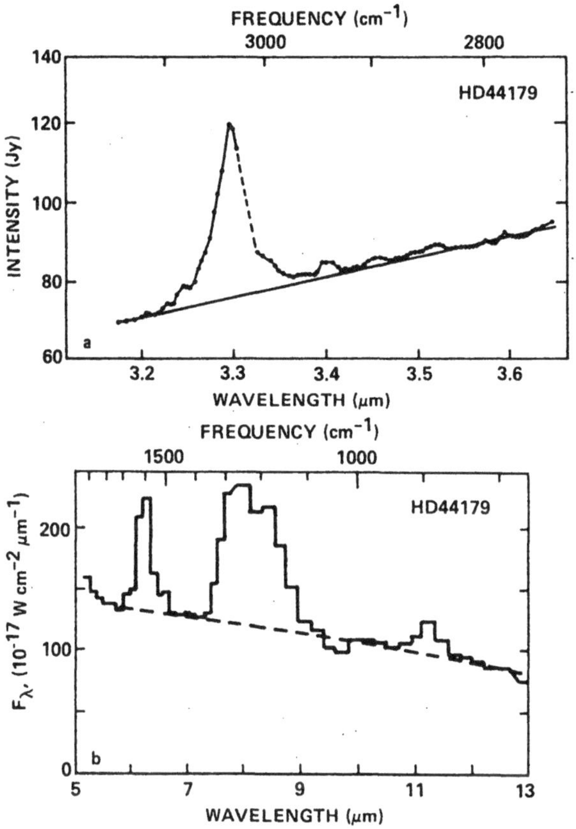

Fig. 2. The infrared emission spectrum from the reflection nebula HD 44179, the Red Rectangle (**a** from Ref. [44], **b** from Ref. [15].)

Good examples of nearly complete mid-IR spectra are shown in Figs. 1, 2 and 3 a, b. In addition to the family of bands at 3050, 1610, "1310", 1160 and 890 cm^{-1} (3.3, 6.2, "7.7", 8.6 and 11.3 μm), these figures show that there are striking variations in the underlying continuum and small changes in the "1310" band. Perusal of these figures also shows that there are 3 types of spectral components: major bands, minor (weak) bands and broad features. The broad features span roughly the 3100 to 2750 cm^{-1}, 2000 to 1100 cm^{-1} and 900 to 700 cm^{-1} regions. The band positions, widths and assignments are summarized in Table 1. In each of the following subsections these will all be discussed in this order: the major bands first, the minor bands next and the broad components last.

2.1 The 3200–2700 cm^{-1} Region

2.1.1 The 3050 cm^{-1} Major Band

Figures 1, 2, 3 a and 4 show that this region is dominated by the 3050 cm^{-1} emission band. This frequency is well known among IR spectroscopists as characteristic of

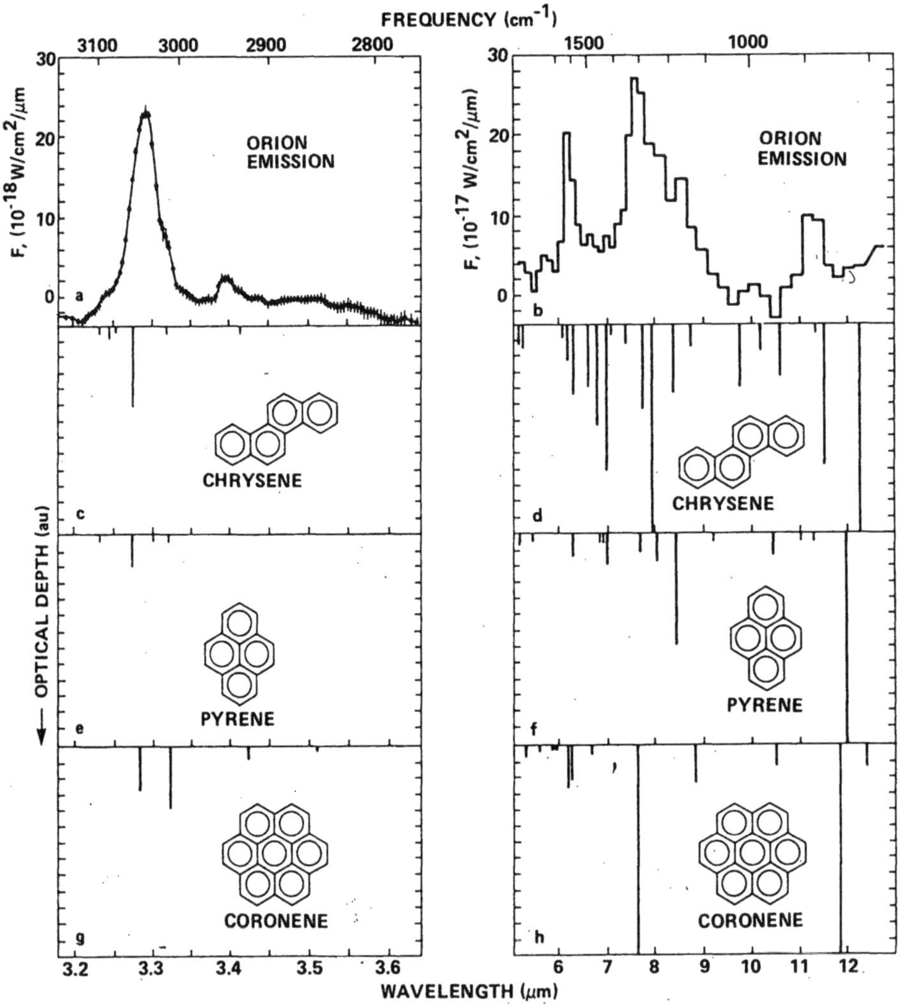

Fig. 3. The 3 to 13 micron emission spectrum from the Orion Bar compared with the absorption spectra of the PAHs chrysene, pyrene and coronene suspended in KBr pellets. (Orion, Ref. [47]; Chrysene, Ref. [38]; Pyrene, Ref. [37]; Coronene, Ref. [39].) A schematic representation for the absorption spectrum is used because the KBr pellet technique alters the spectrum compared to that of a free species

the aromatic CH stretch [35, 36, 37, 38, 39]. This frequency played a strong role in Duley and Williams first suggestion of an aromatic carrier of the interstellar spectra in 1981 [10]. The 30 cm^{-1} FWHH is probably due more to non-radiative vibrational energy redistribution times within an individual molecule rather than due to the overlapping of bands from different PAHs along the line of sight [12, 40]. A homogeneous linewidth of 30 cm^{-1} implies a redistribution time of about 0.2 picoseconds, a value consistent with the vibrational energy distribution time in other large molecules [41, 42]. It has recently been found that the band is narrower than shown in Fig. 2 by nearly a factor of two in the high flux vicinity of the star HD 44179 [43].

Table 1. Emission components: Properties and assignments[1]

v (cm^{-1})	λ (Microns)	FWHH (cm^{-1})	Assignment[2]
The major bands			
3040	3.29	30	Aromatic C–H stretch ($v = 1 \rightarrow v = 0$)
1615	6.2	30	Aromatic C–C stretch
1315–1250	7.6–8.0	70–200	Blending of several strong aromatic C–C stretching bands
1150	8.7	–	Aromatic C–H in-plane bend
885	11.3	30	Aromatic C–H out-of-plane bend for non-adjacent, peripheral H atoms
The minor features			
3085	3.24	–	Overtone and/or combination involving fundamentals in the 1810–1050 cm^{-1} (5.52–9.52 μm) range
2995	3.34	–	Overtone and/or combination involving fundamentals in the 1810–1050 cm^{-1} (5.52–9.52 μm) range
2940	3.4	"20"	Aromatic C–H stretch ($v = 2 \rightarrow v = 1$)
2890	3.46	–	Overtone/combination band involving fundamentals in the 1810–1050 cm^{-1} (5.52–9.52 μm) range, aromatic CH stretch (high v), aliphatic CH stretch,?
2850	3.51	–	Aromatic CH stretch ($v = 3 \rightarrow v = 2$), aliphatic CH stretch, overtone/combination band involving fundamentals in the 1810–1050 cm^{-1} (5.52–9.52 μm) range
2810	3.56	–	Aromatic CH stretch (high v), aldehydic CH stretch, overtone/combination band involving fundamentals in the 1810–1050 cm^{-1} (5.52–9.52 μm) range
1960–1890	5.1–5.3	30	Combination of CH out-of-plane and in-plane bend,?
1785–1755	5.6–5.7	40	Aromatic C–C stretch; overtone of 885 cm^{-1} (11.3 μm) band; carbonyl C = 0 stretch,?
1470–1450	6.8–6.9	30	Aromatic C–C stretch, aliphatic CH deformation
840	11.9	–	C–H out-of-plane bend for doubly adjacent H atoms
790	12.7	–	C–H out-of-plane bend for triply adjacent H atoms
The broad components			
2940	3.5		Overlap of C–H stretching modes, shifted by anharmonic effects,
3115–2740†	3.21–3.65†	"300"	with overtones and combinations of C–C stretch fundamentals in the 1670–1250 cm^{-1} (6–8 μm) region, aliphatic CH stretch?,?
~ 1200	~ 8.5	"400"	Blending of many weak aromatic C–C stretching
1810–1050†	5.52–9.52†		bands
880	12	"160"	Overlap of many aromatic C–H out-of-plane bend-
950–740†	10.5–13.5†		ing modes for non-adjacent as well as doubly and triply adjacent peripheral H-atoms
Red-Near IR Continuum			Electronic transitions between low-lying levels in ionized and complexed PAHs and amorphous carbon particles
Mid-IR Continuum			Quasi-continuum formed by overlapping overtone and combination bands

" " Value estimated from several published spectra. † Rough limits of the feature.

1: When the assignment is not clear, several possible explanations are listed. The first seems most likely.

2: This table is extensively discussed in [16]

2.1.2 The Minor Bands in the 3200–2700 cm^{-1} Region

There are a number of weak bands in this region. Unresolved shoulders straddle the major band at about 3085 and 2995 cm^{-1} (see Fig. 1a, 2a and 3a). These fall in the range generally assigned to overtones and combination bands involving CC stretching fundamentals in the 1810 to 1050 cm^{-1} region and are responsible for most of the weak absorptions in Fig. 3c, e, and g [35, 36, 37, 38, 39]. These features should always

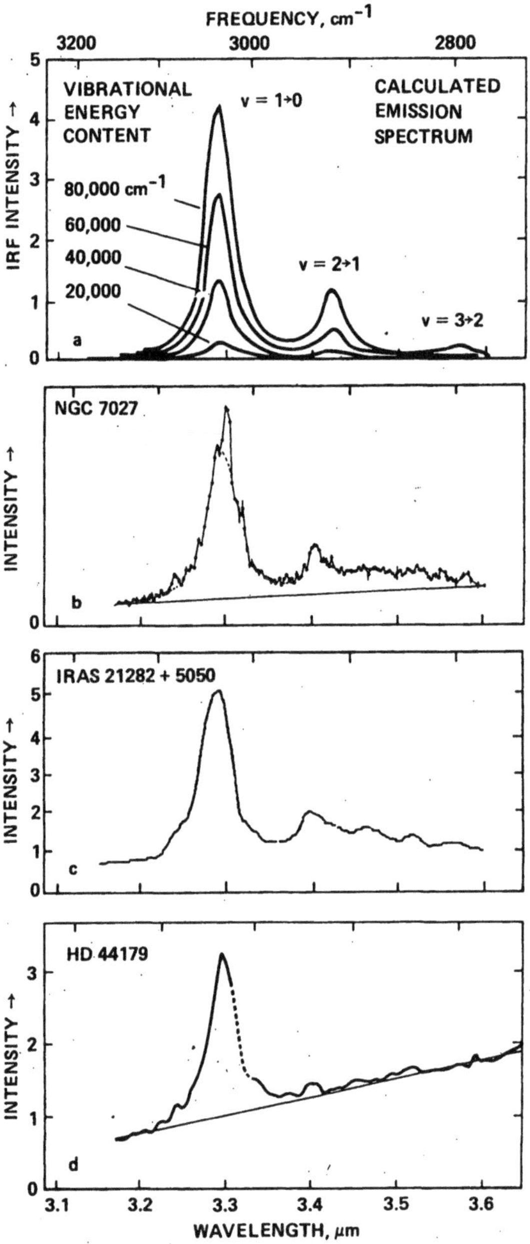

Fig. 4. (a) The calculated emission spectrum for chrysene in the CH stretching region as a function of vibrational energy content. Anharmonicity is assumed to be 120 cm^{-1}. **(b)–(d)** The observed emission spectra of NGC 7027 [61], IRAS 21282 + 5050 [45] and HD 44179 [44] showing how emission from higher vibrational levels depends on the availability of energetic photons. In NGC 7027, where the most energetic pumping photons are available, infrared emision from higher levels is important and produces a prominent $v = 2 \rightarrow v = 1$ band, whereas in the relatively benign HD 44179, emission from $v = 2$ is barely discernible

be present in emission when the 3040 cm^{-1} band is intense because any specie with enough internal energy to populate the fundamental at 3050 cm^{-1} contains enough energy to doubly populate the C–C stretching fundamentals. Fermi resonances between these different modes will certainly contribute as well. Many objects show a clear, weak band at about 2940 cm^{-1} [44], and weaker bands have recently been discovered at 2890, 2845 and 2805 cm^{-1} [45]. We have attributed the 2940 cm^{-1}

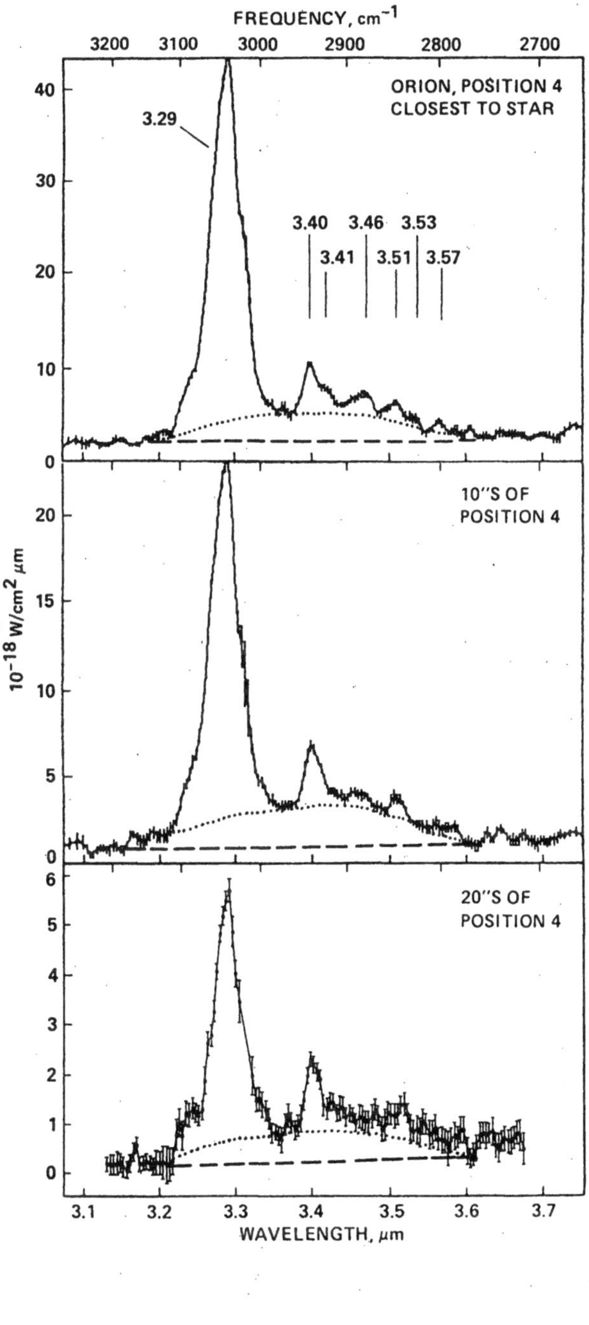

Fig. 5. Spectra in the 3200 to 2700 cm^{-1} range taken in a 5 arcsecond beam at three locations near the ionization ridge in the Orion Nebula. Position 4 is on the ridge between the ionized gas (HII region) and neutral molecular cloud while the positions 10 and 20 arcseconds south are within the molecular cloud. The *dotted lines* indicate the broad component, the *dashed line* the presumed continuum. Wavelengths, in microns, of the features are indicated in the top panel (Figure is from Reference [46])

band and some of the other weak features to the $(v = 1) \leftarrow (v = 2)$ aromatic CH stretch transition [40], although others have argued that it is carried by methyl ($-CH_3$), methylene ($-CH_2-$) and perhaps aldehydic ($-HCO$) sidegroups on PAHs [45]. Within the framework of the anharmonicity picture, Fig. 4 shows that the prominence of the 2940 cm^{-1} band is directly related to the vibrational energy content of the emitter. Thus, anharmonicity can account for the presence of a band at this frequency as well as the observation that it is most prominent in high excitation objects such as NGC 7027 and on the ionization ridge in the Orion Nebula (Fig. 3a). Since the ultraviolet exciting spectrum can be characterized in several astronomical objects, the dependence on energy content can be used to deduce the size range of the species which dominate the emission. Analysis of the observed 3050 to 2940 cm^{-1} intensity ratios show that PAHs containing between 20 and 30 carbon atoms dominate the emission in this region [40]. The corresponding $(v = 2) \leftarrow (v = 3)$ CH stretch transition is expected to fall at about 2800 cm^{-1}. The spectra taken at

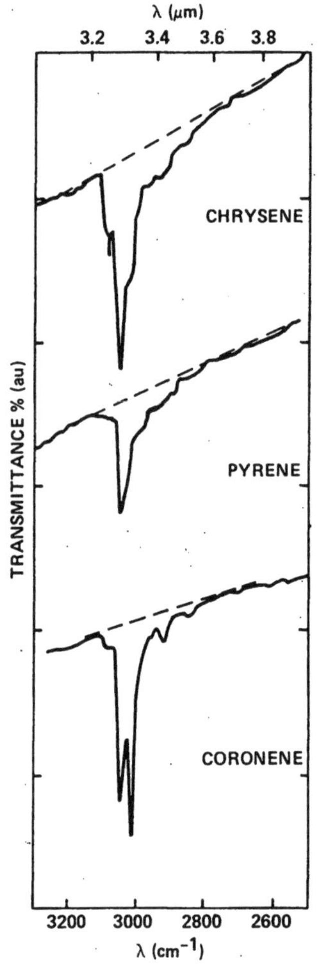

Fig. 6. Infrared absorption spectra in the 3.3 micron region of the PAHs chrysene, pyrene and coronene, suspended in KBr. Spectra courtesy of Drs. Cyvin and Klaeboe, University of Trondheim, Norway (Figure reproduced from Ref. [16])

several positions around the ionization ridge in the Orion Nebula (see. Fig. 5) show that the 2940 and 2850 cm^{-1} band intensities vary with respect to the major band at 3050 cm^{-1} in a way which is consistent with the emission originating from upper CH stretch vibrational levels [46].

Naturally, the bands in this region may well represent a blend of the ($v = 1$) ← ($v = 2$) and ($v = 2$) ← ($v = 3$) aromatic CH stretching transitions with overtones and combinations involving aromatic CC stretches as well as aliphatic CH stretches. Many PAHs which do not have aliphatic side groups show weak absorptions near these frequencies. For example, Fig. 6 shows that chrysene, pyrene and coronene all show substructure on a broad component. Chrysene and coronene show a "peak" at about 2910 and 2845 cm^{-1} while pyrene has a broad (weak) plateau from 2950–2880 cm^{-1}, which is similar to the emission plateau observed from the astronomical object BD + 30°3639 [44]. In the laboratory spectra these are due to overtone and combination bands which have been perturbed sufficiently by solid state effects to absorb weakly [35, 36, 37, 38, 39]. The perturbations within the PAH clusters that are suspended in salt pellets induce IR activity and broaden the individual bands causing them to overlap. In free vibrationally excited PAHs, perhaps Fermi resonances between the overtones and combinations of C–C stretching vibrations with the highly excited C–H modes can sufficiently enhance the intensities of these presumably weak bands to produce the observed intensites.

2.1.3 The Broad Component in the 3200–2700 cm^{-1} Region

Inspection of Fig. 1, 3a, 4 and 5 shows that there is a broad component which stretches from about 3130 cm^{-1} to 2750 cm^{-1}. We have assigned this to a vibrational quasi-continuum produced by the overlapping of the different transitions discussed in the previous section: highly excited aromatic C–H stretching vibrations, overtones and combinations of CC stretching vibrations and perhaps C–H stretches on aliphatic side groups. The spatial behavior of the intensity of this broad component appears to follow the distribution of the major and minor bands in this region, rather than the distribution of the lower frequency, broad features (discussed in Sect. 2.2.3 and 2.3.3) which are due to much larger PAHs and perhaps PAH clusters and amorphous carbon particles [47]. Thus, this broad component is most likely due to molecule-sized free PAHs.

2.2 The 2000–1000 cm^{-1} Region

2.2.1 The 1610, 1350 and 1150 cm^{-1} Major Bands

Léger and Puget pointed out that the frequency of the 1610 cm^{-1} band is as characteristic of polycyclic aromatic hydrocarbons as is the 3050 cm^{-1} band and attributed it to an aromatic C–C stretching fundamental [11]. As with the 3050 cm^{-1} band, the approximately 30 cm^{-1} width probably reflects vibrational energy redistribution times within the molecules contributing to this band rather than a blending of narrower lines from different molecules. Some interstellar spectra show a slight asymmetry on the low frequency side of this feature (e.g., see Fig. 3b and Refs [15] and [48]). This is suggestive of a contribution of emission from higher vibrational levels. As the anharmonicity of the C–C stretch is thought to be smaller than the natural bandwidth (a few cm^{-1} versus 30 cm^{-1}), blending should skew the band to lower frequencies [40].

The peak position and profile of the very broad "1310" cm^{-1} band, on the other hand, vary from object to object and its assignment has been correspondingly less straightforward. We assigned this band to the aromatic CC stretching vibrations for the following reasons [12]. The variation is consistent with PAHs since the frequency range accross the base of the interstellar feature encompasses the frequencies expected for the strongest IR active C–C stretching modes in a mixture of PAHs (e.g., see Fig. 3d, f and h). Unlike the 3050 and 1610 cm^{-1} bands which consistently occur at nearly the same frequency in different aromatic molecules, the precise positions of these other IR active aromatic C–C stretches depend on the molecule. Figure 3 shows that chrysene, pyrene and coronene have several strong C–C stretching bands in the region. As molecule size, degree of ring condensation and symmetry increase, the relative intensities of the 1610 and 1460 cm^{-1} band seem to reverse. Thus, if the interstellar mixture were complex enough, its IR spectrum in this region would show a broad feature, possibly exhibiting some substructure.

In the interstellar medium less stable PAHs will be weeded out, leaving a mixture dominated by the most stable structures [17, 19]. These more stable PAHs are those which are symmetric and generally in the most condensed forms. An example of a mixture comprised largely of the most stable molecular forms is the soot formed in the high temperature combustion of hydrocarbons. A comparison of the Raman spectrum in the 1800 to 1000 cm^{-1} region of auto soot with the interstellar IR emission spectrum from Orion is shown in Fig. 7. The Raman spectrum of soot

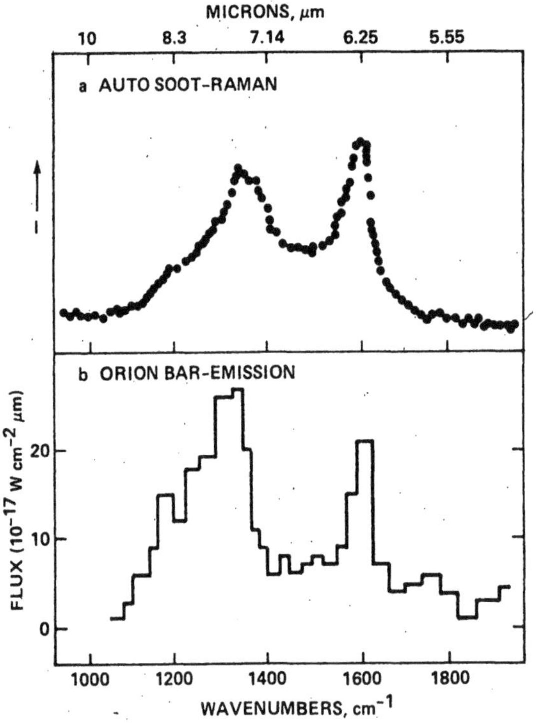

Fig. 7. Comparison of the 5 to 10 micron Raman spectrum of auto soot (a form of amorphous carbon which is rich in the more stable PAHs) with the emission from Orion (soot spectrum, adapted from Ref. [62], Orion from Ref [47]) (Figure from Ref [12])

principally probes the aromatic C–C stretching vibrational frequencies in this material because the Raman scattering cross section for these bonds is very large and they are the most dominant type of bond in the mixture. Of course, for free PAHs with an inversion center of symmetry the infrared and Raman active modes will be mutually exclusive. However, the perturbations in the soot particles break down the symmetry selection rules and the IR and Raman spectra become similar as shown by comparing the IR spectra of high temperature carbonaceous materials in Refs. [22] and [49] with the Raman spectrum shown in Fig. 7. Figure 7 dramatically shows that the distribution of the C–C stretching frequencies in a mixture of the more stable PAHs coincides quite well with the frequency range of the major interstellar emission bands in the 2000–1000 cm^{-1} region [12]. Assignment of the "1310" cm^{-1} interstellar feature to vibrations in symmetric interstellar PAHs is supported by the good agreement of the interstellar emission spectrum from the reflection nebula NGC 2023 with the infrared emission calculated from the measured absorption spectra of four symmetric PAHs shown in Fig. 8 [50].

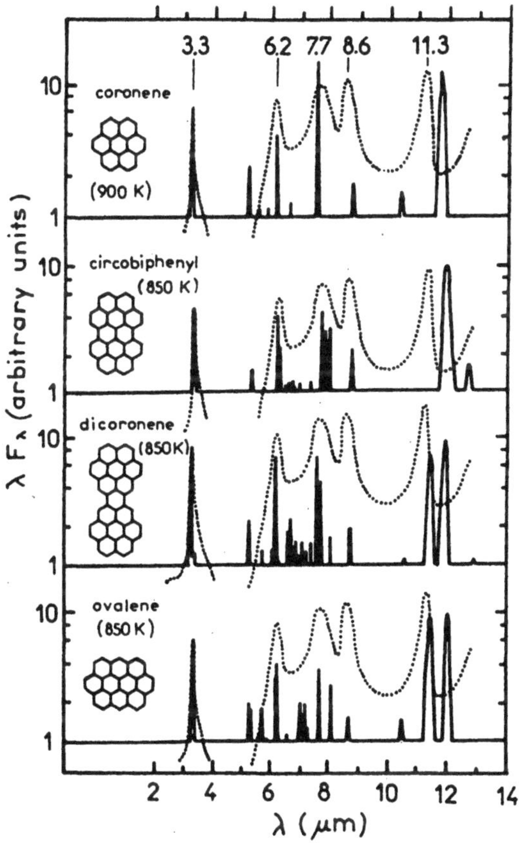

Fig. 8. Emission spectra of several compact PAHs *calculated* from their absorption spectra measured in the laboratory at room temperature compared to the observed spectrum of the reflection nebula NGC 2023 (*dots*) (Figure reproduced from Ref [50])

Finally, a band near 1150 cm^{-1} often appears as a small shoulder to the broad "1310" cm^{-1} feature but is resolved in certain objects (Figs. 1 and 3b). It has been assigned to the in-plane aromatic CH bending vibration [11].

Summing up, the 1610 and "1310" cm^{-1} bands are indicative of aromatic hydrocarbons and have been assigned to the aromatic C–C stretch. The variability of the peak position and profile of the "1310" cm^{-1} feature implies further that symmetric PAHs dominate the interstellar mixture. These are, of course, the most stable members of the PAH family.

2.2.2 The Minor Bands in the 2000–1000 cm^{-1} Region

There are three minor emission bands at about 1920, 1780 and 1460 cm^{-1} which have been associated with the interstellar emission spectrum. Emission in the 1900 cm^{-1} region was searched for because nearly all laboratory PAH spectra show a weak band between 1960 and 1890 cm^{-1}. Based on this expectation, the spectrum of BD + 30°3639 was measured in the 2000 to 1500 cm^{-1} region (see Fig. 9) and a new emission band at about 1900 cm^{-1} was indeed found [51]. Inspection of previous spectra which terminate at about 1900 cm^{-1} show that most start to turn up at about 1850 cm^{-1}, implying that most objects which emit the major features have a band at this position [15, 48]. The confirmation of this feature provides strong support for the PAH hypothesis. Although the mode assignment is not well established, consideration of the relative absorption band strengths in previously published spectra and the symmetry assignments suggest that this is a combination (or overtone?) of the C–H bending modes.

The 1780 and 1460 cm^{-1} band intensities are variable. They were initially recognized in the spectrum of NGC 7027 and attributed to the IR emission band system by Bregman et al [52]. Weak correlation between these and the major bands in several

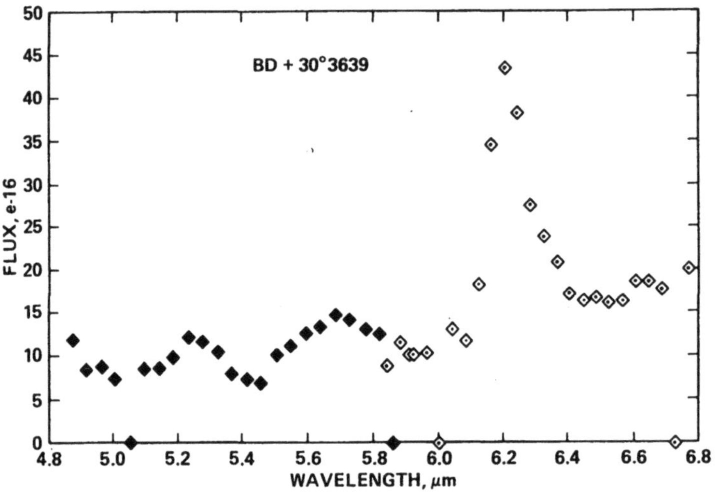

Fig. 9. The 2000 to 1470 cm^{-1} spectrum of the planetary nebula BD + 30°3639 showing the presence of a 1900 cm^{-1} band. This spectrum was measured because nearly all laboratory spectra of PAHs show a band in the 1850 to 2000 cm^{-1} range. The detection of this feature gives strong support to the PAH hypothesis (Figure is from Ref. [51])

objects confirm that they are part of the emission band family [15]. These bands can be seen in Fig. 1 and 9. There are several possible assignments for the 1780 cm^{-1} band. It could correspond to one of the weaker PAH features such as shown in Fig. 3d, f and h, either as a weakly allowed C–C stretching fundamental, or as suggested by Bregman et al [52], it may be an overtone of the major 885 cm^{-1} band. Alternatively, since this frequency is highly characteristic of the C=O stretch, it may arise from a carbonyl attached to a PAH or an interconnecting hydrocarbon chain. A carbonyl in the form of an aldehyde can be ruled out since there has been no detection of a prominent feature attributable to an aldehydic C–H stretch near 2860 cm^{-1} in the objects which show a clear 1760 cm^{-1} band. Until the availability of both higher resolution astronomical spectra (for more than a handful of objects) and lab spectra of suitable candidates, it will not be possible to determine which explanation is more likely.

The 1450 cm^{-1} band probably corresponds to an aromatic C–C stretching vibration (although it is conceivable that it may be an overtone or combination band involving C–H out-of-plane bending modes). Since only some PAHs absorb at this frequency, (Figs. 3d, f, h) emission from a collection of PAHs could produce a relatively weak feature. Since the strength of this band depends strongly on the symmetry and geometry of the molecule, it may be a probe of the chemistry in the emitting region. An aliphatic C–H deformation seems unlikely because, up to now, bands attributable to the aliphatic C–H stretching modes do not seem to correlate with the 1450 cm^{-1} band. Again, more observations are needed.

2.2.3 The Broad Component in the 2000–1000 cm^{-1} Region

Inspection of Figs. 1b and 3b clearly shows the presence of a broad, very roughly triangular-shaped, hump underlying the 1615 and 1300 cm^{-1} bands. This broad component is also evident in the spectra of other objects [15, 48]. As discussed in Sect. 2.2.1 this feature encompasses the region in which the IR active, C–C stretching modes in PAHs are richest. For example, of the 72 normal modes in pyrene, 42 involve C–C vibrations (59%), while in ovalene, 90 of the 132 (68%) are purely C–C in character. While a certain fraction are infrared inactive, or only weakly allowed, and many involve lower frequency bending modes, it is clear that PAHs (and PAH clusters) possess many bands that can overlap and produce a broad feature in this region. In fact, the spectra of amorphous carbon particles (mixtures of aromatic structural units) have a strong feature in this region [22, 32, 49].

The importance of this component in the interstellar emission spectrum depends on both the intensity (flux) and energy of the ultraviolet radiation field. In objects where very energetic photons are plentiful, such as in NGC 7027, larger PAHs and materials composed of several smaller PAHs clustered together (loosely or in small amorphous carbon particles) may be excited sufficiently to emit in this band with intensities comparable to the intensities of the discrete features emitted by isolated PAHs. In regions where the ultraviolet radiation field is less intense and less energetic, such as in many reflection nebulae with much cooler exciting stars, small molecules can survive and their emission features tend to dominate the spectrum. The observed variability in intensity of this band with respect to major band intensities is consistent with an origin in C–C vibrations in larger PAHs, PAH clusters and amorphous carbon particles containing on the order of 400 carbon atoms [16, 47].

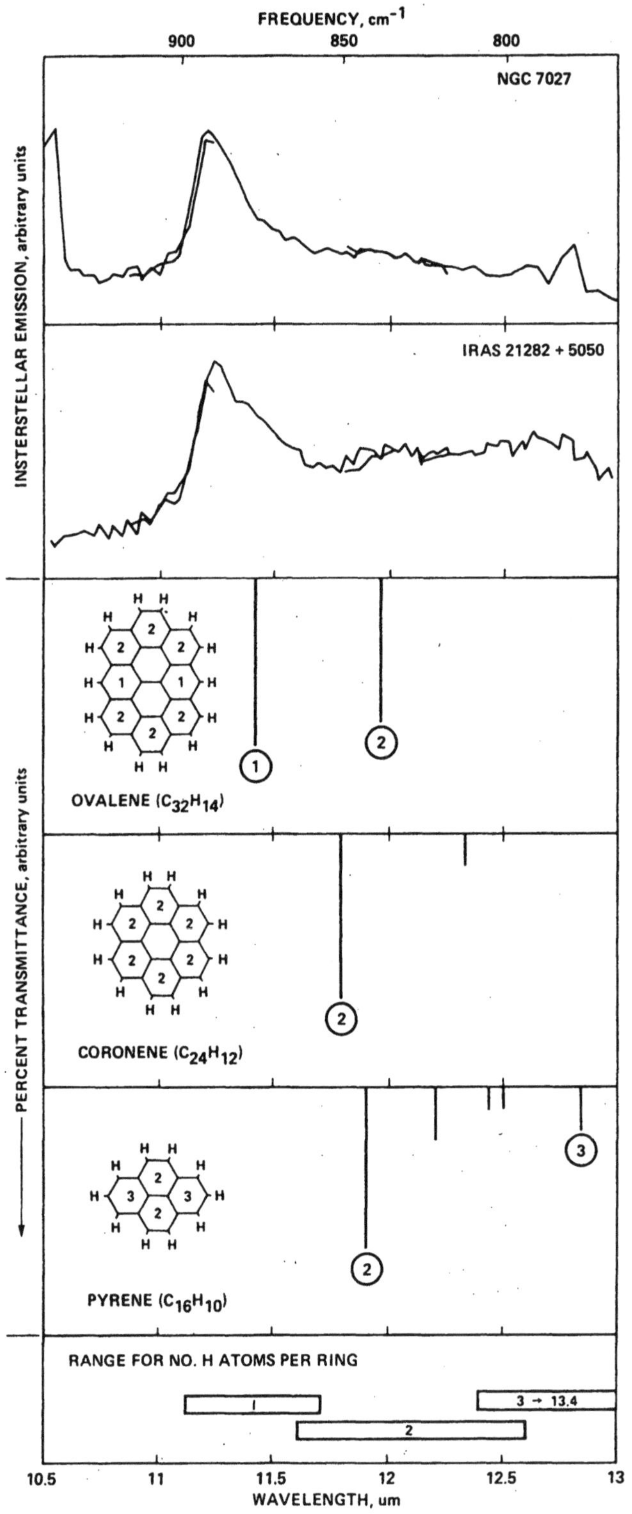
FREQUENCY, cm^{-1}
900
850
800
NGC 7027
IRAS 21282 + 5050
INSTERSTELLAR EMISSION, arbitrary units
PERCENT TRANSMITTANCE, arbitrary units
OVALENE ($C_{32}H_{14}$)
CORONENE ($C_{24}H_{12}$)
PYRENE ($C_{16}H_{10}$)
RANGE FOR NO. H ATOMS PER RING
1
2
3 → 13.4
10.5
11
11.5
12
12.5
13
WAVELENGTH, um

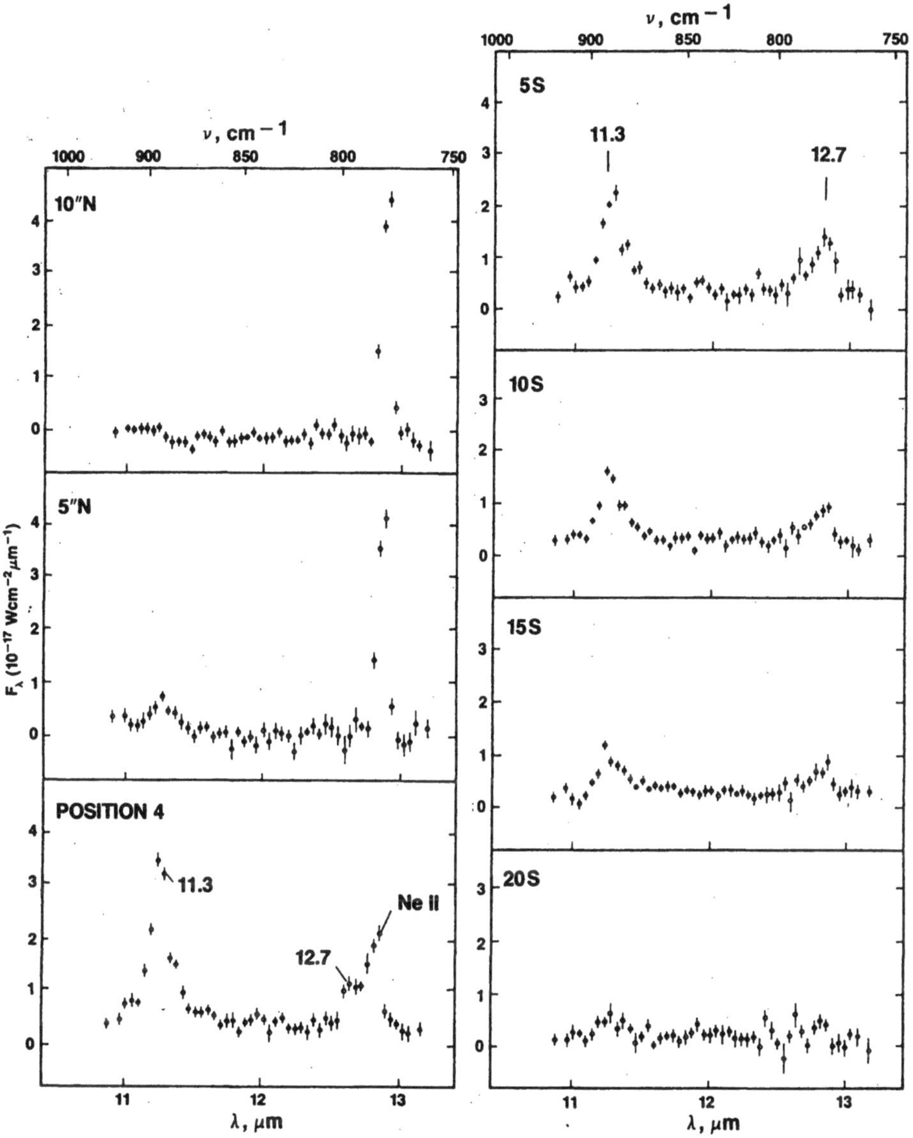

Fig. 11. Spectra between 900 to 770 cm^{-1} from 10 arcseconds north to 20 arcseconds south of position four on the ionization ridge in the Orion Nebula. Positions 4, 10S and 20S are the same as the positions for the spectra shown in Fig. 5 (Figure from Ref. [53])

◀**Fig. 10.** Comparison between the 910 to 770 cm^{-1} emission spectra from NGC 7027 and IRAS 21282 + 5050 with schematic absorption spectra of PAHs in salt pellets. The PAHs shown contain isolated as well as doubly and triply adjacent H atoms on their peripheral rings. (Figure from Ref. [54])

In conclusion, the 2000–1000 cm^{-1} region of the interstellar emission spectrum can have two important contributors – free PAHs and small amorphous carbon particles or clusters of PAHs.

2.3 The 1000–500 cm^{-1} Region

2.3.1 The 890 cm^{-1} Major Band

Until recently, the only feature in this region believed to be associated with the other emission bands was the prominent 890 cm^{-1} band shown in Figs. 1, 2, 3, 10 and 11. This was assigned to the out-of-plane aromatic C–H bending vibration by Duley and Williams in 1981 [10]. This, along with the 3050 cm^{-1} band, provided the original basis for an aromatic explanation of the interstellar emission bands. The 890 cm^{-1} band is skewed to lower frequencies in most emission spectra. This implies that emission from higher vibrational levels contributes to the observed band [40]. For highly vibrationally excited molecules the band will be quite asymmetric, broader, and peaked at slightly lower frequencies than that from molecules containing less vibrational energy. The prominence of the band implied that the aromatic carriers be only partially hydrogenated since IR activity at 890 cm^{-1} is characteristic of aromatic systems with edge rings which contain only isolated, non-adjacent peripheral hydrogen atoms [36]. Fully hydrogenated PAHs possess several strong bands in the 900 to 500 cm^{-1} region. Partial hydrogenation is unexpected because hydrogen is more than 10^7 times more abundant than the proposed PAHs. This apparent inconsistency was resolved by the results described in the next two subsections.

2.3.2 The Minor Bands in the 1000–500 cm^{-1} Region

Quite recently on a weak, resolvable substructure has been detected at about 840 and 780 cm^{-1} [53, 54]. These new bands are just where one would expect to find them for PAHs with doubly and triply adjacent H atoms per edge ring, relieving somewhat the constraint of partial dehydrogenation. As discussed in the next subsection, these new bands can be used to place important constraints on the moleculer geometries of the emitting PAHs. As with some of the minor bands in the 3000–2500 cm^{-1} region, the intensities of the 890 and 780 cm^{-1} bands vary across the ionization ridge in the Orion Nebula (Fig. 11), suggesting that certain PAHs may be particularly favored under the conditions which prevail in that region [53].

The relative intensities of these bands pose several questions. If one assumes that the integrated absorbance per H atom is comparable for single as well as doubly and triply adjacent H atoms per edge ring, then one is still faced with a surprisingly large fraction of partially dehydrogenated PAHs. Detailed analysis of what this means in terms of the interstellar medium must wait until one has reliable band strengths for these different types of peripheral hydrogen atoms in isolated PAHs. Furthermore, the interstellar 950–770 cm^{-1} broad component (discussed in the next section) and 890 cm^{-1} band are generally much less intense than the "1310" cm^{-1} ("7.7" μm) feature whereas the opposite is true in the laboratory spectra. This latter discrepancy has been shown to be the result of a spectral artifact produced by the techniques used to measure most of the PAH spectra published to date. Most laboratory spectra

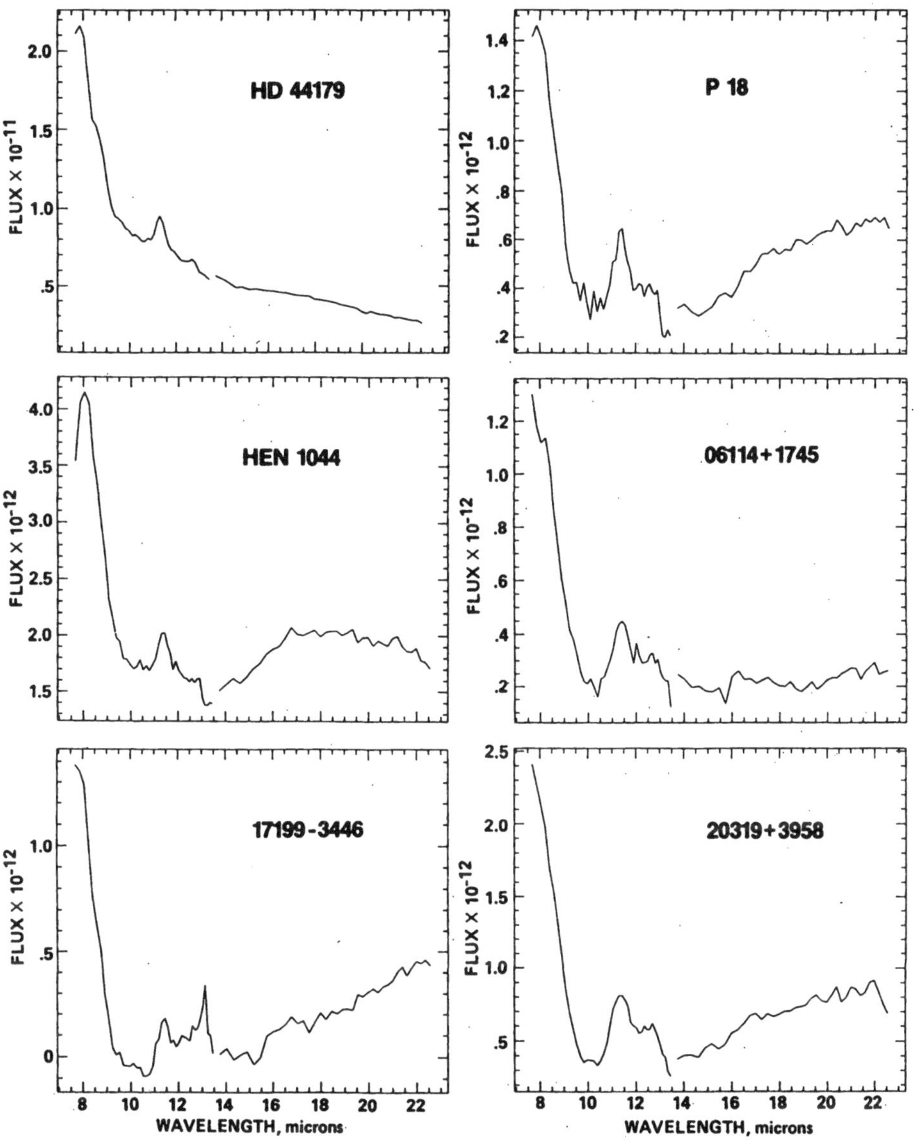

Fig. 12. The LRS (Low Resolution Spectrometer) spectra of six IRAS sources. The short-wavelength data span the range from 1300 to 740 cm^{-1} (7.67–13.45 μm). The long-wavelength data are shown from 730–450 cm^{-1} (13.72 to 22.55 μm). The "1310" (7.7 μm) feature dominates high frequency end of these spectra. The 885 cm^{-1} (11.3 μm) band shows a long-wavelength plateau extending to about 770 cm^{-1} (13.0 μm). Ordinates are flux in W m^{-2} μm^{-1} (Figure from Ref. [56])

are obtained by mixing PAH crystallites with salt and pressing the mixture into a transparent pellet. Individual PAH molecules are not isolated in this way. Molecular interactions between adjacent PAHs produce large band strength enhancements in the 900 to 500 cm^{-1} (11–20 μm) regions [55].

This property is not only important because it reconciles an apparently serious inconsistency with the PAH hypothesis but also because it provides a powerful spectroscopic argument in favor of the well-defined bands arising from free, molecule-sized species rather than particles (in which interactions would be expected to enhance emission in this region).

2.3.3 The Broad Component in the 1000–500 cm^{-1} Region

Figure 12 shows the broad plateau of emission which extends from about 950 to 770 cm^{-1} that is evident in the spectra of objects that produce the IR emission bands. Because of the onset of telluric obfuscation at frequencies below about 800 cm^{-1} this plateau was not recognized until the launch of the Infrared Astronomical Satellite (IRAS) which measured very low resolution spectra from about 1000 to 300 cm^{-1}. This observation relieved the difficulty associated with proposing only partial hydrogenation in exceedingly hydrogen-rich environments because it showed that the edge rings of the PAHs responsible for the features can have non-adjacent as well as 2 or 3 adjacent peripheral H atoms, but not 4 or 5. As mentioned above, this placed strong, important constraints on the molecular geometries of the aromatic hydrocarbons. These constraints are consistent with the conclusion drawn earlier (from the position and profile of the "1310" cm^{-1} band) that the most condensed, stable PAHs dominate the interstellar mixture [56].

2.4 The Far Infrared

Skeletal vibrations such as C–C–C bending and ring puckering modes fall in the far IR. Figure 13 shows that a mixture of PAHs could produce a broad component composed of overlapping individual lines in the 625–330 cm^{-1} ranges. From the limited information available, it seems that these bands are at least a factor of two weaker than the mid-IR bands (although this may be an artifact produced by the steric constraints placed on such collective large molecule vibrations by the solid matrix). Objects showing the PAH mid-IR bands should also have broad, low level emission extending out to about 330 cm^{-1}. At lower frequencies the positions become rather molecule specific. These bands may well be responsible for the emission associated with reflection nebulae and high galactic latitude clouds discovered by the IRAS satellite in broad spectral bands centered at 12 and 25 microns [5, 12, 20, 57].

The intensity of the 12 μm IRAS band drops with respect to that of the 25 μm band in the vicinity of the exciting star in the case of reflection nebulae [58] and close to hot stars in the case of the high galactic latitude clouds [59]. If PAHs and PAH-related materials are responsible for the emission, this behavior can be understood in terms of dehydrogenation rather than grain destruction as previously suggested. As shown in Fig. 13, the 12 μm IRAS band covers the 1250–670 cm^{-1} range which primarily encompasses the C–H in-plane and out-of-plane bending vibrations in PAHs while the 25 μm IRAS band covers the 560–330 cm^{-1} region which encompasses the C–C–C aromatic, plane bending vibrations of PAHs. Consequently, if a PAH is stripped of hydrogen it will lose most of its IR activity in the 1000 to 600 cm^{-1} region but not in the 600 to 10 cm^{-1} region. Small PAHs (15 to 30 carbon atoms) contribute strongly to the infrared emission because they are

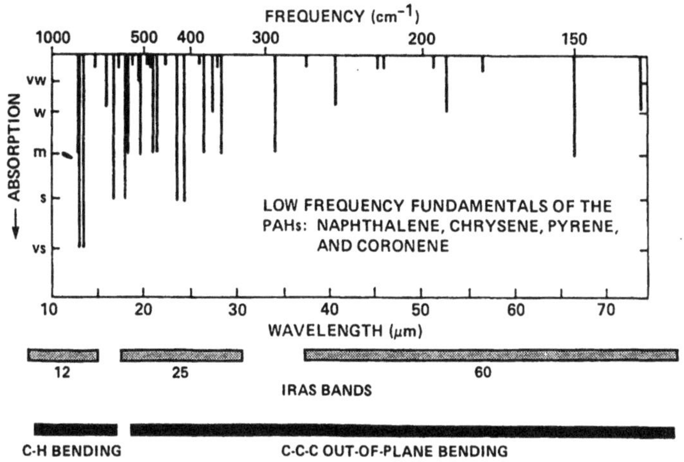

Fig. 13. A schematic composite absorption spectrum for the out-of-plane vibrations in a mixture of the aromatic molecules: naphthalene, chrysene, pyrene and coronene in the 1000 to 130 cm^{-1} range compared to the IRAS bands. The regions in which the CH in- and out-of-plane, and carbon plane bending vibrations fall are also indicated.

the most highly vibrationally excited by the absorption of an ultraviolet photon. These are also the PAHs which will be the first to suffer H atom loss if sufficiently excited [16]. Consequently, if PAHs are indeed responsible for this emission, the drop in intensity of the 12 μm IRAS band with respect to the 25 μm IRAS band can be understood in terms of the dehydrogenation of the most intensely emitting PAHs and not the destruction of the carbon skeleton or of a particular class of grains. As with the other frequency regions, further progress in the far IR will require laboratory experimentation and theoretical development.

3 Conclusions

Many different celestial objects emit an infrared spectrum which has been attributed to a family of benzenoid hydrocarbons. The emission is thought to arise from molecule-sized polycyclic aromatic hydrocarbons (PAHs) and related materials which have been highly vibrationally excited by the absorption of visible or ultraviolet photons. The most intense higher frequency spectral features come primarily from species containing 20 to 50 carbon atoms. Larger species are certainly present as well. Given the same excitation conditions, the larger species probably make increasingly important contributions to the lower frequency bands. There is some controversy over whether the smaller PAHs are free (i.e., in the gas phase) or structural units of amorphous carbon particles and PAH clusters. There is also evidence for the presence of substantially larger PAHs, PAH clusters and amorphous carbon particles containing on the order of 400 carbon atoms.

The spectroscopic assignments are based on a rough similarity between the interstellar spectra and laboratory spectra of PAHs and related materials. As the species in space are most likely in forms little studied in the laboratory, much experimental and theoretical work is called for on aromatic hydrocarbons. In the interstellar medium the proposed mixture includes species which are larger than many previously studied, ionized and dehydrogenated. Until the spectroscopic properties of these types of molecules are characterized, full advantage of the PAH model cannot be realized. This represents an important challenge as PAHs appear to be ubiquitous and abundant throughout the interstellar medium. They are probably more abundant than all of the other known interstellar polyatomic molecules combined. As such, they are potentially excellent probes of conditions in many different objects. They may also be the missing link between the carbon ejected from carbon-rich red giant stars and the carbon particle component of interstellar dust (and perhaps the carbon found in meteorites and interplanetary dust particles).

4 Acknowledgement

I wish to express my sincere appreciation of countless fruitful conversations with John Barker and Xander Tielens on the subject of PAHs in space. I am also indebted to many colleagues with whom I've had the good fortune to collaborate on these problems: J. Bregman, M. Cohen, T. Geballe, S. Leach, S. Sandford, K. Sellgren, M. Werner and F. Witteborn.

5 References

1. Gillett FC, Forrest WJ, Merrill KM (1973) Ap. J. 183: 87
2. Aitken DK (1981) (In: Wynn-Williams CG, cruikshank DP (eds), Infrared astronomy.) D. Reidel, Dordrecht, p 207
3. Willner SP In: Kessler MF, Phillips JP (eds) Galactic and extragalactic infrared spectroscopy. D. Reidel, Dortrecht, p 37 .
4. Phillips MM, Aitken DK, Roche PF (1984) Mon. Not. R. Astr. Soc. 207: 25
5. Low FJ et al. (1984) Ap. J. (Letters) 278: L19
6. Allamandola LJ (1984) In: Kessler MF, Phillips JP Galactic and extragalactic infrared spectroscopy. D. Reidel, Dordrecht, p 5
7. Allamandola LJ, Norman CA (1978) Astron. Astrophys. 66: 129
8. Allamandola LJ, Greenberg JM, Norman CA (1979) Astron. Astrophys. 7: 66
9. Sellgren: K (1984) Ap. J. 277: 627
10. Duley WW, Williams DA (1981) Mon. Not. R. Astr. Soc. 196: 269
11. Léger A, Puget JL (1984) Astron. Ap. 137: L5
12. Allamandola LJ, Tielens AGGM, Barker JR (1985) Ap. J. (Letters) 290: L25
13. Allamandola LJ, Tielens AGGM, Barker JR (1987) In: Léger A, d'Hendecourt LB, Boccara N (eds) Polycyclic aromatic hydrocarbons and astrophysics. D. Reidel, Dordrecht p 255
14. Duley WW (1989) In: Allamandola LJ, Tielens AGGM (eds) Interstellar dust. Kluwer, Dordrecht p 141
15. Cohen M, Allamandola LJ, Tielens AGGM, Bregman JD, Simpson JP, Witteborn F, Wooden D, Rank D (1986) Ap. J. 302: 737
16. Allamandola LJ, Tielens AGGM, Barker JR, (1989) Ap. J. Suppl. (in press)
17. van der Zwet GP, Allamandola LJ (1985) Astr. Ap. 146: 76
18. Léger A, d'Hendecourt LB (1985) Astr. Ap. 146: 81
19. Crawford MK, Tielens AGGM, Allamandola LJ (1985) Ap. J. (Letters) 293: L45
20. Puget JL, Léger A, Boulanger F (1985) Astr. Ap. 142: L19
21. Omont A (1986) Astr. Ap. 164: 159
22. Wdowiak TJ, Flickinger GC, Cronin JR (1988) Ap. J. (Letters) 328: L75
23. d'Hendecourt LB, Léger A (1987) Astr. Ap. 180: L9

24. Lepp S, Dalgarno A (1988) Ap. J. 324: 553
25. Allamandola LJ, Sandford SA, Wopenka B (1987) Science 237: 56
26. Léger A, D'Hendecourt LB, Boccara N (eds.) (1987) Polycyclic aromatic hydrocarbons and astrophysics, D. Reidel, Dordrecht
27. Allamandola LJ, Tielens AGGM, (eds.) (1989) Interstellar dust, Kluwer, Dordrecht
28. Leach S, Eland JHD, Price SD: J. Phys. Chem., (in press)
29. Ruhl E, Price SD, Leach S: J. Phys. Chem. (in press)
30. Sedlmayr E (1989) In: Allamandola LJ, Tielens AGGM (eds) Interstellar dust, Tielens, A.G.G.M. Kluwer, Dordrecht, p 467
31. Frenklach M, Feigelson ED (1989) Ap. J. 341: 372
32. Blanco A, Bussoletti E, Colangeli L (1988) Ap. J. 334: 875
33. Borghesi A, Bussoletti E, Colangeli L (1987) Ap. J. 314: 422
34. Sakata A, Wada S, Onaka T, Tokunaga AT (1987) Ap. J. (Letters) 320: L63
35. Herzberg GH (1968) Infrared and Ramanspectra of polyatomic molecules, D. van Nostrand, Princeton
36. Bellamy LJ (1958) The infrared spectra of complex organic molecules, Wiley, 2nd ed., New York
37. Cyvin SJ, Cyvin BN, Brunvoll J, Whitmer JC, Klaeboe P, Gustavsen JE (1979) Z. Naturforsch. 34a: 876
38. Cyvin SJ, Cyvin BN, Brunvoll J, Whitmer JC, Klaeboe P. (1982) Z. Naturforsch. 37a: 1359
39. Cyvin BN, Klaeboe P, Whitmer JC, Cyvin SJ (1982) Z. Naturforsch. 37a: 251
40. Barker JR, Allamandola LJ, Tielens AGGM (1987) Ap. J. (Letters), 315: L61
41. Oref I, Rabinovitch BS (1979) Acc. Chem. Res. 12: 166
42. Amirav A, Jortner J (1984) J. Chem. Phys. 81: 4200
43. Tokunaga AT, Nagata T, Sellgren K, Smith RG, Onaka T, Nakada Y, Sakata A, Wada S (1988) Ap. J. 328: 709
44. Geballe TR, Lacy JH, Persson SE, McGregor PJ, Soifer BT (1985) Ap. J. 292: 500
45. de Muizon M, Geballe TR, d'Hendecourt LB, Baas F (1986) Ap. J. (Letters) 306: L105
46. Geballe TR, Tielens AGGM, Allamandola LJ, Moorhouse A, Brand PWJL (1989) Ap. J.: 341, 278
47. Bregman JD, Allamandola LJ, Tielens AGGM, Witteborn FC Ap. J. (in press)
48. Cohen M, Tielens AGGM, Bregman JD, Witteborn FC, Rank DM, Allamandola LJ, Wooden DH, de Muizon M 1989 Ap. J.: 341, 246
49. Mortera C, Low MJD (1983) Carbon 21: 283
50. Léger A, d'Hendecourt LB (1987) In: Léger A, d'Hendecourt LB, Boccara N (eds) Polycyclic aromatic hydrocarbons and astrophysics. Reidel, Dordrecht p 223
51. Allamandola LJ, Bregman JD, Sandford SA, Witteborn FC, Wooden DH 1989 Ap. J. (Letters) in press
52. Bregman JD, Dinerstein HL, Goebel JH, Lester DF, Witteborn FC, Rank DM (1983) Ap. J. 274: 666
53. Roche P, Aitken DK, Smith CH (1989), Mon. Not. R. Astr. Soc.: 236, 485
54. Witteborn FC, Sandford SA, Bregman JD, Allamandola LJ, Cohen M, Wooden DH, Graps AL (1989) Ap. J. 341: 270
55. Allamandola LJ, Sandford SA (1988) In: Bailey ME, Williams DA (eds) Dust in the universe Cambridge University Press p 229
56. Cohen M, Tielens AGGM, Allamandola LJ (1985) Ap. J. (Letters) 299: L93
57. Castelaz MW, Sellgren K, Werner MW (1987) Ap. J. 313: 853
58. Sellgren K (1989) In: Allamandola LJ, Tielens AGGM (eds) Interstellar dust, Kluwer, Dordrecht, p 103
59. Boulanger F, Beichman C, Desert FX, Helou G, Perault M, Ryter C (1988) Ap. J. 332: 328
60. Russell RW, Soifer BT, Willner SP (1977) Ap. J. (Letters) 217: L149
61. Geballe TR (1983) Proc. of the workshop on laboratory and observational spectra and interstellar dust period Wolstencroft RD, Greenberg JM (eds) Occasional Reports of the Royal Observatory, Edinburgh p93
62. Rosen H, Novakov T (1978) Atoms. Environ. 12: 923

The Distortive Tendencies of Delocalized π Electronic Systems. Benzene, Cyclobutadiene and Related Heteroannulenes

Philippe Charles Hiberty

Laboratoire de Chimie Théorique,* Bât. 490, Université de Paris-Sud, F-91405 Orsay, France

Table of Contents

---

* Associated with the CNRS, UA 506.

Philippe Charles Hiberty

A Valence Bond theory of delocalization predicts that X_n aggregates of atoms or groups X are generally unstable when X is able to form strongly bound dimers X_2 (e.g. H_2), while X_n may be stable if the X_2 bond is weak (e.g. Li_2). According to this theory, the π systems of aromatic hydrocarbons are expected to be unstable in a regular geometry and to be more stable in an alternated one. To verify this prediction, the driving force responsible for the regular geometry of aromatic π systems is analyzed by means of rigorous ab initio calculations, and decomposed into its π and σ components. The same decomposition is performed for antiaromatic systems. While the σ driving force is, by nature, always symmetrizing, it is shown that the π driving force of a variety of π systems (C_6H_6, N_6, P_6, Si_6H_6, C_4H_4, N_4, Si_4H_4, P_4) is always distortive, and that the π propensity to distort is proportional to the strength of the 2e-π bond, as predicted by the Valence Bond model. Thus, the π system is forced by the σ framework to adopt a regular geometry rather than an alternated one of Kekulé type. The quantum mechanical resonance energy (QMRE), stabilization directly due to the mixing of Kekulé structures, is computed for all the π systems under study. This quantity is distinct from the thermochemical resonance energy (TRE, the usual "resonance energy"). The relationships between TRE, QMRE and π distortivity are examined. The existence of sizeable TREs in aromatic systems is shown not to be contradictory with the distortive propensity of all π electronic systems.

1 Introduction

One of the first successes of Hückel theory has been to rationalize, in simple terms, the special stability of benzene and the unstability of cyclobutadiene with respect to linear polyenes, and one knows that the same theory nicely accounts for the aromaticity vs antiaromaticity of cyclic conjugated molecules according as they bear $(4n + 2)$ or $4n$ π electrons, so that the properties of these hydrocarbons can be predicted by simple topology considerations. Now the dilemma is that the Hückel basis of aromaticity, which is so successful for hydrocarbons, cannot be generalized to other electronic systems sharing the same topologic properties, and thus can hardly be considered as a sound physical explanation of the aromaticity vs antiaromaticity phenomenon. Indeed, the H_6 hexagon is highly unstable [1] with respect to three isolated dihydrogen molecules. On the other hand, the Li_6 hexagon, though not being the most stable Li_6 complex, is more stable than the isolated lithium dimers [2] as is also the square lithium tetramer [3], in spite of its expected antiaromatic properties. Even if we restrict ourselves to π systems, the Hückel rule cannot be applied to second-row compounds like Si_4H_4 or P_4, whose square geometries are not stabilized by a localizing distortion to a rectangular structure [4]. The isoelectronic analogy with hydrocarbons is not more helpful for first-row compounds, since the hexaazabenzene N_6 has very weak, if any, stability, and distorts extremely easily to a Kekuléan D_{3h} structure [5].

It is clear that another theory, more general than the Hückel model, is needed to understand why electrons either tend to delocalize themselves or, on the contrary, tend to form separate localized bonds. We have recently proposed a Valence Bond model [6–8], in which the stability of an X_n complex (X = hydrogen, alkali, halogen, metal, etc.), with equal X-X bond lengths, is shown to correlate with the strength of the X-X bond in the dimer, or more specifically to its singlet-triplet energy gap. The model has been described in detail elsewhere [6] and will not be repeated here; it nicely rationalizes the tendency of weak binders, e.g. alkali atoms, to form stable aggregates while strong binders (H, F) cannot polymerize, and other halogens form marginally stable trimers (Br_3, Cl_3, I_3). Applying now the VB model to π bonds, one observes that the strength and the singlet-triplet gap of an isolated π bond are quite significant, thus putting π electronic systems in the range of unstable delocalized systems [7]. It follows that, considering π bonds only, the D_{6h} geometry of benzene with equal C-C bond lengths corresponds to a transition state between two distorted D_{3h} conformers of Kekulé type, with alternated bond lengths. In other words, the VB model predicts that the resistance of benzene to distortion towards a Kekuléan geometry is composed of two opposing driving forces, one coming from the π bonds, tending to distort the molecule, and a stronger one coming from the σ framework which imposes a D_{6h} geometry with equal bonds, and forces the π electrons to delocalize themselves. It should be noted that such a statement has already been put forward by several workers, on the basis of semiempirical theories [9], or by analyzing the experimental vibrational frequencies of benzene [10]. If this is true, then the π systems find their natural place in a general model for delocalization which also includes clusters of halogens or alkali atoms, and can be used for predicting properties of conjugated molecules containing heteroatoms. If not, one has to assume that the

VB model, though successful for the interactions between monovalent atoms, breaks down when π bonds are considered. The aim of this chapter is to bring a quantitative answer to a question which can be so summarized: What is the nature of the driving force which makes benzene more stable in a D_{6h} geometry than in an alternated D_{3h} geometry of Kekulé type? Exactly the same type of question applies to the allyl radical which will also be investigated and will allow the study of the effects of configuration interaction (CI) and basis set extension.

Another aspect of delocalization, the mixing of Kekulé structures and the energetic gain associated with it, will also be quantitatively investigated, and finally our results will be placed into the context of the thermochemical properties of hydrocarbons and the current definitions of resonance energy.

2 A Simple Computational Experiment

Any standard ab initio computational method is able to reproduce the resistance to distortion of benzene or allyl radical, even at a low level of theory. Thus, if one distorts benzene as in Fig. 1, by alternately compressing one bond by 0.06 Å and stretching the adjacent one by nearly the same quantity (the accurate distortion mode will be specified and justified below), the total energy of benzene goes up by 6.6 kcal/mol (6-31G/SCF), and 7.2 kcal/mol (6-31G + full π-CI) [6]. Similarly, the distortion energy of allyl radical is in the range 3.8–4.4 kcal/mol, depending on the computational level [6]. These values correspond to total energies, in which both σ and π components are included. Now there is a very simple way to estimate the σ component alone, by computing the distortion energy of a state in which the σ bonds still undergo the field of π electrons, the latter being relatively indifferent to a localizing distortion. In the case of an allyl radical, such a state exists; it is the quadruplet state in which the σ frame is left unchanged with respect to the ground state but the π electrons all have parallel spins. Even without accurate calculations, one can estimate the effect of a localizing distortion on the π electrons in this excited state, by analogy with a hydrogen chain having the same multiplicity and the same bond lengths as the C–C bonds in allyl. Thus, the energy of the quadruplet H_3 rises, upon distortion, by only 0.48 kcal/mol [11] at the level of 6-311G [12] + full CI, and this value constitutes a higher limit for the π interactions of quadruplet allyl, since two hydrogenoid atomic orbitals overlap more than two parallel *p* orbitals of carbon at the same interatomic distance. It follows that the high spin state of allyl can be used as a probe for estimating the σ component to the distortion energy of the ground state, as it represents a species in which the π component has been nearly eliminated.

The distortion energies of the quadruplet state of allyl are reported in Table 1 and compared with those of allyl's ground state, as computed in triple-zeta basis set with electron correlation in the π space or in the (σ + π) space. It first appears that the distortion energy of the high spin state is rather insensitive to electron correlation; but the striking result is that the quadruplet allyl resists the distortion *more* than the ground state, by some 0.6–1.0 kcal/mol. Even assuming that the π electrons with parallel spins resist the distortion by 0.5 kcal/mol like the analogous hydrogen chain, which certainly is an overestimation, it remains that the σ component of the distortion energy ranges from 4.3 to 4.5 kcal/mol, and is slightly larger than the total (σ + π)

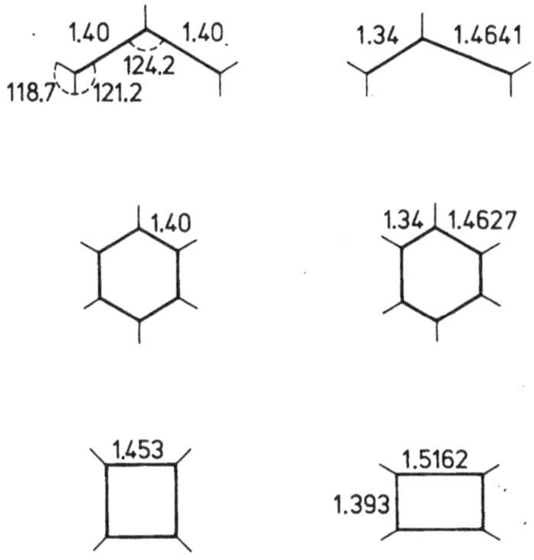

Fig. 1. Symmetric and distorted geometries for allyl radical, benzene, and singlet cyclobutadiene. The distortions keep constant nuclear repulsions between carbons. Bond angles are the same for the distorted and undistorted structures

distortion energy of allyl's ground state. The same computational experiment can be made with the analogous septuplet state of benzene, whose distortion energy is about three times that of quadruplet allyl, and is now much larger than the ground state's resistance to distortion (see Table 1). This means that the geometries of allyl and benzene are governed by two opposing driving forces: the σ bonds which favor a symmetric geometry with equal bond lengths, and the π bonds which favor a Kekuléan alternated geometry. The latter driving force is rather weak in the case of allyl, and stronger in benzene, but in both cases it is outweighed by the σ driving force and a symmetrical geometry results, not because but in spite of the π electrons.

The above computational experiment rests on two assumptions: (i) The interactions between π electrons in the high spin states are not more sensitive to the distortions than they are in the analogous hydrogen chains; (ii) the force constants of the σ bonds do not depend much on the way π electrons are coupled, and can be considered as similar in the ground states and in the high spin states. Both these assumptions will be accurately verified in the following section.

Table 1. Distortion energies (kcal/mol) of allyl radical and benzene in their ground states, and in their high spin states with all π electrons having parallel spins. The distortion transforms a regular geometry into an alternated one, as in Fig. 1

Species, Method	Ground State	High Spin
Allyl, 6-31G/SCF		4.8
Allyl, 6-311G/π-CI	3.8	4.8
Allyl, 6-311G/(π + σ)-CI	4.4	5.0
Benzene, 6-31G/SCF	6.6	14.5

3 Direct Separation of σ and π Driving Forces

The total energy, E, of a conjugated molecule can always be decomposed as in Eq. (1), which is routinely used in π electron calculations [13]:

$$E = \sum_{\pi}^{\mathrm{occ}} h_\pi + R_{\pi\sigma} + R_{\pi\pi} + \sum_{\sigma}^{\mathrm{occ}} h_\sigma + R_{\sigma\sigma} + V_{\mathrm{NN}} \tag{1}$$

Here, h_π and h_σ are the corresponding monoelectronic integrals of the occupied π- and σ-spinorbitals. The R terms stand for electron-electron repulsion of a type that is specified by the subscript, and V_{NN} accounts for nuclear repulsion.

The π electronic energy is readily defined: it is the energy of π electrons in the field of a σ frame constituted of the nuclei and the σ electrons, and involves the two first terms of Eq. (1) (attraction between π electrons and nuclei, partly compensated by repulsion between π electrons and σ electrons), and $R_{\pi\pi}$ (repulsion between π electrons). But the π electronic energy alone does not reflect the energy of a π bond, which is constituted of the interaction between two parallel p orbitals born by two effective positive charges repelling each other, and includes part of the V_{NN} term. To circumvent this difficulty, we have chosen to define the distortion mode as the one which keeps exactly constant the nuclear repulsion between carbon atoms (see Fig. 1). In this manner, the repulsion between positive centers is annihilated in the π component ΔE^π of the resistance to distortion, which can now be defined without ambiguity as

Table 2. Distortion energies (kcal/mol)[a] of allyl radical, benzene, and cyclobutadiene, and their π and σ components separated by means of Eq. (1). In the high spin states, all electrons have parallel spins

Species, Method	ΔE^{tot}	ΔE^π	ΔE^σ
I High spin states			
Allyl, 6-31G/SCF	4.9	0.1	4.8
Allyl, 6-311G/SCF	4.8	0.1	4.7
Allyl, 6-311G/π-CI	4.8	0.1	4.7
Allyl, 6-311G/(π + σ)-CI	5.0		
Benzene, 6-31G/SCF	14.5	0.8	13.7
Cyclobutadiene,6-31G/π-CI	7.7	0.1	7.6
II Ground states			
Allyl, 6-31G/π-CI	3.9	−0.9	4.8
Allyl, 6-311G/π-CI	3.8	−0.9	4.7
Allyl, 6-311G/(π + σ)-CI	4.4		
Benzene, 6-31G/SCF	6.6	−9.7	16.3
Benzene, 6-31G/π-CI	7.2	−9.1	16.3
Cyclobutadiene, 6-31G/π-CI[b]	−3.4	−10.4	7.0

a A negative value means a distortive species.

b The square cyclobutadiene is computed with the Davidson open shell hamiltonian and the distorted one with the standard closed shell hamiltonian. π-CI is then performed over the two species.

the variation of the π *electronic* energy throughout the distortion. The σ component, ΔE^{σ}, of the total distortion energy ΔE^{tot} is defined by Eq. (2):

$$\Delta E^{\text{tot}} = \Delta E^{\sigma} + \Delta E^{\pi} \qquad (2)$$

So defined, ΔE^{σ} is the distortion energy of a bare σ frame devoid of π electrons, whose σ orbitals would have been optimized in the field of the π electrons.

3.1 High Spin States

Let us first use Eq. (1) to estimate the resistance to distortion arising from π interactions in the high spin states. The results are displayed in Table 2, and show a remarkable constancy at the three levels of computations used in the case of allyl. Note that when CI is performed in the π space, the π component ΔE^{π} of the distortion energy is not any more given by the three first terms in Eq. (1), but one can still compute it as the difference between the total distortion energy and its σ component, calculated from the last three terms of Eq. (1). It turns out that ΔE^{π} in the quadruplet allyl is effectively smaller than our higher limit of 0.5 kcal/mol, and does not appear to be sensitive to CI or basis-set extension. Similar results are also observed in benzene and cyclobutadiene, in which the π systems of the high spin states also prove to be rather insensitive to the distortion.

3.2 Ground States

The σ-π partition in the ground states will be used for two aims: (i) to verify that ΔE^{σ} is rougly constant from high spin states to ground states, and (ii) to directly compute ΔE^{π} in the ground states whenever possible. Note that the SCF level is meaningless for the ground state of allyl, because of the doublet instability phenomenon [14], as well as for the ground state of cyclobutadiene, because the square and rectangular geometries respectively correspond to an open shell and a closed shell.

As postulated, the values of ΔE^{σ}, in the case of allyl, are the same in the high spin states and in the ground states: 4.7–4.8 kcal/mol. An additional indication that the σ frame is independent of the π system is provided by the distortion energy of the naked σ frame of allyl, i.e. the cation $(C_3H_5)^{3+}$ devoid of π electrons, which amounts to 4.8 kcal/mol. Projecting allyl's value of 2.4 kcal/mol per σ bond to the benzene σ frame, which involves six σ bonds, one would predict the value 14.1–14.4 kcal/mol for ΔE^{σ} which is calculated to be 13.7 kcal/mol in the septuplet state, vs 16.3 in the ground state and 14.1 in the $(C_6H_6)^{6+}$ cation, at the 6-31G/SCF level. Here the dependence of ΔE^{σ} on the status of σ electrons is not nil, yet it remains too small to alter our previous finding of an unstable π system, all the more as ΔE^{σ} is *larger* in the ground state than it is in the high spin state, indicating that the π system is even more distortive, in benzene, than previously predicted.

In the case of cyclobutadiene, there is a good constancy of ΔE^{σ}. On the other hand, the value of ΔE^{σ} per σ bond cannot be compared with those of allyl and benzene, because the equilibrium bond lengths, in our calculations, are 1.40 Å in the latter vs 1.453 Å in square cyclobutadiene.

The values of ΔE^{π}, as directly computed in the ground states via the σ-π partition of Eq. (1), confirm the previous conclusions arising from considerations of the high spin states: A localizing distortion leading to an alternated geometry of Kekulé type stabilizes the π bonds of benzene by 9.1–9.7 kcal/mol, and those of allyl radical by 0.9 kcal/mol.

An important point to note is that the π system of cyclobutadiene is the most distortive of the three, especially on a per-electron basis. This reflects the well-accepted idea that four-center resonance is less stabilizing than six-center one, a point that will be developed in detail below.

3.3 Effect of σ-Electron Correlation

Do the above results depend on the computational level? To answer this question we computed the distortion energy of both quadruplet and ground states of allyl radical at the level of triple-zeta basis set and correlation of π and σ electrons, by the CIPSI algorithm [15], more than 6×10^6 Slater determinants being included in the CI/MP2 treatment. At this level, the σ-π partition cannot be performed, but one may observe that the resistance to distortion is still larger in the quadruplet than in the ground state, 5.0 kcal/mol vs 4.4 kcal/mol. Assuming the value 0.1 kcal/mol for ΔE^{π} in the quadruplet state, one deduces the value 4.9 kcal/mol for ΔE^{σ}. If one now neglects the dependency of ΔE^{σ} on the type of coupling between the π electrons, one gets the value -0.5 kcal/mol for ΔE^{π}. If, on the other hand, one assumes that ΔE^{σ} varies from the high spin state to the ground state as it does in benzene, one may estimate ΔE^{σ} to be 5.8 kcal/mol, and thus ΔE^{π} to be -1.4 kcal/mol. Therefore, the high computational level confirms the previous conclusions of the respective roles of σ and π driving forces in allyl. One may note anyway that the properties of conjugated hydrocarbons, for example benzene, are well reproduced even with small basis sets and without π-CI [6].

4 Distortive π-Propensity and Resonance Energy

The above computational results show that the π systems of benzene, allyl and cyclobutadiene are stabilized in an alternated geometry of Kekulé type, with respect to fully delocalized systems in symmetrical geometry. This finding may been to be contradictory with the well-established stability of benzene with respect to linear polyenes, or with the known rotational barrier of allyl. What then is the role of π-electron delocalization in the stability of benzene and allyl and in the relative unstability of cyclobutadiene? To answer this question, one must investigate in detail the phenomenon of resonance energy and, first of all, clarify the distinction between three concepts: (i) the distortive propensity of π-electronic systems; (ii) the stabilization resulting from the mixing of several Kekulé structures, *at a given geometry*. Following Coulson [16], we will call this quantity "Quantum Mechanical Resonance Energy" (QMRE). (iii) The modern definition [17, 18] of the resonance energy, aiming at reproducing the experimental stability cyclic polyenes with respect to linear ones. Hereafter, we will call this quantity "Thermochemical Resonance Energy" (TRE).

4.1 Quantum Mechanical Resonance Energy

Kollmar [19] has developed a procedure to estimate, at the SCF level, the energy of one Kekulé structure of a cyclic polyene. Thus, the QMRE is simply the difference between the latter energy and the energy of the molecule as calculated by the regular SCF method, which allows the electrons to delocalize themselves.

The QMREs of a number of four- and six-membered cyclic polyenes, involving carbons as well as heteroatoms, are collected in Table 3. Several points are noteworthy: (i) all QMREs are positive and significant, even in the four-membered rings.
This reflects the obvious fact that the mixing of two Kekulé structures, *at a given fixed geometry,* is always stabilizing. Thus, although QMRE and TRE are occasionally confused with each other in the literature, these concepts are entirely different since the TRE can be negative as, e.g., in cyclobutadiene. (ii) The QMRE of distorted benzene, i.e. benzene in a geometry of Kekulé type, is lower than that of hexagonal benzene, by 31.7 kcal/mol. Therefore, the QMRE is a driving force which *resists* the localizing distortion and tends to impose the symmetric geometry. (iii) The QMREs of four-membered rings are smaller than their six-membered analogs, even on a per-electron basis. This is the root of the Hückel rule, and is in agreement with the VB theory of aromaticity [20], showing that six-center interactions are more stabilizing than four-center ones.

Now how can these large QMREs coexist with the distortive propensities of the π system? This becomes clear by considering in detail how a π system is affected by the geometric transformation of an alternated geometry to a regular one, and can be illustrated with the example of benzene. The π energy can be considered as the π

Table 3. Computed distortion energies[a] with their π and σ components, QMRE values for a series of annulenes, and singlet-triplet gaps ΔE_{ST} of the corresponding 2e-2c π bonds. The ΔE^{σ} values arise from the direct σ-π partition of Eq. (1) in the ground states of the annulenes. The total distortion energies ΔE^{tot} are all calculated at the 6-31G/π-CI level. All energies are in kcal/mol. For all species, the distortion keeps constant nuclear repulsion between the atoms of the ring. The geometries of hetero-annulenes are described in Ref. [4]

Species	ΔE^{tot}	ΔE^{π}	ΔE^{σ}	QMRE	QMRE per e$^-$	ΔE_{ST}[b]
N_6	0.4	−13.3	13.7	102.5	17.1	108
C_6H_6	7.2	−9.1	16.3	85.2	14.2	99
DB[d]				53.5	8.9	
P_6	1.0	−2.8	3.8	44.1	7.4	47[c]
Si_6H_6	3.2	−2.1	5.3	41.6	6.9	42
N_4	−5.5	−14.7	9.2	44.6	11.2	108
C_4H_4	−3.4	−11.0	7.6	30.2	7.6	99
P_4	0.4	−2.6	2.9	24.8	6.2	47[c]
Si_4H_4	0.6	−2.1	2.7	17.7	4.4	42

a A negative value means a distortive species.
b These are 3-21G values for model dimers (e.g. C_2H_4, N_2H_2, etc.).
c This value is calculated with a 3-21G* basis set.
d *DB* stands for Distorted Benzene

energy E_K^π of a single Kekulé structure, lowered by the QMRE due to mixing with the other one, as in Eq. (3), where the variation of π

$$\Delta E^\pi = \Delta E_K^\pi + \Delta \text{QMRE} \tag{3}$$

energy upon distortion is decomposed. Going from the Kekuléan geometry of the distorted benzene (*DB*) to the symmetric one inevitably raises the energy of the most stable Kekulé structure, by an amount $\Delta E_K^\pi = 41.4$ kcal/mol [21], at the 6-31G/SCF level.

$$DB \xrightarrow[\substack{\Delta E_K^\pi = 41.4\,\text{kcal/mol} \\ \Delta\text{QMRE} = 31.7\,\text{kcal/mol}}]{} \text{Benzene} \tag{4}$$

This is *only partly* compensated by a gain of 31.7 kcal/mol in QMRE. So the net result is that the π system is more stable in the alternated geometry, despite the stabilizing effect of electronic delocalization.

4.2 The Origin of the Thermochemical Resonance Energy

In modern definitions of the resonance energy [18, 19], the linear polyenes are assumed to have zero resonance energy, as their π-energies are nearly additive. Their QMREs nearly compensate for the repulsive interactions between ethylenic units, which amount to about 9.5 kcal/mol in butadiene, twice as much in hexatriene, and so on. Thus, the TRE of benzene is calculated by comparing its energy with that of a reference state *R*, an hypothetical cyclohexatriene with just enough QMRE to compensate for three π-bond-π-bond interactions, i.e. 28.5 kcal/mol. Such a reference state has a geometry very close to that of our distorted benzene, but its QMRE is smaller, by 25.0 kcal/mol (see Table 3). Thus, we can decompose the TRE of benzene by a thermodynamic cycle as in Eq. (5):

$$\text{Benzene} \xrightarrow[\substack{\Delta E^\sigma = 16.3\,\text{kcal/mol} \\ \Delta E^\pi = -9.7\,\text{kcal/mol}}]{} DB \xrightarrow[\Delta\text{QMRE} = -25.0\,\text{kcal/mol}]{} R \tag{5}$$

On going from benzene to *DB*, the σ frame is destabilized by 16.3 kcal/mol while the π system is stabilized by 9.7 kcal/mol. Now *DB* has still some significant QMRE and for this reason is more stable than *R*, by 25.0 kcal/mol. The final balance of Eq. (4) yields a TRE of 31.6 kcal/mol, in excellent agreement with other estimations by Hess and Schaad [18c]. This thermodynamic cycle shows that: (i) a sizeable TRE can coexist with the distortive propensity of the π bonds, and (ii) that the QMRE is the root cause for the TRE, the usual "resonance energy". By a similar reasoning [6], one can show that the distortive, or at the very least indifferent π system of allyl can coexist with the rotational barrier.

4.3 The Relationship Between TRE and QMRE

The relationship between TRE and QMRE can be illustrated by another thermodynamic cycle,

$$\text{Benzene} \xrightarrow[\substack{\Delta\text{QMRE} = -31.7\,\text{kcal/mol} \\ \Delta E_K^{tot} = -25.1\,\text{kcal/mol}}]{} DB \xrightarrow[\Delta\text{QMRE} = -25.0\,\text{kcal/mol}]{} R \tag{6}$$

where ΔE_K^{tot} is the *total* ($\sigma + \pi$) energy variation of the lowest Kekulé structure from benzene to *DB*. Distorting benzene to *DB* stabilizes the π bonds of the Kekulé structure by 41.1 kcal/mol and destabilizes its σ frame by 16.3 kcal/mol, so that on the whole the Kekulé structure is stabilized by 25.1 kcal/mol. At the same time the QMRE drops by 31.7 kcal/mol from benzene to *DB*, and by 25.0 kcal from *DB* to *R*. The net balance of the reaction is, of course, 31.6 kcal/mol. It thus appears that the TRE is the sum of two terms, as in Eq. (7), where QMRE(*B*) and QMRE(*R*) are the QMREs of, respectively, benzene and the reference state.

$$\mathrm{TRE} = \Delta E_K^{tot} - \Delta \mathrm{QMRE}_{(B \to R)} \quad (7a)$$

$$\Delta \mathrm{QMRE}_{(B \to R)} = \mathrm{QMRE}(R) - \mathrm{QMRE}(B) \quad (7b)$$

Benzene is aromatic because the QMRE term is the largest one in the right-hand member of Eq. (7a). On the other hand, cyclobutadiene is antiaromatic because now the ΔE_K term is the largest, partly because the QMRE is always weak in four-center interactions (Hückel rule), and partly because the repulsion between two π bonds facing each other at short distances is huge. Note that this last factor must be less important in square P_4 and Si_4H_4 than it is in cyclobutadiene, because of the longer bond lengths, and as a consequence these species do not tend to distort to rectangular structures (see Table 3).

5 Heteroannulenes

To test the generality of the π-distortivity phenomenon and of the Valence Bond model for delocalization, it is of interest to apply the σ-π partition to conjugated molecules other than hydrocarbons, e.g. containing nitrogen, silicon or phosphorus atoms that we have kept in a constrained planar geometry. The total distortion energies, as well as their σ and π components are displayed in Table 3, as calculated at the 6-31G/π-CI level.

5.1 Six-Membered Rings

The qualitative VB model predicts that the π distortive propensity varies like the strength of the two-electron-two-center (2e-2c) π bond, or its singlet-triplet energy gap ΔE_{ST}. This latter value increases from column V to column VI in the periodic table, and decreases from first row to second row. In accord with the VB model, the calculated ΔE^{π} values follow the same tendency. Now the σ driving force also decreases from first row to second row, and from fifth to sixth column. As a consequence, the π distortivity of N_6 is larger than that of benzene while its σ driving force is weaker and opposes little resistance to distortion. The result is that hexagonal N_6 is rather indifferent to distortion, despite its isoelectronic analogy with benzene.

The same phenomenon can be observed in the second row, where now both σ and π driving forces are smaller than in the first row: the total resistance of Si_6H_6 to distortion is smaller than that of benzene, but larger than that of P_6.

5.2 Four-Membered Rings

Again the π-distortivities follow the variations of the ΔE_{ST} values of the 2e-2c π bonds. On the other hand, the σ driving forces of four-membered rings increase from column V to column VI, unlike six-membered rings. As a consequence, the tendency of N_4 to distort to a rectangular structure is not very much larger than that of cyclobutadiene.

Both σ and π bonds are weaker in the second row than in the first one, and this fact alone would lead to the prediction that Si_4H_4 and P_4 have a lesser tendency to distort than their first-row analogs. In addition, because of the rather long bond lengths, the overlap repulsions between the π bonds in the Kekulé structures of square Si_4H_4 and P_4 must not be very large. The striking result is that these two species are nearly indifferent to distortion, despite their expected antiaromatic character.

6 Conclusion

The alternated vs regular geometry of six-membered, four-membered and linear polyenes is the result of a competition between two opposing driving forces: The σ bonds, which always tend to impose equal bond lengths, and the π bonds, which tend to impose an alternated geometry in all the examples investigated here. The distortive propensity of the π systems is shown to vary like the strength of the corresponding 2e–2c π bond, or equivalently like its singlet-triplet energy gap. This property is not contradictory with the known stability of, e.g., benzene, or the unstability of cyclobutadiene (aromaticity vs antiaromaticity) with respect to open polyenes, nor with allyl's rotational barrier.

The electronic delocalization, *at a given fixed geometry,* always brings some stabilization (QMRE) which acts in favor of a rectangular geometry. In six-membered rings, the QMRE is not sufficient for preventing the π system to be distortive, but it weakens the π driving force enough to allow the σ one to win, and a D_{6h} geometry results. Thus, *the π system is forced by the σ frame to adopt a regular hexagonal geometry.* In four-membered rings and open polyenes of the first row, the QMRE is rather weak, either as a consequence of the Hückel rule, or because only one Kekulé structure is available. It results that the π driving force wins over the σ one and imposes an alternated geometry. An additional driving force for distortion is the exchange repulsion energy between the two π bonds of a Kekulé structure and is rather important in C_4H_4 and N_4 but is expected to be weaker in their second row alongs, which are indifferent to distortion.

Thus, the π systems of hydrocarbons and their analogs appear to be similar in nature to aggregates of monovalent atoms like hydrogen, alkali, halogens, etc., and they find their place into a general model for delocalization based on Valence Bond diagrams, whose key parameter is the singlet-triplet gap, or nearly equivalently the strength of the bond, in the related dimer. In this family, the π systems of hydrocarbons are in the range of fairly strong bonds, and in that sense the analogy with hydrogen chains, already suggested by other [1c, 9c], is entirely sound.

7 Acknowledgements

The author is grateful to Professor S. Shaik and to Drs G. Ohanessian, J.-M. Lefour and J.-P. Flament for their collaboration in the previous articles from which the present work is derived.

8 References and Notes

1. (a) Dixon DA, Stevens RM, Herschbach DR (1977) Faraday Discuss. Chem. Soc. 62: 110; (b) Haddon RC, Raghavachari K, Whangbo MH (1984) J. Amer. Chem. Soc. 106: 5364; (c) Ichikawa H (1984) J. Amer. Chem. Soc. 106: 6249; (1983) 105: 7467
2. (a) Plavsic D, Koutecky J, Pacchioni G, Bonacic-Koutecky V (1983) J. Phys. Chem. 87: 1096; (b) Pickup BT (1973) Proc. Roy. Soc. A333: 69
3. (a) Wu CH (1983) J. Phys. Chem. 87: 1534; (b) Beckmann HO, Koutecky J, Bonacic-Koutecky V (1980) J. Chem. Phys. 73: 5182
4. Ohanessian G, Hiberty PC, Lefour J-M, Flament J-P, Shaik S S (1988) Inorg. Chem. 27: 2219
5. Saxe P, Schaefer HF (1983) J. Amer. Chem. Soc. 105: 1760
6. Shaik SS, Hiberty PC, Lefour J-M, Ohanessian G (1987) J. Amer. Chem. Soc. 109: 363
7. Shaik SS, Hiberty PC, Ohanessian G, Lefour J-M (1988) J. Phys. Chem. 92: 5086
8. Shaik SS, Hiberty PC (1985) J. Amer. Chem. Soc. 107: 3089
9. (a) Longuet-Higgins HC, Salem L (1959) Proc. Roy. Soc., A251: 172; (b) Salem L (1972) The Molecular Theory of Conjugated Systems, Benjamin, Reading, MA, p 103–106, 494–505; (c) Paldus J, Chin E (1983) Int. J. Quant. Chem. 24: 373; (d) Epiotis ND (1984) Nouv. J. Chim. 8: 11; (e) For a discussion, see Ref [6]
10. Berry RS (1961) J. Chem. Phys. 35: 29, 2253
11. Shaik SS, Hiberty PC, Ohanessian G, Lefour J-M (1985) Nouv. J. Chim. 9: 385
12. This basis set is of triple-zeta quality, and specially devised for calculation of correlation energies. See: Krishnan R, Binkley J S, Seeger R, Pople J A (1980) J. Chem. Phys. 72: 650
13. Bishop DM (1973) Group Theory and Chemistry, Clarendon, Oxford, p 204
14. McKelvey JM, Berthier G (1976) Chem. Phys. Lett. 41: 476
15. (a) Huron B, Malrieu J-P, Rancurel P (1973) J. Chem. Phys. 58: 5745; (b) Malrieu J-P (1982) Theor. Chim. Acta 62: 163; (c) Evangelisti S, Daudey J-P, Malrieu J-P (1983) Chem. Phys. 75: 91
16. Coulson CA, Altman S L (1952) Trans. Faraday Soc. 48: 293
17. (a) Dewar MJS, de Llano C (1969) J. Amer. Chem. Soc. 91: 789; (b) Dewar MJS, Gleicher GJ (1965) J. Amer. Chem. Soc. 87: 685, 692
18. For recent papers on the thermochemical stability of benzene and instability of cyclobutadiene, see (a) Haddon RC (1979) J. Amer. Chem. Soc. 101: 1722; (b) Haddon RC, Raghavachari K (1985) J. Amer. Chem. Soc. 107: 28; (c) Hess BA, Schaad LJ (1983) J. Amer. Chem. Soc. 105: 7500
19. Kollmar H (1979) J. Amer. Chem. Soc. 101: 4832
20. (a) van der Hart WJ, Mulder JJC, Oosterhoff LJ (1972) J. Amer. Chem. Soc. 94: 5724; (b) Malrieu J-P, Maynau D (1982) J. Amer. Chem. Soc. 104: 3021; (c) Kuwajima S (1984) J. Amer. Chem. Soc. 106: 6496
21. This value is simply estimated by applying Eq. (3)

The Spin-Coupled Valence Bond Description of Benzenoid Aromatic Molecules

David L. Cooper[1], **Joseph Gerratt**[2] **and Mario Raimondi**[3]

[1] Department of Chemistry, University of Liverpool, P.O.Box 147, Liverpool L69 3BX, U.K.
[2] Department of Theoretical Chemistry, University of Bristol, Cantocks Close, Bristol BS8 1TS, U.K.
[3] Dipartimento di Chimica Fisica ed Elettrochimica, Università di Milano, Via Golgi 19, I-20133 Milano, Italy

Table of Contents

Clear theoretical evidence is presented to show that the π-electron systems of benzenoid aromatic molecules are described well in terms of localized, non-orthogonal, singly-occupied orbitals. The characteristic properties of molecules such as benzene or naphthalene arise from a profoundly quantum mechanical phenomenon, namely the mode of coupling of the electron spins, rather than from any supposed delocalization of the orbitals. Other systems considered include azobenzenes, such as pyridine, five-membered rings, such as furan, and inorganic heterocycles, such as borazine ("inorganic benzene").

This description is furnished by spin-coupled theory, which represents the modern development of valence bond theory. The approach incorporates from the outset the chemically most important effects of electron correlation, but it retains a simple, clear-cut visuality.

1 Introduction

Most organic chemists are familiar with two very different and conflicting descriptions of the π-electron system in benzene: molecular orbital (MO) theory with delocalized orthogonal orbitals and valence bond (VB) theory with resonance between various canonical structures. An attitude fostered by many text books, especially at the undergraduate level, is that the VB description is much easier to understand and simpler to use, but that MO theory is in some sense "more fundamental".

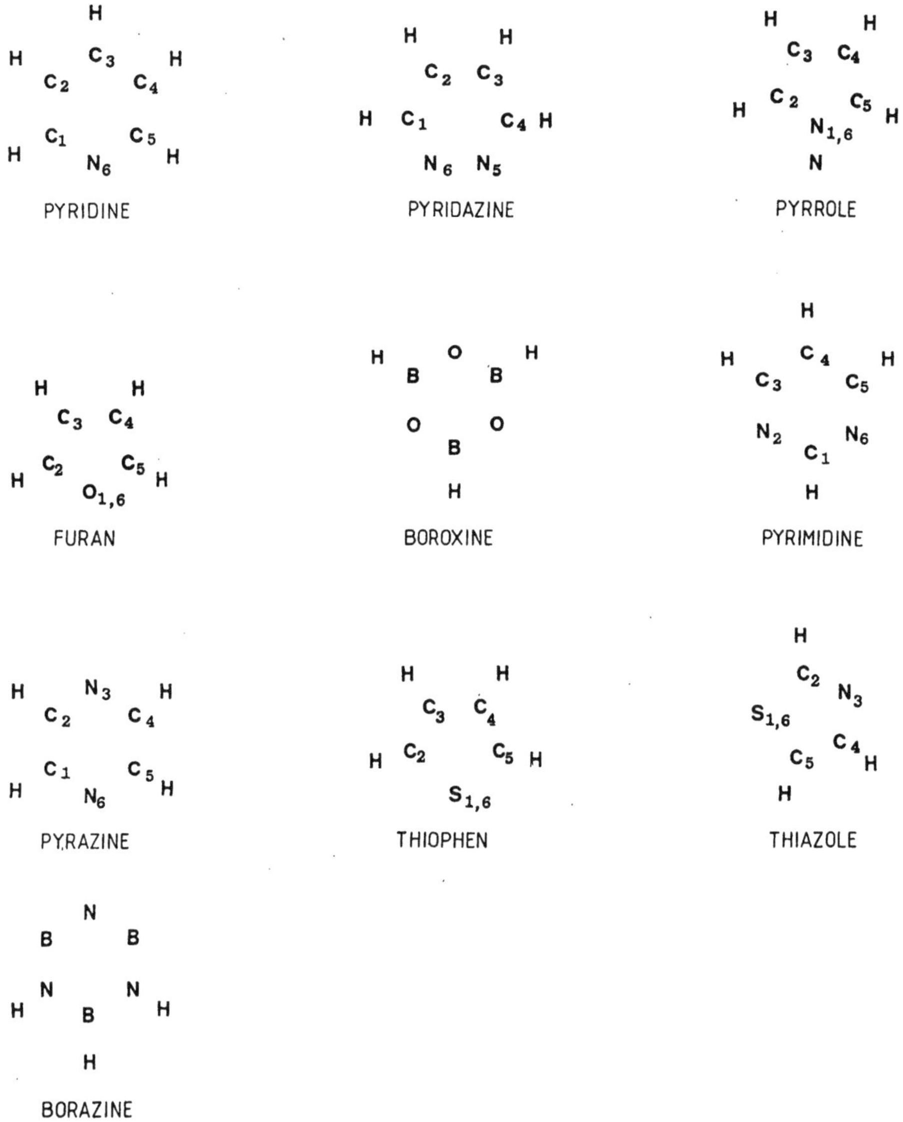

Fig. 1. Structural formulae of heterocyclic molecules, drawn approximately to scale

The purpose of this review is to discuss the main conclusions for the electronic structure of benzenoid aromatic molecules of an approach which is much more general than either MO theory or classical VB theory. In particular, we describe some of the clear theoretical evidence which shows that the π electrons in such molecules are described well in terms of localized, non-orthogonal, singly-occupied orbitals. The characteristic properties of molecules such as benzene arise from a profoundly quantum mechanical phenomenon, namely the mode of coupling of the spins of the π electrons. This simple picture is furnished by spin-coupled theory, which incorporates from the start the most significant effects of electron correlation, but which retains a simple, clear-cut visuality. The spin-coupled representation of these systems is, to all intents and purposes, unaltered by the inclusion of additional electron correlation into the wavefunction.

It is not appropriate here to present a detailed account of the formalism of spin-coupled valence bond theory, and of its computational implementation. Instead, we provide a brief, almost entirely qualitative overview of our method, concentrating on the most interesting results for benzenoid aromatic molecules. Further details may be found in the literature cited and in various recent reviews [1, 2, 3].

The spin-coupled method has now been applied to a large number of aromatic systems: benzene and naphthalene; azobenzenes, such as pyridine, pyridazine, pyrimidine and pyrazine; five-membered rings, such as furan, pyrrole, thiophen, and thiazole; and inorganic heterocycles, such as borazine ("inorganic benzene") and boroxine, for which we find little evidence of aromaticity. Structural formulae are collected in Fig. 1. For all of these molecules we have included the effects of electron correlation for the π electrons but not for the σ framework. This σ–π separation is an approximation whose utility rests upon the chemistry of aromatic systems – to abandon it would be to ignore this entire body of experience. Furthermore, very extensive calculations [4] have demonstrated that 'π-electron only' correlation affords an excellent description of ground and excited states of benzene.

The π orbitals lie to a considerable extent within the space of the σ orbitals and not well outside it, as is commonly assumed. This would seem to bring into question the fundamental concept of σ–π separation. However, the π orbitals are a great deal more *polarizable* than the σ orbitals, and consequently, to a good approximation, they provide almost all of the response of the system to chemical and other influences. As a result, much of the chemistry of aromatic systems can be understood on the basis of the properties of the π electrons alone.

2 Spin-Coupled Theory

We start with a description of the spin-coupled wavefunction for the general case in which electron correlation is included for all of the electrons. We return later to the question of σ–π separation. In the spin-coupled approach to molecular electronic structure, an N-electron system is described by N orbitals, all of which are allowed to be distinct and non-orthogonal. One consequence of the non-orthogonality of these singly-occupied orbitals is that there is usually more than one way of coupling together the spins of the individual electrons so as to achieve the required overall resultant S. In general, it is important to include all of the modes of spin coupling, especially when considering the making or breaking of chemical bonds.

The spin-coupled wavefunction may be written in the form [5]

$$\Psi_{SM} = \sum_{k=1}^{f_S^N} c_{Sk} \mathscr{A}(\varphi_1 \varphi_2 \ldots \varphi_N \Theta_{SM;k}^N) \tag{1}$$

where the c_{Sk} are called spin-coupling coefficients, $\mathscr{A}$ is the antisymmetrizer, $\varphi_\mu(r_\mu)$ is one of the spatial orbitals, and $\Theta_{SM;k}^N$ is an N-electron spin function. The spin functions (and thus the total wavefunction) are eigenfunctions of $\hat{S}^2$ and of $\hat{S}_z$, with eigenvalues $S(S+1)\hbar^2$ and $M\hbar$, respectively. The particular mode of coupling the electron spins is labeled by the index k, where the total number of allowed couplings for each value of M is

$$f_S^N = \frac{(2S+1)\,N!}{(\frac{1}{2}N + S + 1)!\,(\frac{1}{2}N - S)!} \tag{2}$$

For a system consisting of six electrons with a net spin of zero, $f_S^N = 5$. For our present purposes, it is probably most useful to consider the modes of spin coupling in terms of the traditional basis of Rumer functions used in classical VB theory. For a discussion of different spin functions, and of the relationships between them, see for example the book by Pauncz [6]. The (complete) set of Rumer functions may be visualized easily in terms of the diagrams in Fig. 2, where an arrow $i \rightarrow j$ signifies

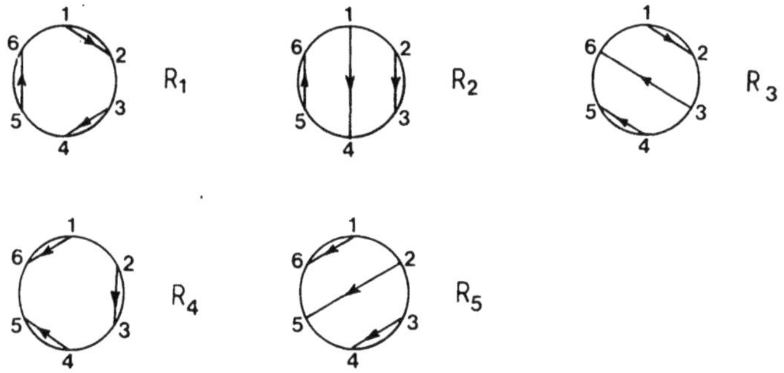

Fig. 2. Rumer diagrams for $N = 6$ and $S = 0$

that the spins of electrons i and j are singlet coupled, and this corresponds to the occurrence in the spin function of a factor of

$$(1/\sqrt{2})\,(\alpha(i)\,\beta(j) - \beta(i)\,\alpha(j)) \tag{3}$$

The orbitals are expanded in a basis set of atom-centered basis functions χ_p in the form

$$\varphi_\mu = \sum_p c_{\mu p} \chi_p \tag{4}$$

much as in MO theory. The coefficients $c_{\mu p}$ and the spin-coupling coefficients c_{Sk} are all optimized simultaneously so as to minimize the total energy. It is important to note that the spin-coupled wavefunction is not invariant to arbitrary linear transformations of the orbitals, whose form is a unique outcome of the variational procedure. Furthermore, the orbitals are allowed to be as localized or as delocalized as is necessary to achieve the lowest energy, without constraints on the overlaps between them.

For the benzenoid aromatic molecules considered here, we wish to include electron correlation for the N electrons of the π-electron system but not for the $2n$ σ electrons. This is achieved most readily if we first carry out a standard Hartree-Fock MO theory calculation, yielding a set of orthogonal self-consistent field (SCF) molecular orbitals ψ_i. The spin-coupled wavefunction may then be written

$$\Psi_{SM} = \sum_{k=1}^{f_S^N} c_{Sk} \mathscr{A}(\psi_1^2 \psi_2^2 \ldots \psi_n^2 \varphi_1 \varphi_2 \ldots \varphi^N \Theta_{0,0;f}^{2n} \Theta_{S,M;k}^N) \tag{5}$$

in which the spin function $\Theta_{0,0;f}^{2n}$ corresponds to singlet coupling of the two electrons in each of the n doubly-occupied 'core' orbitals. The N spin-coupled orbitals φ_μ are then optimized as completely general linear combinations of all the occupied and unoccupied (or 'virtual') SCF MOs of π symmetry. This is strictly equivalent to expanding the spin-coupled orbitals in terms of all the atomic basis functions of π symmetry. More sophisticated calculations have also been carried out in which the doubly-occupied σ orbitals are allowed to relax, in a self-consistent manner, in the field of the correlated π electrons.

3 Benzene [7]

For the equilibrium geometry of benzene (D_{6h} symmetry) we used bond lengths C–C 1.3964 Å and C–H 1.0831 Å, taken from vibration-rotation spectroscopy [8]. The basis set, consisting of (10*s* 5*p*/4*s*) Gaussian functions contracted to [3*s* 2*p*/2*s*], was taken directly from the tables of Dunning and Hay [9], without any scaling of the hydrogen exponents. A slightly different geometry was used in a preliminary account of this work [10].

Spin-coupled calculations were then performed for the six π electrons, with the thirty-six electrons of the σ framework accommodated in doubly-occupied MOs. The form of the spin-coupled wavefunction

$$\Psi_{00} = \sum_{k=1}^{5} c_{0k} \mathscr{A}(\sigma_1^2 \sigma_2^2 \ldots \sigma_{18}^2 \varphi_1 \varphi_2 \varphi_3 \varphi_4 \varphi_5 \varphi_6 \Theta_{0,0;f}^{36} \Theta_{0,0;k}^6) \tag{6}$$

encompasses a very wide range of possible solutions with varying degrees of delocalization. The SCF solution is just one such possibility.

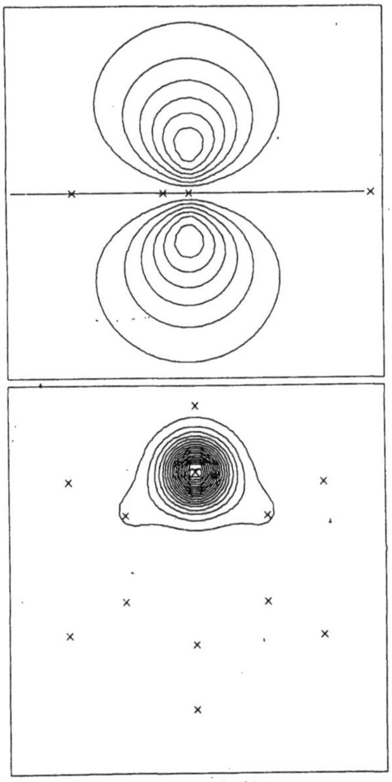

a

b

Fig. 3. Spin-coupled orbital φ_1 in benzene – throughout this review we present contour plots of the square modulus of the orbitals, $|\varphi_\mu(r)|^2$. Five other orbitals are related by symmetry to this one. **(a)** at one of the C atoms in a σ_v mirror plane, **(b)** in the plane 1 bohr above the σ_h molecular plane (the projections of the nuclear coordinates onto this plane are denoted with crosses)

We find that the converged spin-coupled wavefunction for benzene consists of six distinct, singly-occupied, highly localized, non-orthogonal orbitals. They all possess an identical energy and shape, and they can be transformed into one another by successive $\hat{C}_6$ rotations. One of the orbitals is depicted in Fig. 3 as contour plots of $|\varphi_1|^2$ in a σ_v mirror plane (at a C atom) and in the plane one unit of atomic distance (1 bohr, $\approx 5.3 \times 10^{-11}$ m) above the σ_h molecular plane. Each of the orbitals closely resembles a C(2p) function, pointing perpendicular to the ring, but is distorted symmetrically towards the carbon atoms on either side.

It is important to bear in mind that each spin-coupled orbital contains just one electron. Thus, for example, if orbital φ_1 is localized about nucleus A and φ_2 is localized about nucleus B, then this is equivalent to saying that the electrons tend to "keep out of each other's way": their Coulombic repulsions are reduced, and so the state becomes considerably more stable.

The spin-coupled wavefunction provides an improvement over the SCF energy of 199 kJ mol^{-1} (0.0758 hartree), which is considerable. This arises from the effects of electron correlation in the π-electron system. The distortion of the spin-coupled orbitals for benzene occurs because of the small amount of ionic character needed to describe the C–C π bonds. Ionic structures in spin-coupled theory are those in which one or more of the orbitals is allowed to be doubly occupied. Allowing for the different numbers of allowed spin functions, and including the spin-coupled configuration, a total of 175 VB structures can be generated in this way. A non-orthogonal CI calculation using these 175 structures (a "full-valence VB" calculation) gives a further energy lowering of 18 kJ mol^{-1}. The relative magnitude of this figure suggests that the small distortion effects in the spin-coupled orbitals, which are no larger than those we have seen for the C−C π bonds in alkenes, already take into account the most important of the ionic effects. The same would not be true in classical VB theory, where ionic structures are essential even to match the SCF energy. If the full-valence VB or spin-coupled wavefunctions were projected onto a basis of VB structures constructed from strictly localized orbitals, then the many contributions from ionic structures would reappear [11]. The spin-coupled configuration is overwhelmingly dominant in the full-valence VB wavefunction and so we can claim with confidence that the inclusion of spin-coupled ionic structures does not change the essential physical picture. One of the main conceptual objections to a VB-type of description is thus removed, namely the apparent importance of ionic structures even in obviously covalent situations.

The relative importance of the different modes of spin coupling may be examined most easily in the Rumer basis described above. For benzene, with the orbitals ordered φ_1 to φ_6 around the ring, we find that the spin couplings corresponding to Kekulé structures (R_1 and R_4) contribute 40% each and that those corresponding to *para*-bonded or Dewar structures (R_2, R_3 and R_5) contribute $6\frac{1}{2}$% each. These values are very close to those given many years ago by Pauling [12] in his original, and much simplified, classical VB calculation, and discussed by Coulson [13] in his text book.

The difference in energy between one of the Kekulé-type structures (R_1) and the full spin-coupled wavefunction, with all five Rumer structures, can reasonably be termed the resonance energy. In this way, we obtain a value of 88 kJ mol^{-1}. If we allow also for the spin-coupled ionic structures, this value is increased to 106 kJ mol^{-1}.

A second series of calculations for benzene was rather more elaborate. A much larger basis set of triple-zeta plus polarization (TZVP) quality was used, consisting of (10*s* 6*p*/5*s*) primitive Gaussians contracted to [5*s* 3*p*/3*s*] [14], and augmented with polarization functions with exponents $p_H = 1.0$ and $d_C = 0.72$ (six components). We first carried out a complete active space multiconfiguration SCF (CASSCF) calculation which included all 175 configurations which may be generated by distributing the six 'active' electrons in six orthogonal MOs. Both the form of the (orthogonal) orbitals and the weights of the different configurations were fully optimized. The final set of six active π orbitals is that which gives the lowest energy for this list of 175 configurations, where the doubly-occupied σ orbitals for the 'inactive' electrons are allowed to 'relax', in a self-consistent manner, in the field of the improved description of the π electrons. Spin-coupled calculations were then carried out with the electrons of the σ framework accommodated in the doubly-occupied 'inactive' CASSCF orbitals. As before, the spin-coupled calculations involved the six π electrons, which were expanded in all the basis functions of π symmetry.

For a system with N electrons, a spin-coupled calculation followed by a full-valence VB calculation produces a total wavefunction which is formally equivalent to a CASSCF description with all allowed distribution of N electrons in N orthogonal orbitals. Robb and co-workers [15] have shown how in principle one might transform between these two descriptions. The interpretation of the CASSCF wavefunction for benzene is not straightforward, as many of the 175 configurations make significant contributions. Furthermore, the natural orbital occupation numbers are not very meaningful. Although our full-valence VB wavefunction has the same number of structures and is formally identical, it is overwhelmingly dominated by the spin-coupled wavefunction, so that it is very easy to interpret.

Table 1. Overlap integrals between the spin-coupled orbitals for benzene

	φ_1	φ_2	φ_3	φ_4	φ_5	φ_6
φ_1	1	0.52	0.03	−0.16	0.03	0.52
φ_2		1	0.52	0.03	−0.16	0.03
φ_3			1	0.52	0.03	−0.16
ψ_4				1	0.52	0.03
φ_5					1	0.52
φ_6						1

We found that these more sophisticated spin-coupled calculations, which used larger basis sets with polarization functions on all of the atoms and which allowed the σ orbitals to relax, produced a picture of bonding in the π-electron system of benzene which is practically identical to that described earlier. As before, we found six equivalent spin-coupled orbitals which are transformed into one another by successive $\hat{C}_6$ rotations. The overlaps between the orbitals, ordered φ_1 to φ_6 around the ring, are reported in Table 1. In this case, the electron correlation effects incorporated in the spin-coupled model provide an energy improvement over the SCF description of 170 kJ mol^{-1}, with a further lowering of 20 kJ mol^{-1} on including spin-coupled ionic structures.

The Kekulé description of benzene, as expressed in the classical VB form, appears to be much closer to reality than is a description in terms of delocalized molecular orbitals, and it provides a clear picture of the behavior of *correlated* electrons in this molecule. The special properties of benzene arise fundamentally from the mode of coupling of the electron spins around the carbon ring framework. Except for small but crucial distortions of the orbitals, the spin-coupled and classical VB descriptions of this molecule are very similar.

4 Pyridine and Diazobenzenes [16]

Spin-coupled calculations have been performed for various azobenzenes, in a manner analogous to our first series of calculations on benzene. Bond lengths and angles were taken from the volumes edited by Katritzky and Rees [17], but where we were unable to find precise C−H bond lengths we took values from related molecules.

We found that the replacement of one C−H unit in benzene by a nitrogen atom resulted in only small changes in the form of the spin-coupled orbitals, in the relative importance of the different modes of spin coupling, and in our estimate of the resonance energy (92 kJ mol^{-1}). The weights of the different Rumer functions for pyridine are collected in Table 2, with the orbitals ordered according to the labelling of the atoms in Fig. 1. Analogous calculations on the diazobenzenes show that these systems also have very similar descriptions. Occupation numbers in the Rumer basis are listed in Table 2. The spin-coupled orbitals for one of these molecules, pyrimidine, are shown in Fig. 4 as contour plots of $|\varphi_\mu|^2$ in the plane 1 bohr above the molecular plane. As before, these are clearly distorted atomic $2p$ functions, pointing perpendicular to the molecular plane. For all of these systems, the orbitals localized on symmetry-related centres are themselves transformed into one another by symmetry operations of the molecular point group, and the relative distortion of orbitals on C and N appears to reflect the difference in electronegativity of these two atoms.

Table 2. Occupation numbers of the different spin couplings in the Rumer basis for six-membered heterocycles. The Rumer diagrams (R_i) are numbered as in Fig. 2, and the orbitals as in Fig. 1 (and as described in the text).

	Occupation numbers (expressed as percentages)				
	R_1	R_2	R_3	R_4	R_5
Benzene	40	$6\frac{1}{2}$	$6\frac{1}{2}$	40	$6\frac{1}{2}$
Pyridine	40	$7\frac{1}{2}$	$5\frac{1}{2}$	40	$7\frac{1}{2}$
Pyridazine	$20\frac{1}{2}$	$8\frac{1}{2}$	$8\frac{1}{2}$	54	$8\frac{1}{2}$
Pyrimidine	40	$8\frac{1}{2}$	6	40	6
Pyrazine	40	8	$4\frac{1}{2}$	40	8
Borazine	91	3	3	0	3
Boroxine	$95\frac{1}{2}$	$1\frac{1}{2}$	$1\frac{1}{2}$	0	$1\frac{1}{2}$

The occupation numbers of the different Rumer functions for the diazobenzenes appear to be ordered according to simple rules:

(a) The most important contributions are those from the Kekulé-type structures.
(b) The next most important structures involve two C–N singlet couplings. It is often the case that the first two rules describe the same structures.
(c) The third most important group of structures consists of those with one C–N singlet coupling.
(d) All other structures make smaller contributions.

The reasons behind this preference for singlet coupling between orbitals on carbon and nitrogen are not clear, although it is worth noting that the energy difference between single and double bonds is larger in the case of C–N than it is for C–C. Exceptions to these rules may well appear later, but they are consistent with all the heterocyclic systems studied near equilibrium geometry.

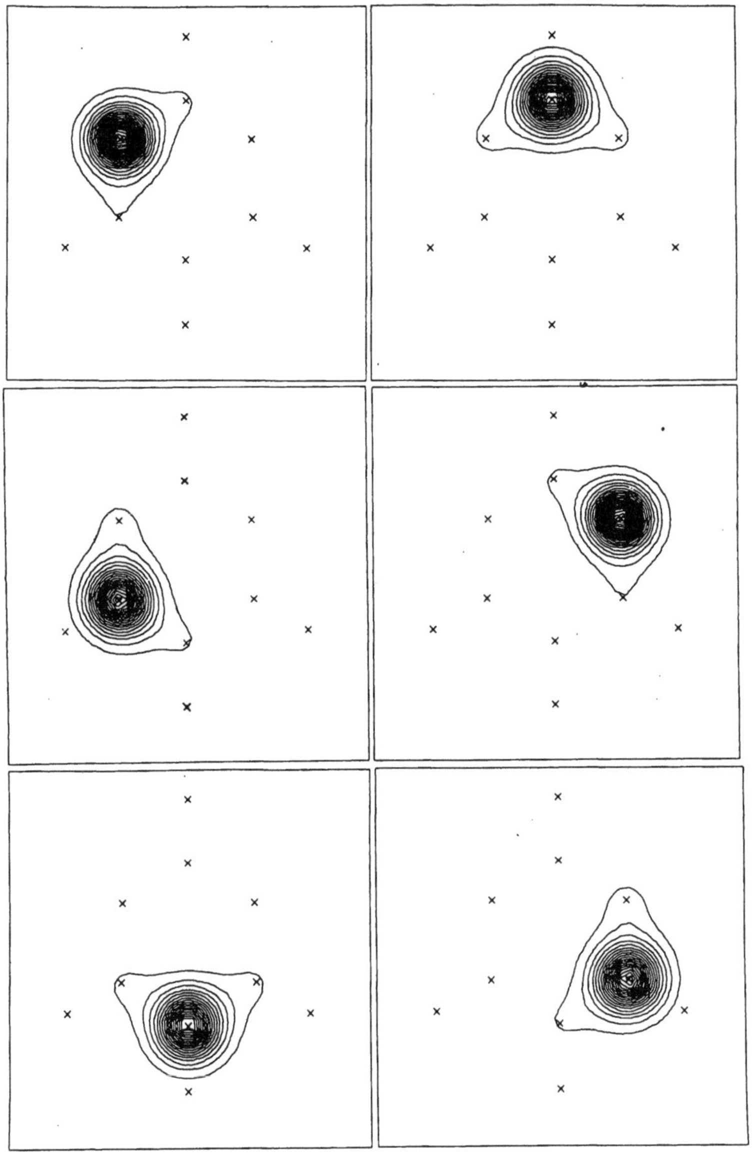

Fig. 4. Spin-coupled orbitals for pyrimidine. The view corresponds to that in Fig. 3(**b**), and the molecule is orientated as in Fig. 1

5 Five-Membered Heterocycles [18]

The spin-coupled description of the π-electron systems in the heteroaromatic molecules furan, pyrrole, thiophen and thiazole is also based on localized but non-orthogonal orbitals. The spin-coupled orbitals for pyrrole are shown in Fig. 5. There is one orbital localized on each carbon atom and one on the heteroatom (φ_1). The sixth orbital

(φ_6) is also associated with the heteroatom, but is very significantly distorted and shows delocalization onto the neighbouring carbon atoms.

We find that the only important Rumer structures are R_4 and R_5, which correspond to singlet coupling of the two electrons on nitrogen. In a description based on orbitals

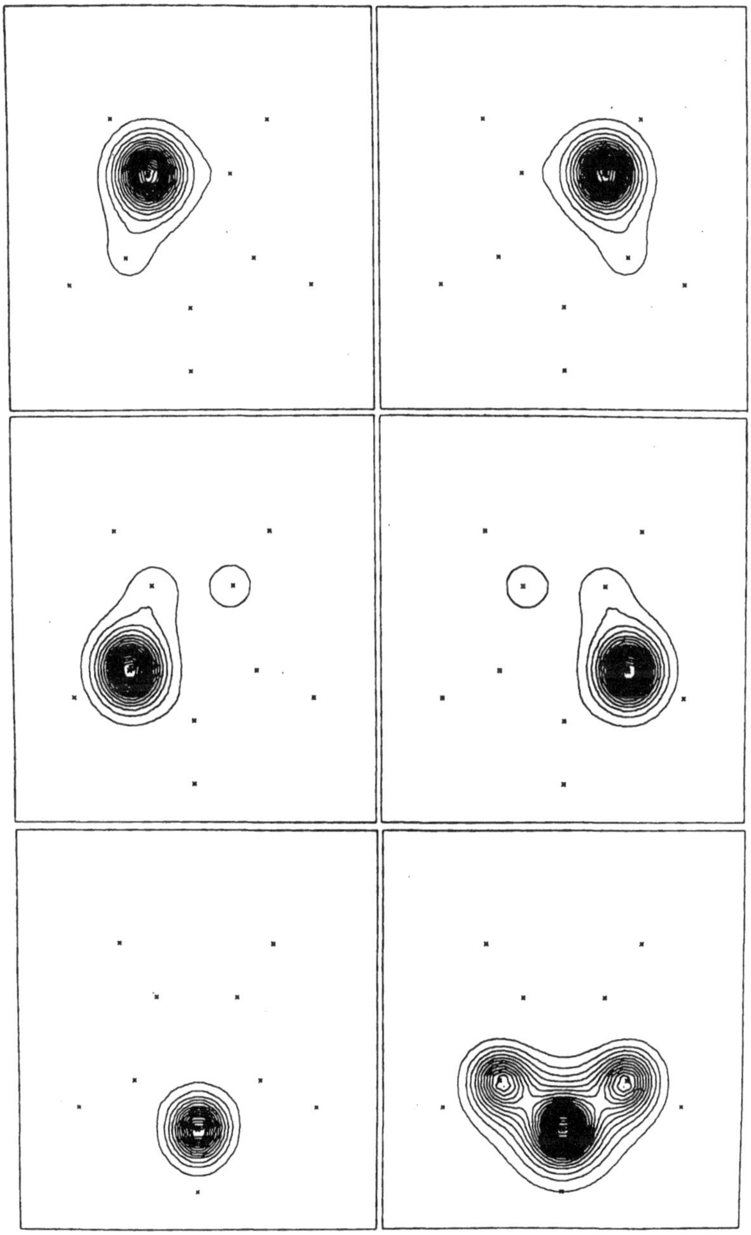

Fig. 5. Spin-coupled orbitals for pyrrole. The view corresponds to those in Figs. 3(**b**) and 4, and the molecule is orientated as in Fig. 1

fully localized on each atom, we can identify the Rumer functions R_4 and R_5 with the classical valence bond structures

and

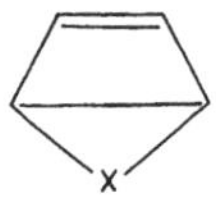

I II

respectively. When the orbitals cease to be completely localized, then this interpretation ceases to be so precise. However, with this reservation, structures I and II for pyrrole contribute roughly 80% and 20%, respectively. The distortion of the orbitals on carbon is consistent with the dominance of VB structure I. Using the same orbitals, the energy of structure R_1 is 31 kJ mol^{-1} higher than the full solution, and we can identify this difference as the resonance energy. It is this, rather than delocalized orbitals, that distinguishes aromatic systems from other molecules.

Similar descriptions emerge for the other five-membered rings we have studied, and occupation numbers in the Rumer basis are collected in Table 3. The main difference is in the delocalization of the second orbital stemming from the heteroatom (φ_6), the extent of which decreases with increasing electronegativity of the heteroatom (sulphur in the case of thiazole).

Table 3. Occupation numbers of the Rumer structures for five-membered heterocycles. Further details are given in the text.

	R_4	R_5
Pyrrole	80%	20%
Furan	85%	15%
Thiophen	88%	12%
Thiazole	85%	11%

One of the great strengths of MO theory is the guidance it provides for the assignment of photoelectron spectra by means of Koopmans' theorem. Using physical assumptions which closely parallel those of Koopmans' theorem, we have used the spin-coupled orbitals to examine simple valence bond estimates of the ionization potentials. We find that the results for the lowest ionization potentials are at least as good as those derived from Koopmans' theorem, while the higher potentials appear to be considerably more reliable.

6 Inorganic Heterocycles: Borazine and Boroxine [7]

There now exists a vast literature on inorganic heterocycles, and much of this has been reviewed in the two recent volumes edited by Haiduc and Sowerby [19]. The most extensively studied of the boron-nitrogen heterocycles, and probably the system

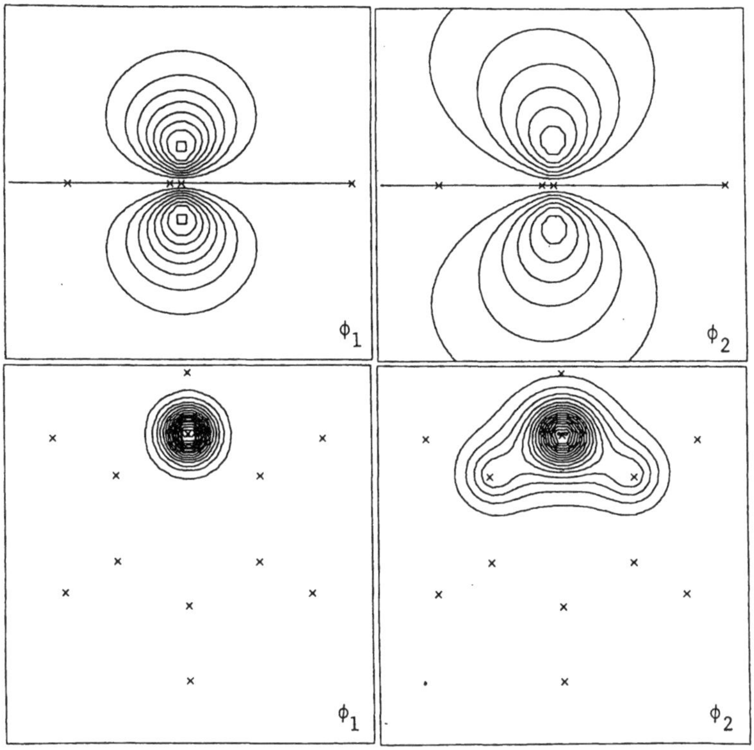

Fig. 6. Spin-coupled orbitals φ_1 and φ_2 in borazine. Two other orbitals are related by symmetry to each of these. **Top row:** at one of the N atoms in a σ_v mirror plane. **Bottom row:** in the plane 1 bohr above the σ_h molecular plane

of most interest in the current context, is borazine ($B_3N_3H_6$). This molecule has frequently been termed 'inorganic benzene' because of the possibility of aromatic stabilization in the π-electron system. Boroxine ($B_3O_3H_3$) is of interest for much the same reason.

Equilibrium geometries are available from electron diffraction studies of borazine [20] and of boroxine [21], both of which are planar in the gas phase with D_{3h} symmetry. For borazine we used bond lengths B–N 1.4355 Å, B–H 1.258 Å and N–H 1.050 Å, and angles NBN 118.3° and BNB 121.7°. For boroxine we used B–O 1.375 Å, and B–H 1.192 Å with angles of 120°.

Spin-coupled calculations, analogous to those for the organic heterocycles, have been carried out for both molecules. Three of the spin-coupled π orbitals for borazine are related to each other by successive $\hat{C}_3$ rotations and they take the form of fairly localized, slightly distorted N(2*p*) functions. The other three orbitals are also transformed into one another by successive $\hat{C}_3$ operations. These are also based on N(2*p*) functions but clearly show significant delocalization onto neighboring boron centres. Nonetheless, most of the electron density remains on N.

Starting at one of the *N* atoms, and proceeding around the ring, we refer to the more localized orbitals as φ_1, φ_3 and φ_5, and to the less localized orbitals as φ_2, φ_4 and

φ_6. Orbitals φ_1 and φ_2 are shown in Fig. 6: the top row of views corresponds to a σ_v mirror plane (at one of the N atoms), while the bottom row is analogous to Figs. 3(b), 4 and 5. Despite the difference in the degree of delocalization onto B atoms, the overlap between φ_1 and φ_2 is 0.81. A very similar picture emerges for boroxine, although there is slightly less delocalization onto neighboring boron centres.

The weights of the different Rumer structures are collected in Table 2. Structure R_1, which corresponds to singlet coupling of the spins of the two electrons in the orbitals associated with each N or O atom, is dominant in both cases. The contributions from the other structures are very small, and we find that the resonance energies are negligible.

Aromaticity is a concept that has proven to be of great utility at a qualitative level, and is well understood by organic chemists in terms of the chemical stability of ring systems. However, it proves to be very difficult to formulate a more precise definition based on a clear set of physical, chemical or theoretical criteria [22]. Certainly, experimental energy differences are not conclusive, particularly as different reaction schemes can produce widely differing estimates of stabilization energies [23]. The large anisotropy in the observed diamagnetic susceptibility of benzene, which has been interpreted as evidence for 'ring currents', is difficult to extend to molecules other than benzene. Indeed, there is clear theoretical evidence which shows that the proton deshielding is in fact due to *local* effects arising in the adjacent carbon atom and in the C–H bond [24]. It is also hard to produce convincing chemical criteria. Nonetheless, we believe that it is fair to say that there is very little evidence for the supposed aromaticity of borazine.

At least in the gas phase, borazine is fairly similar to benzene in terms of physical properties and structure. However, $B_3N_3H_6$ is much more reactive than C_6H_6 and readily undergoes addition reactions rather than electrophilic substitution. The boron-nitrogen ring thus appears to lack the characteristic stability associated with aromatic systems. Similarly, boroxine is stable in the gas phase in the presence of a diluting inert gas, but disproportionates at room temperature in condensed phases to yield B_2O_3 and B_2H_6. The fact that our spin-coupled calculations suggest that neither of these heterocycles has any significant aromatic character thus accords well with these observations. A consistent picture emerges from more sophisticated calculations, analogous to the second series of calculations for benzene.

7 Conclusions

Aromatic systems play a central role in organic chemistry, and a great deal of this has been fruitfully interpreted in terms of molecular orbital theory: that is, in terms of electrons moving more-or-less independently of one another in delocalized orbitals. The spin-coupled model provides a clear and simple picture of the motion of *correlated* electrons in such systems. The spin-coupled and classical VB descriptions of benzene are very similar, except for the small but crucial distortions of the orbitals. The localized character of the orbitals allows the electrons to avoid one another. Nonetheless, the electrons are still able to influence one another directly because of the non-orthogonality of the orbitals.

The form of the spin-coupled orbitals for pyridine and for the diazobenzenes, and the weights of the Rumer structures, generally conform to the pattern found for benzene, but with the added proviso of a marked preference for C=N bonds, which seems entirely reasonable. The five-membered rings are also very similar, except for the marked tendency of one of the orbitals stemming from the heteroatom to delocalize onto neighboring centers. A similar tendency is observed for the inorganic heterocycles borazine and boroxine, but we find no evidence for significant aromatic character in either of these molecules. We have shown recently that the spin-coupled description of larger aromatic molecules, such as naphthalene [25], is very similar to the one we have described here for the π-electron system in benzene.

A simple, consistent and highly-visual model emerges for the behavior of correlated π electrons in all of the aromatic systems that we have studied. The special stability of such molecules arises from the mode of coupling of the electron spins, as shown by the magnitudes of the resonance energies, and not from any supposed delocalization of the orbitals. We have thus developed a theoretical framework on which to build a coherent picture of the structure and properties of aromatic molecules.

8 References

1. Cooper DL, Gerratt J, Raimondi M (1987) Adv. in Chem. Phys. 69: 319
2. Cooper DL, Gerratt J, Raimondi M (1988) Int. Rev. Phys. Chem. 7: 59
3. Gerratt J, Cooper DL, Raimondi M In: Klein DJ, Trinajstić N (eds) Valence bond theory. Elsevier, Amsterdam (in press)
4. Matos JMO, Roos BO, Malmqvist P-A (1987) J. Chem. Phys. 86: 1458
5. Gerratt J (1971) Adv. Atomic and Molec. Phys. 7: 141
6. Pauncz R (1979) Spin eigenfunctions, Plenum, New York
7. Cooper DL, Wright SC, Gerratt J, Hyams PA, Raimondi M (1989) J. Chem. Soc., Perkin Trans. 2: 719
8. Cabana A, Bachaud J, Giguerè J (1974) Can. J. Phys. 52: 1949
9. Dunning TH, Hay PJ (1977) In: Schaefer HF (ed) Methods of electronic structure theory. Plenum, New York
10. Cooper DL, Gerratt J, Raimondi M (1986) Nature 323: 699
11. Hiberty PC, Cooper DL (1988) J. Mol. Struct. (Theochem) 169: 437
12. Pauling L (1933) J. Chem. Phys. 1: 280
13. Coulson CA (1961) Valence, 2nd end, Clarendon, Oxford Ch. 9
14. Dunning TH (1971) J. Chem. Phys. 55: 716
15. McDouall JJW, Robb MA (1987) Chem. Phys. Lett. 142: 131
16. Cooper DL, Wright SC, Gerratt J, Raimondi M J (1989) Chem. Soc., Perkin Trans. 2: 255
17. Boulton AJ, McKillop A (1984) In: Katritzky AR, Rees CW (eds) Comprehensive heterocyclic chemistry, Vol 2, Sect 2.01, Pergamon, Oxford, Johnson; CD: ibid., Sect. 2.04
18. Cooper DL, Wright SC, Gerratt J, Raimondi M (1989) J. Chem. Soc., Perkin Trans. 2: 263
19. Haiduc I, Sowerby DB (eds) (1987) The Chemistry of inorganic homo- and heterocycles, vols 1 & 2. Academic, London
20. Harshbarger W, Lee G, Porter RF, Bauer SH (1969) Inorg. Chem. 8: 1683
21. Chang CH, Porter RF, Bauer SH (1969) Inorg. Chem. 8: 1689
22. Garratt PJ (1986) Aromaticity, Wiley, New York Ch 11
23. Haddon RC (1982) Pure Appl. Chem. 54: 1129
24. Lazzeretti P, Rossi E, Zanasi R (1982) J. Chem. Phys. 77: 3129; Lazzeretti P, Zanasi R (1983) Chem. Phys. Lett. 100: 67
25. Sironi M, Cooper DL, Gerratt J, Raimondi M (1989) J. Chem. Soc., Chem. Comm 1989: 675

Semiempirical Valence Bond Views for Benzenoid Hydrocarbons

Douglas J. Klein

Department of Marine Sciences, Texas A&M University at Galveston, Galveston, Texas 77553-1675, USA

Table of Contents

Because the π-networks of benzenoid hydrocarbons are the classical prototypical example of delocalized bonding, they provide a crucial test for chemical-bonding theories. Here there is revealed a systematic organization for valence-bond views to describe such π-bonding. With an initiation near the ab initio realm a sequence of semiempirical valence-bond models is identified and characterized. The refinement from one model to the next proceeds via: either a (perturbative) restriction to a smaller model space or orthogonalization of a suitable natural basis for the model space. The known properties of the models are indicated, and possible methods of solution are mentioned. A great diversity of work is outlined, related, systematized and extended. New research is suggested.

1 Introduction

A key area in the theory of benzenoid hydrocarbons is the semiempirical valence-bond (VB) view, which was begun over 50 years ago by Pauling and Wheland [1] as a quantum-theoretic realization and extension of classical chemical-bonding ideas. In these earlier classic ideas, initiating with Kekulé [2] in 1865 and modified up through the work of Armit and Robinson [3] in 1922, each benzenoid had been described in terms of its collection of Kekulé structures. It had been realized [4] that such benzenoids were *not* equilibrium mixtures of different conformations (or tautomers) corresponding to each individual Kekulé structure, because the aromatic properties and reactivities of typical benzenoids were not consonant with strictly localized structures. Moreover, it had been suggested by Thiele [5] that lesser influences were manifested by longer bond structures. These and other general classical ideas were rather neatly validated and extended with the development of our present subject: semiempirical VB views for benzenoid hydrocarbons.

Here we skim over the field of semiempirical VB theory of the π-systems of benzenoids. Primary focus is on a systematic derivational development of a hierarchical sequence of VB models. Different VB-based models are addressed in different sections (2, 3, 5, 6) here, and the overall development is summarized in the diagram at the conclusion of Sect. 7. Section 4 serves as an interlude on quantum chemical computational methods, with emphasis on the VB basis and its relationship to chemical structure – this being crucial for the following sections. Along the way we indicate some of the history and general characteristics of the models. The unifying view which emerges not only incorporates many aspects of past work but reveals avenues for future research.

As a tool, which is useful in relating the theory to chemical structure and classical chemical-bonding ideas, it is of advantage to utilize a few graph-theoretic ideas, described by Trinajstić [6] and also by others, especially Cyvin and Gutman. Thus we speak of the *molecular graph* G of the system as corresponding to the molecule's π-network. Familiarity is presumed with the terms: subgraph, spanning subgraph, connected graph, and Kekulé structure (or subgraph). A *component* of a graph is understood to be a maximal connected subgraph. What little additional is needed will be explained along the way.

Overall the present article seeks to meld chemical graph-theoretic (chemical-bonding) ideas with conventional quantum-chemical approaches, all within the framework of traditional VB theory here extended to encompass more recent results and models. Thence use is made of some quantum-chemical nomenclature, which, however, is standard fare in any of a number of quantum chemistry texts, though they seldom seriously discuss VB models for π-network systems. Some effort is made to incorporate solid-state theoretic results on one of the models which has arisen with different applications in mind. As such, the present article offers a novel global perspective which (as is so often the case) emphasizes the author's own work in the area.

2 Primitive VB Model

2.1 Formal Presentation

With the advent of quantum mechanics in the mid 1920s the stage was set for a fantastic theoretical development of molecular theory. Heitler and London [7] had quickly applied the theory to describe the covalent bond in the Hydrogen molecule. Sometime around 1930 the $\sigma - \pi$ separability for conjugated hydrocarbons was realized, for it was utilized by several workers; see, e.g., refs. [1] and [8].

In the VB approach [1, 9] attention is focused on the covalent structures, with one π-electron associated to each conjugated carbon, because of the cost of intraatomic Coulomb repulsion for ionic structures. Thence the subspace of configurations to be considered is based on a simple orbital product over (non-orthogonal) atomic π-orbitals (AO's)

$$\Phi \equiv \prod_{i=1}^{N} \chi_i(i) \tag{2.1.1}$$

where $\chi_i(j)$ would be for electron j in the orbital of site i. But also there is an assignment of spins to each electron, and the associated N-particle spin function

$$\Theta \equiv \prod_{i=1}^{N} \sigma_i(i) \tag{2.1.2}$$

where σ_i is the ith electronic spin (α or β). The overall VB configurations are antisymmetrized

$$|\Phi \wedge \Theta\rangle \equiv \mathscr{A}(\Phi \times \Theta) \tag{2.1.3}$$

Here $\mathscr{A}$ is the (unnormalized) antisymmetrizer

$$\mathscr{A} \equiv \sum_{P} (-1)^P P^{SF} \times P^{\sigma} \tag{2.1.4}$$

where the sum is over permutations of electron indices, with P^{SF} acting on the electrons spatial (spin-free) coordinates and P^{σ} acting on the spin coordinates, and $(-1)^P$ is the parity of P.

The primitive VB model is defined in terms of overlap and Hamiltonian matrix elements over the basis states of eqn. (2.1.3). For fixed Φ there are 2^N possible spin-product functions so that this gives the dimension of the model's space. Indeed (though not originally formulated in this manner) the model may be mathematically represented entirely in spin space, despite the fundamental spin-free nature of the interactions. One may introduce a spin-space overlap operator by integrating out the spin-free coordinates

$$\mathscr{S} \equiv \frac{1}{N!} \langle\Phi| \mathscr{A}^+ \mathscr{A} |\Phi\rangle = \langle\Phi| \mathscr{A} |\Phi\rangle = \sum_{P} (-1)^P \langle\Phi| P^{SF} |\Phi\rangle P^{\sigma} \tag{2.1.5}$$

Likewise a spin-space Hamiltonian operator is introduced

$$\mathscr{H} \equiv \frac{1}{N!} \langle\Phi| \mathscr{A}^+ H^{SF} \mathscr{A} |\Phi\rangle = \sum_{P} (-1)^P \langle\Phi| P^{SF} H^{SF} |\Phi\rangle P^{\sigma} \tag{2.1.6}$$

where H^{SF} is the basic spin-free Schroedinger Hamiltonian.

The many-body matrix elements appearing in $\mathscr{S}$ and $\mathscr{H}$ factor. First, $\langle\Phi| P^{SF} |\Phi\rangle$

is a product of 1-electron overlaps $\langle \chi_i | \chi_j \rangle$, with $i = Pj$ the index to which P carries j. If the χ_i instead of being chosen as bare AO's are (symmetrically) orthogonalized, then one is conventionally [10] led at least initially to Parisier-Parr-Pople-like models, and thence typically to MO-based solution schemes and pictures. A *simplest* presumption avoiding orthogonal orbitals is that for $j \neq i$ only nearest-neighbor $\langle \chi_i | \chi_j \rangle$ are nonzero. Further simplification assumes independence of nearestneighbor pairs,

$$\langle \chi_i | \chi_j \rangle = \begin{cases} 1, & i = j \\ s, & i \sim j \\ 0, & \text{otherwise} \end{cases} \tag{2.1.7}$$

where $i \sim j$ indicates a nearest-neighbor pair of sites in the molecular graph G. Then $\langle \Phi | P^{SF} | \Phi \rangle$ is 0 if P does not move edges of G about amongst themselves; otherwise it is $s^{n(P)}$ where $n(P)$ is the number of sites P moves to an adjacent site in G. Another way to view this is to express P as a disjoint product of cyclic permutations C, whence

$$\langle \Phi | P^{SF} | \Phi \rangle = \prod_{C}^{P} \langle \Phi | C^{SF} | \Phi \rangle \tag{2.1.8}$$

where the factors $\langle \Phi | C^{SF} | \Phi \rangle$ are nonzero (equal to $s^{n(C)}$) only if C also corresponds to a cycle in G.

The Hamiltonian matrix elements also factor in a manner related to that for the overlaps. The matrix element over $P^{SF}H^{SF}$ modifies only one or two cycle factors at a time in the corresponding overlap, over P^{SF}. That is, the analysis of the next section reveals

$$\langle \Phi | P^{SF} H^{SF} | \Phi \rangle = \langle \Phi | P^{SF} | \Phi \rangle \left\{ \sum_{C}^{P} \varepsilon(C) + \sum_{C_1, C_2}^{P} \varepsilon(C_1 \times C_2) \right\} \tag{2.1.9}$$

where the energies $\varepsilon(C)$ and $\varepsilon(C_1 \times C_2)$ involve interactions amongst the electrons and/or associated sites of the cycle C or of the disjoint cycles C_1 and C_2. The most important ε-terms are those that are associated to the fewest interchanges of sites and to shorter range interactions. Up through near-neighbor interactions with no more than two interchanges, there are four ε-terms: $\varepsilon(i)$ is a site energy; $\varepsilon(i, j)$, for the 2-site cycle (ij), corresponds to an "exchange" energy; $\varepsilon(i \times j)$, for two 1-site cycles, corresponds to a "Coulomb" energy; and $\varepsilon(ij \times k)$ involves an electron interchange on sites i and j whilst there is interaction between the electrons of this pair and that of site k not affected by the permutation. Through appropriate choice of reference energy $\varepsilon(i)$ may be identified as 0, as in the next section. A typical nearest-neighbor $\varepsilon(i, j)$ we write as $\varepsilon(\bullet\!\!\bigcirc\!\!\bullet)$, and typical $\varepsilon(i \times j)$ and $\varepsilon(ij \times k)$ for $i \sim j \sim k \neq i$ we take equal and represented by $\varepsilon(\cdot\,\cdot)$.

The model then may be expressed in terms of these parameters. The overlap operator is

$$\mathscr{S} = \sum_{P}^{G} (-1)^P s^{n(P)} P^{\sigma} \tag{2.1.10}$$

where the sum is over all permutations which carry no site index to any other site more distant than the nearest neighbor in G, and $n(P)$ is the number of site indices that P moves to nearest neighbors. The Hamiltonian operator, up through the 2-site contributions, is

$$\mathscr{H} = \sum_{i\sim j}^{G} \sum_{P}^{G_{(ij)}} (-1)^P \mathscr{s}^{n(P)} \{\varepsilon(\cdot\,\cdot) - \mathscr{s}^2 \varepsilon(\bullet\!\!\bigcirc\!\!\bullet)\,(ij)^{\sigma}\}\, P^{\sigma}$$

$$- \mathscr{s}^2 \varepsilon(\cdot\,\cdot) \sum_{i\sim j\sim k}^{G} \sum_{P}^{G_{(ijk)}} (-1)^P \mathscr{s}^{n(P)} (ij)^{\sigma}\, P^{\sigma} \qquad (2.1.11)$$

where (ij) is the permutation transposing indices i and j, and the P-sums are much as in eqn. (2.1.10) but now for graphs $G_{(S)}$ with sites of set S deleted.

2.2 Digression on Matrix Elements

A closer look may be taken at the few-site ε matrix elements of the preceding section. To do this the Hamiltonian is written as a sum of 1- and 2-electron terms

$$H^{SF} = \sum_{i} H_i(i) + \sum_{i<j} H_{ij}(i,j) \qquad (2.2.1)$$

Here the subscript indices i, j refer to sites, and the indices in parentheses refer to electrons. One choice for the 1- and 2-site operators is

$$H_i(i) \equiv -\frac{\hbar^2}{2m} \nabla^2(i) - \frac{e^2}{r_i(i)} - \varepsilon_i$$

$$H_{ij}(i,j) \equiv e^2 \left\{ \frac{1}{r_{ij}} - \frac{1}{r_i(j)} - \frac{1}{r_j(i)} + \frac{1}{r(i,j)} \right\} \qquad (2.2.2)$$

where ∇^2 is the usual Laplacian associated to the kinetic energy, $\hbar$ is $h/2\pi$ with h Planck's constant, m is the electronic mass, e is the electronic charge, and the various r-variables are interparticle distances. In "improved" choices some of the Coulomb potentials might be replaced by various other sorts of (e.g. core) potentials. The reference orbital energies are

$$\varepsilon_i \equiv \langle \chi_i | \left\{ -\frac{\hbar^2}{2m} \nabla^2 - \frac{e^2}{r_i} \right\} | \chi_i \rangle \qquad (2.2.3)$$

as will prove convenient.

Now the single-cycle ε parameters of the preceding section are

$$\varepsilon(C) = \langle \Phi | C^{SF} \left\{ \sum_{i}^{C} H_i(i) + \sum_{i<j}^{C} H_{ij}(i,j) \right\} | \Phi \rangle / \langle \Phi | C^{SF} | \Phi \rangle \qquad (2.2.4)$$

where the sums are solely over electron indices in the cycle C. Clearly with the definition of (2.2.3) the 1-site cycles give $\varepsilon(i) = 0$. In fact, for general C the $H_i(i)$ should make little contribution since the χ_i are presumed similar to bare atomic orbitals which themselves would be eigenvectors to $H_i(i)$ and give 0. Thence with the neglect of these at most minor contributions from $H_i(i)$, the 2-site single cycle (ij) gives

$$\varepsilon(i,j) = \langle \chi_j(1)\,\chi_i(2)|\,H_{ij}(1,2)\,|\chi_i(1)\,\chi_j(2)\rangle/\mathcal{s}^2 \tag{2.2.5}$$

which is of the form of a "normalized" exchange term.

The two-cycle ε-parameters are

$$\varepsilon(C_1\times C_2) = \langle\Phi|\,C_1^{SF}C_2^{SF}\sum_i^{C_1}\sum_j^{C_2} H_{ij}(i,j)\,|\Phi\rangle/\langle\Phi|\,C_1^{SF}C_2^{SF}\,|\Phi\rangle \tag{2.2.6}$$

For the simplest (2- and 3-site) cases, we have

$$\begin{aligned}\varepsilon(i\times j) &= \langle\chi_i(1)\,\chi_j(2)|\,H_{ij}(1,2)\,|\chi_i(1)\,\chi_j(2)\rangle\\ \varepsilon(ij\times k) &= \langle\chi_j(1)\,\chi_i(2)\,\chi_k(3)|\,\{H_{ik}(1,3)+H_{ik}(2,3)\\ &\quad + H_{jk}(1,3)+H_{jk}(2,3)\}\,|\chi_i(1)\,\chi_j(2)\,\chi_k(3)\rangle/\mathcal{s}^2\end{aligned} \tag{2.2.7}$$

which are of the form of Coulomb interactions.

With the choice of interactions as in Eq. (2.2.2) the terms of (2.2.5) and (2.2.7) may be given a classical interpretation. To do this first introduce charge densities

$$\begin{aligned}\varrho_i(\vec r) &\equiv -e\{\chi_i(\vec r)\}^2 + e\delta(\vec r,\vec R_i)\\ \varrho_{i\to j}(\vec r) &\equiv -e\chi_i(\vec r)\,\chi_j(\vec r)\,\frac{1}{s} + e\delta(\vec r,\vec R_i)\\ \varrho_{i-j}(\vec r) &= -e\,\chi_i(\vec r)\,\chi_j(\vec r)\,\frac{1}{s} + \frac{e}{2}\,\delta(\vec r,\vec R_i) + \frac{e}{2}\,\delta(\vec r,\vec R_j)\end{aligned} \tag{2.2.8}$$

where $\delta(\vec r,\vec R_i)$ is a Dirac delta function on site i located by the vector $\vec R_i$. Overall these charge densities are electrically neutral (in that they integrate to 0). Then using (2.2.2) in (2.2.5) and (2.2.7), one may show that

$$\begin{aligned}\varepsilon(i,j) &= \iint \varrho_{i\to j}(\vec r)\,\frac{1}{|\vec r-\vec r'|}\,\varrho_{j\to i}(\vec r')\,d\vec r\,d\vec r'\\ \varepsilon(i\times j) &= \iint \varrho_i(\vec r)\,\frac{1}{|\vec r-\vec r'|}\,\varrho_j(\vec r')\,d\vec r\,d\vec r'\\ \varepsilon(ij\times k) &= \iint \varrho_k(\vec r)\,\frac{1}{|\vec r-\vec r'|}\,\varrho_{i-j}(\vec r')\,d\vec r\,d\vec r'\end{aligned} \tag{2.2.9}$$

That is, these parameters are represented as classical Coulombic interactions between two pairs of overall electrically neutral charge distributions. For $\varepsilon(i,j) = \varepsilon(\cdot = \cdot)$ the interaction is between two interpenetrating counter-oriented dipolar distributions. For $\varepsilon(i \times j)$ or $\varepsilon(ij \times k)$ the interaction is between two partially overlapping parallel quadrupolar distributions; they seem to be roughly the same size $\varepsilon(\cdot\,\cdot)$ for $i \sim j \sim k$. From the forms here $\varepsilon(\cdot = \cdot)$ is positive. Presumably $\varepsilon(\cdot\,\cdot)$ is positive and larger than $\varepsilon(\cdot = \cdot)$.

2.3 Discussion

The primitive model of Sect. 2 may be used to obtain the (approximate ground-state) wavefunction. This is found by solving the eigenproblem

$$\mathscr{H}\Psi' = E'\Psi' \tag{2.3.1}$$

where the eigenstate Ψ' is a linear combination of the various spin functions Θ. Though with modern quantum-chemical computational techniques the treatment of this model is only slightly more difficult than for either that with $\mathscr{s} = 0$ or that of Sect. 3, it has seldom been treated in a quantitative fashion. This is unfortunate since Fisher and Murrell [11] and also Mulder and Oosterhoff [12] have shown how $\mathscr{s} \neq 0$ in such models is crucial for two (related) qualitative features, namely: first, the Hückel $4n$- and $(4n + 2)$-rule; and second, the Woodward-Hoffmann rules. These aspects have been reiterated and discussed extensively by Epiotis [13].

The value of $\mathscr{s}$ and whether it can reasonably be taken to be zero has long been a point of some confusion. With the coefficient $-\mathscr{s}^2\varepsilon(\cdot = \cdot)$ negative (for $i \sim j$) as in Sect. 2.2 and $\mathscr{s}$ elsewhere set to 0, one obtains [14] an "inverted" eigenvalue spectrum, with highest spin multiplicities lowest. But if instead $-\mathscr{s}^2\varepsilon(\cdot = \cdot)$ is treated parametrically and taken as positive, the resulting eigenvalue spectra are much more reasonable. In fact a rough explanation of why this latter model should yield spectra comparable to the "proper" one with $\mathscr{s}\varepsilon(\cdot = \cdot) > 0$ and $\mathscr{s} > 0$ was given in 1936 by Van Vleck [15]. But still there remain occasional references to this (and related computational problems) as a supposedly unresolved *nonorthogonality catastrophe.* This is returned to in the next section.

With the choice of $\varepsilon(\cdot = \cdot)$ as a positive exchange parameter Epiotis [13] has addressed what the value of $\mathscr{s}$ should be. He argues that it should be larger than that for a nearest-neighbor pair of bare (untransformed) AO's. More particularly, he argues that the preferred AO's should be antiorthogonalized, as indicated in Fig. 1 below, showing schematic orbital amplitudes χ_j along bond axes to neighboring sites. Figures 1a, b, and c respectively show χ_j for a bare AO, an orthogonalized AO (as typically used in Hückel MO theory), and an antiorthogonalized AO. These ideas are in fact consonant with much "standard wisdom". That is, it is widely recognized [10] that orthogonalization enhances electron hopping at the expense of exchange, thereby leading to Parisier-Parr-Pople or Hückel-MO models. Antiorthogonalization naturally has the opposite effect (of enhancing exchange and covalency). This view finds further ab initio support in the recent and widely acclaimed quantum-chemical calculations of Cooper et al. [16] on benzene where variationally optimized nonor-

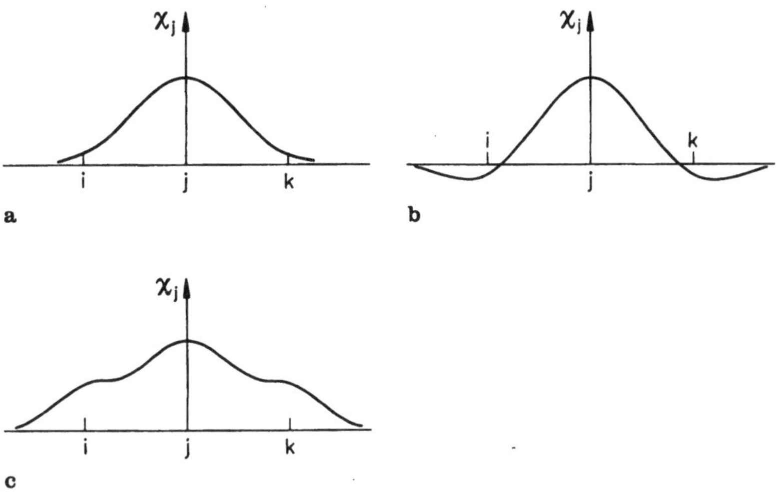

Fig. 1. Sketches of amplitudes for the three types of atomic orbitals (AOs) for a π-center j and how they spread out to neighboring π-centers i and k. These AO types are bare, orthogonalized, and antiorthogonalized, in **(a)**, **(b)** and **(c)** above

thogonal orbitals in a spin-coupled wavefunction are found to take roughly the antiorthogonalized form. With the enlarged $\mathscr{s}$, the semiempirical approach has enhanced next-neighbor and cyclic interactions, which are especially of relevance for rings of size 3 or 4.

The fundamental graph-theoretic nature of the primitive model is of some note. Each permutation P arising in the sums of Eqs. (2.1.9) and (2.1.10) are in correspondence with a special class of spanning subgraphs, each of whose components is either a single site, a pair of sites joined by an edge, or a cycle. (The practice of using such graphs to represent the various ε-matrix elements is fairly common, even with several electrons per site, as in Ref. [17], though often the tight correspondence to the molecular graph has not been emphasized.) Then $n(P)$ is a graph-invariant for such a spanning subgraph; it simply is the number of sites which are not single components in the spanning subgraph. Such an identification offers the possibility of using various standard graph-theoretic techniques in treating and further investigating this model.

3 Pauling-Wheland Valence-Bond Model

3.1 First Orthogonalization

The overlap operator of the primitive VB model may be rigorously transformed away. Indeed, the eigenvalue problem of (2.2.1) is seen to be equivalent to

$$(\mathscr{S}^{-1/2}\mathscr{H}\mathscr{S}^{-1/2})\,\mathscr{S}^{1/2}\Psi' = E'\mathscr{S}^{1/2}\Psi' \qquad (3.1.1)$$

Because $\mathscr{S}$ is positive definite and Hermitean there is a natural choice [18] for the square-root $\mathscr{S}^{1/2}$, taking it also to be positive definite and Hermitean. Thus also $\mathscr{S}^{-1/2}$ is defined as the (unique) inverse to $\mathscr{S}^{1/2}$. Equation (3.1.1) then suggests a transformed Hamiltonian into which we also incorporate an energy shift E_0,

$$H \equiv \mathscr{S}^{-1/2}(\mathscr{H} - E_0\mathscr{S})\,\mathscr{S}^{-1/2} \tag{3.1.2}$$

The associated orthonormal eigenstates then are

$$\Psi = \mathscr{S}^{1/2}\Psi' \tag{3.1.3}$$

That is, the new eigenproblem

$$H\Psi = E\Psi \tag{3.1.4}$$

now appears in a more customary form. The energies E and E' differ only by the additional reference shift, conveniently now taken as

$$E_0 = \varepsilon(\cdot\,\cdot)\,|\mathscr{E}(\mathrm{G})| \tag{3.1.5}$$

where $|\mathscr{E}(\mathrm{G})|$ is the order of the edge set of G (i.e., the number of edges in G).

A problem to determine $\mathscr{S}^{-1/2}$ now emerges. One way to compute the already mentioned "natural choice" for $\mathscr{S}^{-1/2}$ is [18] to: first transform via a unitary matrix U to diagonal form $\mathbf{U}\mathscr{S}\mathbf{U}^\dagger = \boldsymbol{\Delta}$; second, take the inverses of the positive square roots of the diagonal of $\boldsymbol{\Delta}$ to obtain $\boldsymbol{\Delta}^{-1/2}$; and third compute $\mathscr{S}^{-1/2} = \mathbf{U}^\dagger\,\boldsymbol{\Delta}^{-1/2}\mathbf{U}$.

Another approach [18] is to compute $\mathscr{S}^{-1/2}$ and thence also H via a perturbative expansion in powers of the overlap parameter $\mathfrak{s}$. That is, one writes

$$\begin{aligned}
\mathscr{S} &= \mathbf{1} + \sum_{m\geqq 1} \mathfrak{s}^m \mathscr{S}^{(m)} \\
\mathscr{S}^{-1/2} &= \mathbf{1} + \sum_{m\geqq 1} \mathfrak{s}^m (\mathscr{S}^{-1/2})^{(m)} \\
\mathscr{H} &= \sum_{m\geqq 0} \mathfrak{s}^m \mathscr{H}^{(m)} \\
H &= \sum_{m\geqq 0} \mathfrak{s}^m H^{(m)}
\end{aligned} \tag{3.1.6}$$

where $\mathbf{1}$ is the identity operator. Then one finds, for the primitive VB model,

$$\begin{aligned}
\mathscr{S}^{(1)} &= 0 \\
\mathscr{S}^{(2)} &= -\sum_{i\sim j}^{\mathrm{G}} (ij)^\sigma \\
\mathscr{S}^{(3)} &= 0 \\
&\vdots \\
\mathscr{H}^{(0)} &= E_0\mathbf{1} \\
\mathscr{H}^{(1)} &= 0
\end{aligned} \tag{3.1.7}$$

$$\mathscr{H}^{(2)} = -\sum_{i\sim j}^{G} \{\varepsilon(\cdot\!\bigcirc\!\cdot) - \varepsilon(\cdot\,\cdot)\,[|\mathscr{E}(G)| - 1)]\}\,(ij)^{\sigma}$$

$$\mathscr{H}^{(3)} = 0$$

$$\mathscr{H}^{(4)} = \frac{1}{2}\sum_{i\sim j}^{G}\sum_{k\sim l}^{G(ij)} \{2\varepsilon(\cdot\!\bigcirc\!\cdot) - \varepsilon(\cdot\,\cdot)\,[|\mathscr{E}(G)| - 2]\}\,(ij)^{\sigma}\,(kl)^{\sigma}$$

$$\vdots$$

where the possibility of rings of sizes 3 or 4 has not been admitted. The factors $|\mathscr{E}(G)| - n$ appear in counting near-neighbor Coulomb interactions $\varepsilon(\cdot\,\cdot)$ between sites other than the $n = 1$ or 2 exchanging pairs. In $\mathscr{H}^{(4)}$ the factor of 1/2 arises because the summations count each disjoint pair of bonds twice, while $\varepsilon(\cdot\!\bigcirc\!\cdot)$ appears with a factor of 2 because either bond may be involved in the exchange interaction. Next following through standard but tedious Taylor series expansions to obtain the $(\mathscr{S}^{-1/2})^{(n)}$ and then the $H^{(n)}$, one finally obtains

$$\begin{aligned} H^{(0)} &= H^{(1)} = H^{(3)} = 0 \\ H^{(2)} &= \mathscr{H}^{(2)} \\ H^{(4)} &= \{\varepsilon(\cdot\!\bigcirc\!\cdot) - \varepsilon(\cdot\,\cdot)\}\,\{|\mathscr{E}(G)| + \frac{1}{2}\sum_{i\sim j\sim k} [(ij)^{\sigma}\,(jk)^{\sigma} + (jk)^{\sigma}\,(ij)^{\sigma}]\} \end{aligned} \tag{3.1.8}$$

Notably in $H^{(4)}$ the disjoint (or "unlinked") sum of $\mathscr{H}^{(4)}$ over pairs of edges has been cancelled by independent sums arising from products of $\mathscr{H}^{(2)}$ and $\mathscr{S}^{(2)}$. Such cancellations continue to higher orders, leaving a *size-extensive* effective Hamiltonian. Indeed we might anticipate this if the total energy is imagined to be Taylor-series expandable in s while also being size-extensive.

The resultant effective Hamiltonian may be given in a simplified form. First the 3-site permutations simplify on spin space

$$(ij)^{\sigma}\,(jk)^{\sigma} + (jk)^{\sigma}\,(ij)^{\sigma} = (ij)^{\sigma} + (jk)^{\sigma} + (ki)^{\sigma} - 1 \tag{3.1.9}$$

Thence next-nearest neighbor "exchange" interactions appear, as well as modifications to the nearest-neighbor interactions. Additional structure-dependent modifications to the nearest-neighbor interactions occur if the $\varepsilon(i \times j)$ and $\varepsilon(ij \times k)$ parameters are distinguished. Such modifications we pass over here and use simply a second-order nearest-neighbor interaction parameter

$$J = s^{2}\{\varepsilon(\cdot\,\cdot) - \varepsilon(\cdot\!\bigcirc\!\cdot)\} \tag{3.1.10}$$

Then through second order

$$H = J \sum_{i\sim j}^{G} (ij)^{\sigma} \tag{3.1.11}$$

which is our currently desired effective Hamiltonian.

3.2 The Model

The second-order result of Eq. (3.1.11) is essentially the Pauling-Wheland VB model. Formally this differs from the $\mathscr{s} \to 0$ primitive model (with $-\mathscr{s}^2\varepsilon(\bullet\bigcirc\bullet)$ fixed) mentioned in Sect. 2.3 only in having a different parameter multiplying the Hamiltonian. In particular, even with estimates of $\mathscr{s}$, $\varepsilon(\bullet\bigcirc\bullet)$ and $\varepsilon(\cdot\,\cdot)$ appropriate to bare AO's the sign of this new multiplying parameter J is positive, as desired. With $\mathscr{s}$ even larger, as we noted it should be in Sect. 2.3, this circumstance is only more so engendered. In fact, by a rather different (but perfectly reasonable) argument Simpson [19] obtained much the same second-order expressions. Other still different approaches beyond second-order have also been given [20, 21], sometimes starting from a primitive VB model somewhat more general than that presented in Sect. 2.

The Pauling-Wheland model may also be derived starting with orthogonalized AO's and proceeding through the Parisier-Parr-Pople model, or some relative threof. These approaches can be perturbative [22, 23] or a little "better" [23, 24]. Again there results the Pauling-Wheland VB model with the correct sign for J.

A point of some confusion is that there are different representations of the Pauling-Wheland VB model. In fact, Pauling and Wheland [1] did not represent it in the form of Eq. (3.1.10), but rather they presented it as a matrix on the Rumer basis (mentioned in Sect. 4.2 here). The appearance of (3.1.10) may be further modified through the use of the Dirac identity [25]

$$(ij)^{\sigma} = 2\vec{s}_i \cdot \vec{s}_j + \frac{1}{2} \tag{3.2.1}$$

where $\vec{s}_i$ and $\vec{s}_j$ are conventional spin operators for the spins of sites i and j. Then except perhaps for an additive scalar we have

$$H = J \sum_{i \sim j}^{G} 2\vec{s}_i \cdot \vec{s}_j \tag{3.2.2}$$

(The supressed scalar serves only to further shift all the eigenvalues.)

In the form of Eq. (3.2.2) this model Hamiltonian is recognized to be (mathematically) identical to the Heisenberg spin Hamiltonian of solid-state theory. This seems to have been first noted by Van Vleck and Sherman [26]. It is to be emphasized [17] that the present model describes what are fundamentally spin-free interactions, with the representation on spin space allowed because of the restriction to covalent structures. Indeed it is a point of bemusement that Heisenberg [27] originally represented the model in terms of spin-free permutations, the representation of Eq. (3.2.2) being effected (perhaps first) by Dirac [25]. Another difference in Heisenberg's work is that he wished to treat rather different underlying systems where J would take the negative "ferromagnetic" sign. Much of the work in physics up to about 1962 has been nicely reviewed by Herring [21].

3.3 Theorematic Results

Though the exact solution of the Pauling-Wheland VB model (or the positive-J Heisenberg spin Hamiltonian) is generally a nontrivial matter, there are a number of rigorous results that may be established. These are primarily for *alternant* systems with bipartite graphs G and sites partitionable into *starred* and *unstarred* subsets such that sites of one set have neighbors solely from the other set. We denote these subsets of sites (or vertices) by $\mathscr{V}_*$ and $\mathscr{V}_0$, we specify orders of these sets by $|\mathscr{V}_*|$ and $|\mathscr{V}_0|$, and we choose $|\mathscr{V}_*| \geqq |\mathscr{V}_0|$. The theorems to be quoted all share a common hypothesis:

Hypothesis – The connected system graph G is alternant with exchange parameters $J_{ij} \neq 0$ only for $i \sim j$ whence they are positive. Then the theorems are:

Theorem A – The ground-state spin is $S = (|\mathscr{V}_*| - |\mathscr{V}_0|)/2$.

Theorem B – The ground-state is nondegenerate beyond the spin-multiciplicity degeneracy implied by theorem A.

Theorem C – The ground-state spin densities ϱ_i (i.e., the expectation values of s_i^z for the maximum-M_s ground state) are 0 iff $S = 0$ and otherwise are >0 or <0 as $i \in \mathscr{V}_*$ or $\mathscr{V}_0$.

Theorem D – The ground-state expectation value of $\vec{s}_i \cdot \vec{s}_j$ is positive or negative as i and j are in the same or different subsets, $\mathscr{V}_*$ and $\mathscr{V}_0$.

Theorem E – If there is an involutary permutation T commuting with H, then it corresponds to a spin-free point-group symmetry of $+$ or $-$ $(-1)^{|\mathscr{V}_*|}$ as T either leaves invariant $\mathscr{V}_*$ (and $\mathscr{V}_0$) or interchanges the sites of $\mathscr{V}_*$ and $\mathscr{V}_0$.

Theorems A and B (along with an underlying "working" theorem) were first proved by Lieb and Mattis [28]. Theorems C, D and E were [29] established a little later.

All of these theorems are of chemical interest. Theorem A (and B) evidently accurately predict [30] the ground-state spins of general conjugated hydrocarbons. Theorems C and D give qualitative characteristics of certain ground-state expectation values, of relevance [31] in ESR and NMR. These characterizations of theorem C seem to be consistent with a large number of unrestricted Hartree-Fock calculations, as, e.g. in Refs. [31] and [32]. Theorem D may be used [33] to understand relative stabilities of related species. Theorem E validates [29] part of the reasoning that goes into Craig's rules [34] in understanding aromatic and antiaromatic tendencies from the VB viewpoint. The results of theorems A and D were independently surmised [35] as reasonable.

There are extensions of all these theorems. The interaction strengths between neighboring sites need not all be equal, as is relevant in dealing with corrections, due to unequal bond lengths. Further interactions between sites of the same set ($\mathscr{V}_*$ or $\mathscr{V}_0$) are permissible if the associated exchange coupling parameters are $\leqq 0$. In fact this extension is [36] crucial in (very successfully) dealing with conjugated hydrocarbons that also include one or more carbene centers. Finally these theorems extend to the situation where the site spins are other than 1/2 (as can occur especially in inorganic crystals).

4 Representations and Chemical Structure

4.1 Computational Methodology

The solution of the eigenproblem for H of Sect. 3 appears to increase rapidly in difficulty as the size of the system increases. That is, the dimension of the space upon which H is defined is 2^N with N the number of sites, so that the matrix representation becomes larger quite rapidly with increasing N. There is some aid in block-diagonalizing the matrix upon utilizing the spin (and point-group) symmetries. But the blocks of interest generally remain formidable: they associate to subspaces of dimensions $\sim 2^N/N$ if one takes into account the always available spin-symmetry; or if suitably high point-group symmetry is available, they may associate to subspaces of dimensions as small as $\sim 2^N/N^2$. Standard brute-force quantum-chemical unitary-group configuration-interaction techniques then enable [37] one to deal with systems of up to $N = 24$ sites. Further refinement of this technique for the particular model at hand might allow $N = 26$ to be done where the blocks of interest involve about a million configurations. Indeed this size of system has been managed [38] to be treated via a related "Rumer-state" based scheme so specially adapted. More readily foreseen near-term improvements in computers might enable N to be increased by 10%. Though these values of N include many benzenoids, many more are beyond this reach. In addition more maneagable and more clearly comprehendable descriptions of the ground-state wavefunction are desirable.

There are some exact ground-state wavefunctions and energies available. However, this so-called "Bethe *Ansatz*" solution scheme [39] has several difficulties: first, (so far) it seems applicable only to cyclic rings (with equal-strength near-neighbor interactions); second, it is exceedingly complicated, so that only very few expectation values have been evaluated; and third, it yields only a fraction of all the states, though the ground state has been shown [40] to be included. Some much simpler ground states are known [41] for other Heisenberg spin Hamiltonians with nonnearest-neighbor interactions of certain prescribed (usually rather strong) interaction-strengths.

Because of the difficulty and complexity of the calculations of the preceding two paragraphs, "suitable" approximation schemes are of interest. Here "suitable" most desirably means both simply maneagable and easily comprehendable (with physico-chemical content). For alternants one such scheme involves the Nèel state [42]

$$\Theta_0 \equiv \prod_i^{\in \mathcal{V}_*} \alpha(i) \prod_j^{\in \mathcal{V}_0} \beta(j) \tag{4.1.1}$$

Its energy is readily evaluated to be

$$E(\text{Nèel}) = \langle \Theta_0 | H | \Theta_0 \rangle = -\frac{1}{2} |\mathscr{E}(\mathrm{G})| J \tag{4.1.2}$$

Because of the Rayleigh-Ritz principle E(Nèel) is an upper bound to the exact ground-state energy. Moreover, a perturbation expansion based upon the Nèel state may be developed, as apparently first done by Hartmann [43] for chemical applications, and later pursued by others [44]. The resulting expressions yield an energy expansion in terms of "local" graph-theoretic invariants, and thence is of much interest. Some other Nèel-state based approaches (e.g., the antiferromagnetic spin-wave development, the Jordan-Wigner transformed SCF approximations, or Kasteleyn's wavefunction *Ansatz*) are reviewed elsewhere [45]. There is much additional (mostly more recent) work; see, e.g., refs. [46–48]. Particularly the work of Kaplan et al. [47] and of Huse and Elser [48] suggest possibilities of very high accuracy results with reasonable effort.

4.2 VB Structures

The traditional way to strengthen the ties of the VB model with classical chemical-bonding ideas is through the use of its representation on a suitable basis. A step toward this end involves the VB structure idea wherein it is noted that a product over disjoint pairs of singlet coupled sites yields an overall singlet. Thus for four sites and two such pairs one has three possibilities, corresponding to three VB *diagrams* indicating the pairs,

$$\Theta_1 = [1, 2]\,[3, 4] \quad \text{with}$$

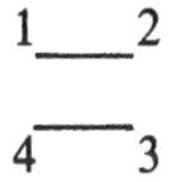

$$\Theta_2 = [1, 3]\,[2, 4] \quad \text{with} \tag{4.2.1}$$

$$\Theta_3 = [1, 4]\,[2, 3] \quad \text{with}$$

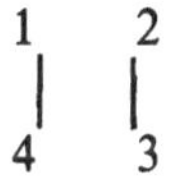

where we have abbreviated the (unnormalized) singlet pair functions

$$[i, j] \equiv \alpha(i)\,\beta(j) - \beta(i)\,\alpha(j) \tag{4.2.2}$$

Rumer [49] noted that these three structures are linearly dependent, and proposed that in general a linearly independent set would be obtained from only those structures represented by a diagram with sites situated about the periphery of a circle and paired by noncrossing straight lines. Soon thereafter this was proved [50].

Pauling [51] developed matrix element formulas between pairs of VB structures. They are graphical in nature involving *superposition* diagrams obtained by superimposing the appropriate pair of VB structures. Thus the superposition diagrams corresponding to the pairs of Rumer structures from Eq. (4.2.1) are

for Θ_1, Θ_1

$$\text{for} \quad \Theta_1, \Theta_3 \quad \text{or} \quad \Theta_3, \Theta_1 \tag{4.2.3}$$

for Θ_3, Θ_3

The components (maximal connected subgraphs) of these diagrams are termed *islands*. The number of *little* islands with 2-sites in a superposition diagram S is denoted by $i(S)$, and the number of larger *big* islands is denoted by $I(S)$. Then Pauling showed that

$$\langle \Theta_a | \Theta_b \rangle = \pm 2^{i(S_{ab}) + I(S_{ab})} \tag{4.2.4}$$

where S_{ab} is the superposition diagram corresponding to Θ_a and Θ_b. Further the interaction matrix elements over $\vec{s}_i \cdot \vec{s}_j$ were found to be 0 if i and j are in different islands of S_{ab}, and otherwise

$$\langle \Theta_a | \vec{s}_i \cdot \vec{s}_j | \Theta_b \rangle = \pm \frac{3}{4} \langle \Theta_a | \Theta_b \rangle \tag{4.2.5}$$

In fact Pauling's formulas were slightly different in that he dealt with normalized VB structures and with transpositions (or more general permutations) rather than $\vec{s}_i \cdot \vec{s}_j$. McWeeny and Cooper [52] and Sutcliffe [53] give systematic formulas for more general operators with VB structures of general spin multiplicities.

Finally some points about the phase of the VB structures and the signs appearing in (4.2.4) and (4.2.5) need to be cleared up. In a VB diagram many workers append arrows to a pairing bond between sites i and j to distinguish between singlet pair functions $[i, j]$ and $[j, i]$. Thence the signs in (4.2.4) and (4.2.5) are associated to the arrangement of arrows. Another approach, for the overall singlet case, is to view the set of sites as partitioned into two equinumerous sets, such as $\mathscr{V}_*$ and $\mathscr{V}_0$ in dealing with even alternants. Then we may choose all (independent) VB structures to pair sites from $\mathscr{V}_*$ with sites in $\mathscr{V}_0$ in such a manner that for a pair $[i, j]$ we take $i \in \mathscr{V}_*$ and $j \in \mathscr{V}_0$ (rather than the other way around). Then the sign in (4.2.4) is always + and that in (4.2.5) is + or − as i and j are in the same or different sets.

One way to use all this VB structure technology is to set up configuration interaction Hamiltonian and overlap matrices to be treated much as in the previous section. Indeed this seems to have been a major rationale from the early days, as is witnessed in Sherman's (precomputer) treatment [54] of the naphthalene molecule with 42 singlets, though point-group symmetry adaptation reduced the ground-state block to a size of 13. The next section discusses another use of VB structures thereby making closer contact with classical chemical-bonding ideas.

5 Pauling-Wheland Resonance Theory

5.1 The Model

Some VB structures are more "favored" in exhibiting a lower energy expectation for the VB model Hamiltonian H of Sect. 3. Indeed from the matrix element formula of (4.2.5), along with the ensuing comments on signs, one sees that the more favored VB structures are those wherein as many spin-pairings $[i, j]$ as possible occur along bonds of the molecular graph G. If all the spin-pairings can be so accomodated then the structure is termed a *Kekulé* structure. Pauling and Wheland [1] were thus motivated to make a simple ground-state approximation by diagonalizing H on the subspace of Kekulé structures. Of course not all polyhex graphs G exhibit a Kekulé structure, e.g. as discussed by Hosoya [55], but as noted empirically by Fries [56] it appears this may be a requirement for stability.

Because of the close correspondence to classical chemical-bonding structures, the use of the Kekulé structures could be suggested without a detailed consideration of matrix elements, as invoked in the preceding paragraph. Thence Rumer [49] was first intrigued by this close correspondence. Both Pauling [57] and Wheland [9] wrote books broadly advocating and illustrating this idea. Basically since each spin-pairing when between nearest neighbors represents a Heitler-London covalent bond one anticipates a maximal number of such near-neighbor spin-pairings to be preferred. An even more sophisticated view is that of Simpson [58] who views Kekulé structures as coherent combinations of an equal number of the lower-lying primarily covalent states. This view however leads to a model differing from the simple restriction of H of Sect. 3 to the Kekulé subspace, and will be discussed further in Sect. 6.

There are indications that the consequent resonant-theoretic model leads to better agreement with experiment on benzenoids than might be expected from a first-order degenerate perturbation argument (such as we presented in the first paragraph of this section). That is, the higher-order corrections to the model of Sect. 3 seem [59] to stabilize Kekulé structures somewhat, while "frustrating" (or destabilizing) the Nèel states. Another piece of numerical evidence is that resonance energies of benzenoids, for the VB model of Section 3 have been found [37] to correlate closely with those from the Hückel MO model, while those from the present Pauling-Wheland resonance-theoretic model correlate [60, 61] more closely with the resonance-energies from the conjugated circuits model – and the SCF PPP-like resonance energies of Dewar and deLlano [62] – which have been judged [63] to be very good. It is to be emphasized that the circumstance is entirely different for non-benzenoids with $4n$-membered (especially $n = 1$) rings, whence the higher-order corrections destabilize the Kekulé structures.

There is yet another rather unusual motivation of the present model, starting from the Hückel MO, Hubbard or Parisier-Parr-Pople models. This is Živković's so-called "bond-orbital resonance-theoretic" approximation [64] for solving these latter models. He makes a wavefunction *Ansatz*

$$\Psi = \mathscr{A}(\Psi_\alpha \cdot \Psi_\beta) \tag{5.1.1}$$

where Ψ_σ is to be a function optimally defined on the subset of electron indices associated to spin σ (which can be α or β). Instead of breaking each up into orbital products as in SCF theory, he seeks each Ψ_σ as a linear combination of configurations (involving only σ-electrons). In his "resonance theoretic" *Ansatz* for alternants he takes Ψ_σ as a linear combination of states $|\sigma K\rangle$ labelled by Kekulé-structure diagrams

$$|\sigma K\rangle = \prod_{i}^{\in \mathscr{V}_\sigma} \{\chi_i(i) + \chi_{i(K)}(i)\}\, \sigma(i) \tag{5.1.2}$$

where we take $\mathscr{V}_\alpha = \mathscr{V}_*$, $\mathscr{V}_\beta = \mathscr{V}_0$, and $i(K) \in \mathscr{V}_{\bar{\sigma}}$ is the site to which i is paired in K. The optimal linear combinations of the $|\sigma K\rangle$ in Ψ_σ are independent of σ in alternants.

For benzenoids one obtains essentially the same [64] Hamiltonian and overlap matrices as for the Pauling-Wheland resonance-theoretic model. That is the energies are the same except for a proportionality and possibly a shift (the shift being zero for "resonance energies"). For non-benzenoids the matrices and resultant energies differ notably, leading in Živković's approach to Hückel $4n$-rule type results.

One approach to a complete solution of the Pauling-Wheland resonance-theoretic model is to treat it as a quantum-chemical configuration-interaction problem, now defined on a space of reduced size (corresponding only to those VB structures which are Kekulé structures). However this space still has a size that often increases exponentially with the size of G. Thence one (ultimately) wishes further simplifications in dealing with the Hamiltonian and overlap matrixes **H** and **S**.

5.2 Resonance-Theoretic Wavefunction *Ansatz*

The resonance theory of Pauling and Wheland is brought into readier contact with classical ideas with the introduction of the resonance-theoretic wavefunction *Ansatz* appropriate for alternant systems. This is simply an equal-weighted sum over Kekulé structures

$$|\Psi\rangle = \sum_{K}^{\mathrm{G}} |K\rangle \tag{5.2.1}$$

also proposed by Pauling and Wheland [1]. One motivation for this *Ansatz* is that when expanded in terms of simple spin-products as in (2.1.2) the signs of their coeffficients agree with those that are proved [28] to apply for the exact wavefunction. Also empirically in a number of cases, the *Ansatz* appears [65] to be fairly close to the variationally determined optimal combination of Kekulé structures.

To use the wavefunction Ψ of (5.1.1) one needs to evaluate its matrix elements. One way to do this is to compute representation matrices on the "basis" of Kekulé structures, then sum the elements of these matrices. But graph-theoretic representations for the overall matrix element of Ψ may also be obtained. Using (4.2.4) with the sign convention possible for alternants, one obtains

$$\langle\Psi\,|\,\Psi\rangle = \sum_{K,K'}^{\mathrm{G}} 2^{i(K\cup K')+I(K\cup K')} \tag{5.2.2}$$

where $K \cup K'$ denotes the superposition diagram arising from K and K'. Now to limit attention solely to superposition diagrams one need note that a particular superposition diagram may arise in more than one way. In particular each big island arises in two ways with every other edge possibly coming from either bra or ket (K or K'). Thence upon restriction to summation over superposition diagrams, we have

$$\langle\Psi\,|\,\Psi\rangle = \sum_{S}^{\mathrm{G}} 2^{i(S)}4^{I(S)} \tag{5.2.3}$$

This form first noted in Ref. [66] has been again obtained by Sutherland [67]. It appears much in the form of a statistical mechanical partition function, so that a variety of statistical mechanical, or graph-theoretic, techniques might be used in the evaluation of it.

For alternants the Hamiltonian matrix element over Ψ may similarly be obtained. This utilizes Eq. (4.2.5) and the associated sign discussion to give

$$\langle\Psi| H |\Psi\rangle = -\frac{3}{2} J \sum_{S}^{G} n(G, S)\, 2^{i(S)} 4^{I(S)} \qquad (5.2.4)$$

where $n(G, S)$ is the number of edges of G joining sites in the same island of S.

These formulas (5.2.3) and (5.2.4) have been used [60, 66, 68] in a number of investigations, especially for polymeric systems. There a powerful (statistical mechanical) transfer matrix technique [60] has been used for the evaluation of the sums in (5.2.3) and (5.2.4), applicable to either finite or infinite chains. Moreover, in these studies a type of long-range ordering of spin-pairing structures emerges, and it is argued [69] to give rise to bond localization and the possibility of novel "solitonic" excitations. More commonly these last two topics have received much attention [70] within the Hückel MO framework, where bond localization is connected with the so-called Pierels distortion ideas.

6 The Herndon-Simpson Model

6.1 Second Orthogonalization

Another approach to treating the Pauling-Wheland resonance-theoretic model of Sect. 9 is to transform the Kekulé structure basis to eliminate the overlap. This may be done following the "first orthogonalization" approach of Sect. 3, whence we are to develop an expansion of the inverse square root $\mathbf{S}^{-1/2}$ of the overlap matrix $\mathbf{S}$ for the Pauling-Wheland resonance-theoretic model of Sect. 5. To manage this in a systematic fashion it is of advantage to introduce local operators on the space of Kekulé structures. We let $|K)$ denote the orthonormalized Kekulé structure corresponding to $|K\rangle$. Further we let γ denote a connected subgraph of the parent molecular graph G such that γ itself admits Kekulé structures denoted e.g. by $\varkappa$. Then we define operators

$$\Lambda_\gamma(\varkappa \mid \varkappa') \equiv \sum_{K}^{\supseteq \varkappa} \sum_{K'}^{\supseteq \varkappa'} |K)(K'| \qquad (6.1.1)$$

where e.g. the first sum is over all Kekulé structures of G such that $\varkappa$ on γ is a substructure of K. That is, the operator $\Lambda_\gamma(\varkappa \mid \varkappa')$ changes a Kekulé-structure state $|K')$ to $|K)$ iff K' and K are identical except (possibly) on γ where $K \cap \gamma = \varkappa$ and $K' \cap \gamma = \varkappa'$. To make full use of these fundamental operators their multiplication rules need to be deduced. In fact, this has already been done, to reveal a *Kekulé-space* algebra. Using such ideas to compute $\mathbf{S}^{-1/2}$ in powers of a local overlap $s = 1/4$ (arising in the analysis of the "island structure" in **S** and **H**), then computing $\mathbf{S}^{-1/2}\mathbf{H}\mathbf{S}^{-1/2} = \mathscr{H}$ up through third-order, it has been found [61] that

$$\mathscr{H} = -\frac{3}{2}J\{|\mathscr{E}(G)|\,\mathbf{1} + \frac{51}{64}\Lambda(6) - \frac{3}{16}\Lambda(6') + \frac{3}{16}\Lambda(10) - \frac{3}{128}\Lambda(10') + \frac{3}{64}\Lambda(14a) + \frac{3}{128}\Lambda(14b) + \frac{7}{64}\Lambda(14c)\} \tag{6.1.2}$$

where the $\Lambda(n)$ are sums over $\Lambda_\gamma(\varkappa|\varkappa')$ with subgraphs γ having n sites. For instance, both $\Lambda(6)$ and $\Lambda(6')$ involve sums over 6-membered ring graphs γ and associated $\varkappa$, $\varkappa'$ with $\varkappa \neq \varkappa'$ in $\Lambda(6)$ and $\varkappa = \varkappa'$ in $\Lambda(6')$. The operators $\Lambda(6)$, $\Lambda(10)$, $\Lambda(14a)$, $\Lambda(14b)$, $\Lambda(14c)$ interchange pairs of Kekulé structures differing only in their pattern of conjugation around a single 6-, 10- or 14-cycle, with the a, b, c identifying the three possible shapes for 14-cycles on the hexagonal (benzenoid) lattice,

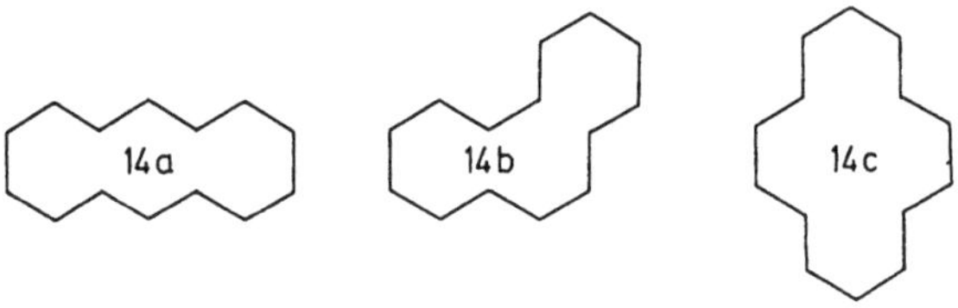

The operator $\Lambda(10')$ interchanges Kekulé structures differing only within a single naphthalenic cell: that with a π-bond in the center is interchanged with the other two. Resonance energies are obtained by referencing the ground-state energy of $\mathscr{H}$ against the single-structure energy $\langle K|\,\mathbf{H}\,|K\rangle / \langle K\,|\,K\rangle$.

In fact, via different reasoning much this same sort of model was proposed sometime ago by Simpson [58]. He noted that e.g. for benzene the + and − combinations of the ground $^1A_1$ and excited $^1B_2$ states of benzene were in some ways similar to the conventional Kekulé structures. Thence he proposed that these and similar coherent combinations for other benzenoids could be interpreted as *exact orthogonal* Kekulé structures. With such structures given, Simpson and later Herndon [71] imagined the simplest interaction mixing such structures to be a linear combination of $\Lambda(6)$ and $\Lambda(10)$, though they did not give it in the operator form noted e.g. in Eq. (6.1.2). This model may be treated exactly via matrix diagonalization on the space of Kekulé structures, but this has been little done, even with the simplified models involving just $\Lambda(6)$ and $\Lambda(10)$. These matrices can of course increase exponentially in size with system size.

6.2 Randić's Conjugated Circuits

An approriate wavefunction *Ansatz* simplifies ground-state computations. In analogy to the Pauling-Wheland resonance-theoretic *Ansatz* of Sect. 5.2, Herndon [71] proposed

$$|\Psi) = \sum_{K}^{G} |K) \tag{6.2.1}$$

whence the resonance energy may be expressed [72] as a combination of graph-theoretic invariants. To describe this, consider a Kekulé structure K for a graph G. Then a *conjugated circuit* in K is [73] identified to be a cycle of G corresponding to alternating single and double bonds in K. The invariants of interest are the numbers $\#_\xi(G)$ of conjugated circuits of a given type ξ summed over all Kekulé structures of G. These invariants arise because for the wavefunction of (6.2.1),

$$(\Psi|\,\Lambda(\xi)\,|\Psi) = \#_\xi(G)\,, \qquad \xi = 0, 6, 10, 14a, 14b, 14c \tag{6.2.2}$$

where $\#_0(G)$ is the number of Kekulé structures of G. Further it may be seen that

$$\begin{aligned} (\Psi|\,\Lambda(6')\,|\Psi) &= \#_6(G) \\ (\Psi|\,\Lambda(10')\,|\Psi) &= 2\,\#_{10}(G) \end{aligned} \tag{6.2.3}$$

Then with the use of $\mathscr{H}$ of (6.1.2) one obtains a resonance energy

$$\begin{aligned} \mathrm{RE}(G) = &+ \frac{3}{2} J \left\{ \frac{39}{64}\,\#_6(G) + \frac{9}{64}\,\#_{10}(G) \right. \\ &+ \left. \frac{3}{64}\,\#_{14a}(G) + \frac{3}{128}\,\#_{14b}(G) + \frac{1}{64}\,\#_{14c}(G) \right\} \Big/ \#_0(G) \end{aligned} \tag{6.2.4}$$

all in terms of simple graph-theoretic invariants.

In fact Randić [73] proposed expressions like that of (6.2.4) independently of a development from the Herndon-Simpson model. His idea may have been to graph-theoretically quantify the qualitative classical "aromatic-sextet" ideas of Clar [74]. The coefficients of $\#_6(G)$, $\#_{10}(G)$ and that of a particular combination of the conjugated 14-circuit invariants were treated as empirical parameters. The derived parameters of Eq. (6.2.4) compare to only within a factor of ~50% to Randić's parameters. But the comparison especially for the two larger parameters is even closer (to ~20%) and is really somewhat "remarkable" considering that Randić's values were obtained by fitting to SCF-MO results [62].

A number of computational approaches to the $\#_\xi(G)$ have been developed and there have been widespread applications of the conjugated-circuits model, motivated both from Herndon's and from Randić's approaches. The applications extend even much beyond benzenoids. This is reviewed elsewhere by Randić' et al. [76].

7 Overview

7.1 Synthesis

From the survey here it is seen that activity in working on semiempirical VB theory has continued over the years with much widening interest during the last few years. It is seen that there are several VB-type models all interrelated, as indicated now in a summary fashion in Diagram 1. In the first column are indicated models and their interrelation, while in the second column are indicated simplifying means of approximate solution. Associated sections of our survey are also indicated. In this diagram the top model and bottom two models have not yet been discussed to any great extent in the literature (or here). The top simply includes the ionic terms, occupying the orbitals of the primitive VB model with varying numbers (0, 1 or 2) electrons. Particularly to deal with ions (or many heteroatom species, or "ionic" excited states even of homoatomic neutral species) this model is relevant. Some brief promising mention of this model is made by Fischer and Murrell [11] and Epiotis [13]. The bottom two models are based upon *Clar* structures, which graphically are spanning subgraphs C of G with components that are hexagons or edges such that no 3 disjoint edges of C are contained in a 6-cycle of G. In terms of quantum states $|C\rangle$ (for the model next from the bottom) would be a sum over all $|K)$ such that $K \subseteq$ C. Thence $|C\rangle$ would be nonorthogonal (though the $|K)$ are orthogonal), since two different $|C\rangle$ could share a common $|K)$. Transformation by an appropriate inverse square root would lead to the bottom model with orthogonalized Clar structures $|C)$. Indeed Herndon and Hosoya [77] have built upon Clar's qualitative ideas [74] and suggested this model; the fit of the solutions obtained by matrix diagonalizations to the data of Dewar and deLlano [62] seems encouraging. A natural wavefunction cluster-expansion *Ansatz* would be

$$|\Psi) = \sum_{C}^{G} x^{h(C)} |C) \tag{7.1.1}$$

where x is a variational parameter and h(C) is the number of hexagonal "aromatic sextets" in C. But this is as yet untested. The work of Živković et al. [78] might possibly be interpreted to arise from some such approximation. Of course, Clar's qualitative development [74] almost purely from experimental observations argues strongly for further study of these bottom two models. The question marks (?) in several of the solution boxes of Diagram 1 indicate a dearth of work in these areas. Finally both above and below the extreme models of Diagram 1 there should be additional models. In the next position above, the model might include various sigma-electron configurations. In the next position below, attention might be restricted to the most preferred Clar structures with a maximum number of aromatic sextets.

The derivations of the models as outlined here should be amenable to improvement. Basically the contractions to smaller subspaces were developed here within a first-order degenerate perturbation-theoretic framework, so that further improvements might be obtained in proceeding to higher orders. This then further "renormalizes" the parameters appearing in these models. Work in this area has been little developed.

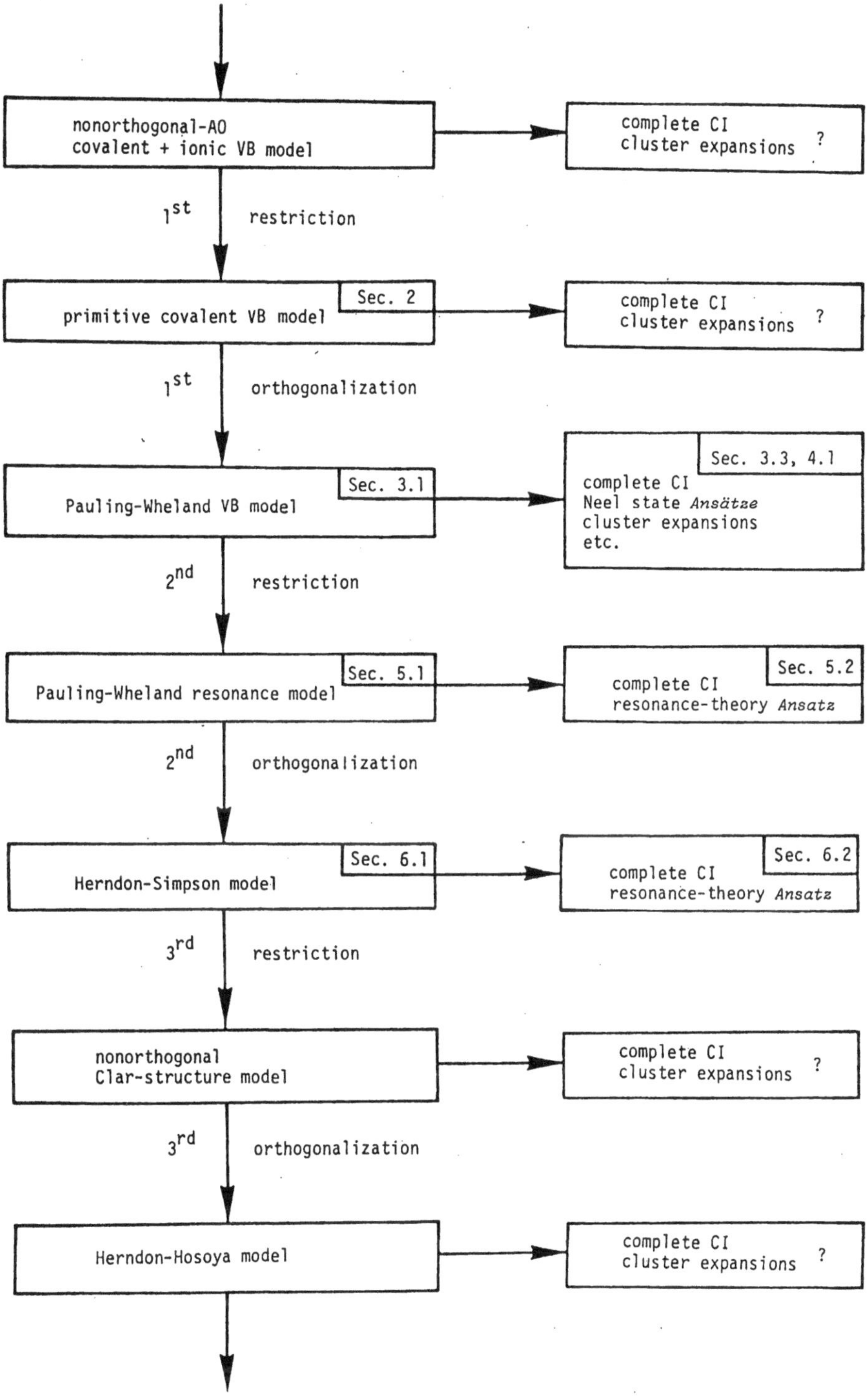
nonorthogonal-AO
covalent + ionic VB model
complete CI
cluster expansions ?
1st restriction
Sec. 2
primitive covalent VB model
complete CI
cluster expansions ?
1st orthogonalization
Sec. 3.1
Pauling-Wheland VB model
Sec. 3.3, 4.1
complete CI
Neel state Ansätze
cluster expansions
etc.
2nd restriction
Sec. 5.1
Pauling-Wheland resonance model
Sec. 5.2
complete CI
resonance-theory Ansatz
2nd orthogonalization
Sec. 6.1
Herndon-Simpson model
Sec. 6.2
complete CI
resonance-theory Ansatz
3rd restriction
nonorthogonal
Clar-structure model
complete CI
cluster expansions ?
3rd orthogonalization
Herndon-Hosoya model
complete CI
cluster expansions ?

Diagram 1

Many misperceptions about VB theory are clarified with our ideas going into Diagram 1. Criticisms concerning orthogonalization of AO's and the sign of the exchange parameter J in the Pauling-Wheland VB model are misplaced, as long ago argued by Van Vleck [15] and further elaborated on in Sect. 3 here. The orthogonalization step is properly viewed as an orthogonalization of Slater determinants made up of AO's (as emphatically distinguished from an orthogonalization of AO's); thence there is no sign problem (for J). A number of resolutions or ways around the notorious "nonorthogonality catastrophe" have been indicated. Criticisms concerning the adequacy of the Pauling-Wheland VB theory for cyclo-butadiene and other non-benzenoids is revealed as neglect of the next corrections, for which there are estimates [12, 37, 79]. But even as it stands this VB model makes some (especially spin-related) predictions [30, 33, 36] much more reliably than (at least) the simplest MO theories. A discussion of these and a number of further criticisms has been given elsewhere [33].

It is of some note that many of the models may be (and often were) obtained by-passing the derivational approach here. Basically each model may be viewed as represented by the first terms in a graph-theoretic cluster expansion [80]. Once the space on which the model to be represented is specified, the interactions in the orthogonal-basis cases are just the simplest "additive" few-site operators possible. For the nonorthogonal bases the overlaps are just the simplest "multiplicative" operators possible, while the associated Hamiltonian operators are the simplest associated "derivative" operators. These ideas lead [80] to proper size-consistency and size extensivity. Similar sorts of ideas apply in developing wavefunction *Ansätze* or ground-state energy expansions for the various models.

A further type of approximation depending only on counts of Kekulé (or perhaps Clar) structures may be argued to be suggested by VB ideas. Configuration mixing and consequent resonance energies evidently increase with the number of Kekulé (or Clar) structures, other things being "fixed". Thence one might seek understanding simply in terms of these counts. Much of the consequent work is indicated in the recent book [81] of Cyvin and Gutman. Similar sorts of ideas based upon counts of other relevant spanning subgraphs (e.g., superposition graphs, weighted Clar graphs, or suitably weighted longer bond VB structures) seem to have been little investigated.

7.2 Outlook

The global view attained here suggests many avenues of further work. Rather than having achieved a conclusive stage it seems that VB theory is still very active, with much yet to be done and many as yet unanswered questions. Notably most of the work in the literature has here been identified only through a few lead references [9, 13, 21, 45, 57, 76, 81 and 82] each of which might be traced to 100 additional references. Several of the models, their derivations, their parameterizations, solution procedures to them, their strengths, their limitations, and most especially their applications to particular systems have here only been cursorily investigated. Nevertheless, beyond the ground-state energy of benzenoids, there remains much work to be done, concerning properties (polarizabilities, susceptibilities, reactivities, etc.), excitation spectra (and transition probabilities), as well as extensions to ions, nonbenzenoids, nonalternants, heteroatoms, etc. Some current development beyond conjugated systems, even into the ab initio realm, is surveyed elsewhere [82].

Especially much of the recent semiempirical work on benzenoids has been phrased in terms of graph-theoretic concepts, which aid in binding the developments closely to chemical structure and to classical empirically developed chemical-bonding ideas, such as those of Armit and Robinson [3] and of Clar [74]. A stronger complementary role to MO theory might be anticipated – perhaps even a type of merging may result. One might envision a broad hierarchical network of schemes ranging from ab initio to semiempirical models, such that interrelations between and qualifications at each level of simplification are understood. In addition to the interrelations between VB schemes as focused on in the present survey, interrelations to MO models would be available and possibly new hybrid models would emerge.

8 Acknowledgements

Discussions with numerous researchers are acknowledged but most extensively with W. A. Seitz, T. G. Schmalz, N. Trinajstić, M. Randić, M. A. García-Bach, T. P. Živković, W. C. Herndon, F. A. Matsen, and R. D. Poshusta. Indeed much research in the area has been done in collaboration with these able scientists. Brief discussions with Profs. R. McWeeny, N. H. March and L. Pauling have inspired much of the work. Finally the support of the Welch Foundation of Houston, Texas is gratefully acknowledged.

9 References

1. Pauling L, Wheland GW (1933) J. Chem. Phys. 1: 362
2. Kekulé A (1865) Bull. Acad. Roy. Belg. 19: 557; (1866) Ann. 137: 129
3. Armit JW, Robinson R (1922) J. Chem. Soc. 38: 827
4. Erlenmeyer E (1901) Ann. 316: 57
5. Thiele FKJ (1899) Liebs Ann. Chem. 306: 89, 125
6. Trinajstić N (1983) Chemical graph theory, CRC Press, Boca Raton, Florida
7. Heitler W, London FW (1927) Zeit. Phys. 44: 455
8. Hückel E (1930) Zeit. Phys. 60: 423
9. Wheland GW (1955) Resonance in organic chemistry, Wiley, New York
10. Fischer-Hjalmars I (1965) J. Chem. Phys. 42: 1962; Freed KF (1983) Acct. Chem. Res. 16: 137
11. Fischer H, Murrell JN (1963) Theor. Chim. Acta 1: 463
12. Mulder JJC, Oosterhoff LJ: Chem. Comm. 1970: 305, 307
13. Epiotis ND (1982) Unified valence-bond theory, Springer, Berlin Heidelberg New York
14. Inglis DR (1934) Phys. Rev. 46: 135
15. Van Vleck JH (1936) Phys. Rev. 49: 232
16. Cooper IL, Gerratt J, Raimondi M (1986) Nature 323: 699
17. Matsen FA, Klein DJ, Foyt DC (1971) J. Phys. Chem. 75: 1866
18. Löwdin PO (1950) J. Chem. Phys. 18: 365; (1970) Adv. Quantum Chem. 5: 185
19. Simpson WT (1956) J. Chem. Phys. 25: 1124
20. Carr WT (1953) Phys. Rev. 92: 28; Mullin WC (1964) Phys. Rev. A136: 1126
21. Herring C (1963) In: Rado GT, Suhl H (eds) Magnetism 2B, Academic, New York, pp 1–181
22. Linderberg J, Ohrne Y (1986) J. Chem. Phys. 49: 716; (1979) Brandow BH: Intl. J. Quantum Chem. 15: 207
23. Buleavski LN (1966) Zh. Eksp. Teor. Fiz. 51: 230
24. Klein DJ, Foyt DC (1973) Phys. Rev. A8: 2280; Girardeau MD (1976) J. Math. Phys. 17: 431; Economou EN, Mihas P (1977) J. Phys. C10: 5017; Malrieu JP, Maynau D (1982) J. Am. Chem. Soc. 104: 3021; Poshusta RD, Klein DJ (1982) Phys. Rev. Lett. 48: 1555;

Maynau D, García-Bach MA, Malrieu JP (1986) J. Physique 47: 207
25. Dirac PAM (1930) The principles of quantum mechanics, Oxford University Press, Clarendon
26. Van Vleck JH, Sherman A (1935) Rev. Mod. Phys. 7: 167
27. Heisenberg W (1928) Zeitz. Physik 49: 619
28. Lieb EH, Mattis DC (1962) J. Math. Phys. 3: 749
29. Klein DJ (1982) J. Chem. Phys. 77: 3098; Klein DJ, Alexander SA (1987) In: King RB, Rouvray DH (eds) Graph theory and topology in chemistry, Elsevier, Amsterdam, pp 404–419
30. Klein DJ, Nelin CJ, Alexander SA, Matsen FA (1982) J. Chem. Phys. 77: 3101; Koutecký J, Döhnert D, Wormer PES, Paldus J, Čížek J (1984) J. Chem. Phys. 80: 2244
31. Carrington A, McLachlan AD (1967) Introduction to magnetic resonance, Harper and Row, New York
32. Tyutyulkov N, Schuster P, Polansky OE (1983) Theor. Chim. Acts 63: 291; Karafiloglou P (1983) Intl. J. Quantum Chem. 25: 293; Yamaguchi K, Fukui H, Fueno T: Chem. Lett. 1986: 625; Fukutome H, Takahashi A, Ozaki M (1987) Chem. Phys. Lett. 142: 181
33. Klein DJ (1983) Pure & Appl. Chem. 55: 299
34. Craig DP: J. Chem. Soc. 1951: 3175
35. McConnell HM (1959) J. Chem. Phys. 30: 126; McLachlan AD (1959) Mol. Phys. 2: 223; Ovchinikov AA (1978) Theor. Chim. Acta 47: 297
36. Alexander SA, Klein DJ (1987) J. Am. Chem. Soc. 110: 3401; Iwamura H, Izuoka A: J. Chem. Soc. Japan 1987: 595
37. Alexander SA, Schmalz TG (1987) J. Am. Chem. Soc. 109: 6933
38. Ramasesha S, Soos ZG (1984) Intl. J. Quantum Chem. 25: 1003
39. Hulthén L (1938) Arkiv. Mat. Astron. Fys. A26: 1; Takhtadzhan LA, Fadeev LD (1979) Russ. Math. Surveys 34: 11; Sogo K, Waditi M (1982) Prog. Theor. Phys. 68: 85
40. Yang CN, Yang CP (1966) Phys. Rev. 150: 221
41. Majumdar CK (1970) J. Phys. C3: 911; van den Broek PM (1980) Phys. Lett. 77A: 261; Klein DJ (1982) J. Phys. A15: 661; Shastry BS, Sutherland B (1981) Physica 108B: 1069; Caspars WJ, Magnus W (1983) Physica 119A: 29; Affleck I, Kennedy T, Lieb EH, Tasaki H (1987) Phys. Rev. Lett. 59: 799
42. Nèel L (1948) Ann. Phys. Paris 3: 137
43. Hartman H (1947) Z. Naturforsch. A2: 259
44. Davis HL (1960) Phys. Rev. 120: 787; Parrinello M, Scirè M, Arai T (1973) Lett. Nuovo Cim. 6: 138; Maynau D, Said M, Malrieu JP (1983) J. Am. Chem. Soc. 105: 5244; Karafiloglou P (1985) J. Chem. Phys. 82: 3728
45. Jones W, March NH (1973) Theoretical solid state physics, Wiley-Interscience, New York; Mattis DC (1965) Theory of Magnetism, Harper and Row, New York
46. Vroelant C, Daudel R (1949) Bull. Soc. Chim. France 16: 36; Nebenzahl I (1969) Phys. Rev. 177: 1001; Bartowski RR (1972) Phys. Rev. B5: 4536; García-Bach MA, Klein DJ (1977) Intl. J. Quantum Chem. 12: 273; Klein DJ, García-Bach MA (1979) Phys. Rev. B19: 877; Miyashita S (1984) J. Phys. Soc. Japan 53: 44; Suzuki M (1986) J. Stat. Phys. 43: 883; Cioslowski J (1987) Chem. Phys. Lett. 134: 507; Barnes T, Swanson ES (1988) Phys. Rev. B37: 9405
47. Kaplan TA, Horsch P, Fulde P (1982) Phys. Rev. Lett. 49: 889
48. Huse DA, Elser V (1988) Phys. Rev. Lett. 60: 2531
49. Rumer G: Nachr. Ges. Wis. Gött., Math-physik. Klasse 1932: 337
50. Rumer G, Teller E, Weyl H: Nachr. Ges. Wis. Gött., Math-physik. Klasse 1932: 499
51. Pauling L (1933) J. Chem. Phys. 1: 280
52. Cooper IL, McWeeny R (1966) J. Chem. Phys. 45: 226
53. Sutcliffe BT (1966) J. Chem. Phys. 45: 235
54. Sherman A (1934) J. Chem. Phys. 2: 488
55. Hosoya H (1986) Croat. Chem. Acta 59: 583
56. Fries K (1927) Liebigs Ann. Chem. 545: 121
57. Pauling L (1939) The nature of the chemical bond, Cornell University Press, Ithaca, New York
58. Simpson WT (1953) J. Am. Chem. Soc. 75: 597

59. Klein DJ, Alexander SA, Seitz WA, Schmalz TG, Hite GE (1986) Theor. Chim. Acta 69: 393
60. Klein DJ, Hite GE, Schmalz TG (1986) J. Comp. Chem. 7: 443
61. Klein DJ, Trinajstić N (in press) Pure & Appl. Chem.
62. Dewar MJS, deLlano C (1969) J. Am. Chem. Soc. 91: 787
63. Randić M, Trinajstić N (1987) J. Am. Chem. Soc. 109: 6923
64. Živković TP (1982) Theor. Chim. Acta 61: 363; (1983) Croat. Chem. Acta 56: 29, 525
65. Gomes JANF (1981) Theor. Chim. Acta 59: 333
66. Klein DJ, Schmalz TG, Hite GE, Metropoulos A, Seitz WA (1985) Chem. Phys. Lett. 120: 367
67. Sutherland B (1988) Phys. Rev. B37: 3786
68. Hite GE, Metropoulos A, Klein DJ, Schmalz TG, Seitz WA (1986) Theor. Chim. Acta 69: 369
69. Klein DJ, Schmalz TG, Seitz WA, Hite GE (1986) Intl. J. Quantum Chem. S19: 707
70. Su WP, Schrieffer JR, Heeger AJ (1980) Phys. Rev. B22: 2099; Rice MJ (1979) Phys. Lett. 71A: 152; Heeger AJ (1981) Comments Sol. State Phys. 10: 53; Chien JCW (1984) Polyacetylene, Academic, New York; Bozovic I (1985) Mol. Cryst. Liq. Cryst. 119: 475; Hoffman R (1988) Rev. Mod. Phys. 60: 601
71. Herndon WC (1973) J. Am. Chem. Soc. 95: 2404; Herndon WC, Ellzey ML Jr (1974) J. Am. Chem. Soc. 96: 6631
72. Schaad LJ, Hess BA Jr (1982) Pure & Appl. Chem. 54: 1097
73. Randić M (1975) Chem. Phys. Lett. 38: 68; (1977) Tetrahedron 33: 1905
74. Clar E (1972) The aromatic sextet, Wiley, New York
75. Randić M, Nikolić S, Trinajstić N (1987) In: King RB, Rouvray DH (eds) Graph theory and topology in chemistry, Elsevier, Amsterdam, pp 429–447
76. Nicolić S, Randić M, Klein DJ, Plavšić D, Trinajstić N (1989) J. Mol. Struct. (Theochem) 198: 223
77. Herndon WC, Hosoya H (1984) Tetrahedron 40: 3987
78. Živković TP, Trinajstić N, Randić M (1981) Croat. Chem. Acta 54: 309
79. Kuwajima S (1984) J. Am. Chem. Soc. 106: 6496; Poshusta RD, Schmalz TG, Klein DJ (1989) Mol. Phys. 66: 317
80. Klein DJ (1986) Intl. J. Quantum Chem. S20: 153
81. Cyvin SJ, Gutman I (1988) Kekulé structures in benzenoid hydrocarbons, Springer, Berlin Heidelberg New York
82. See numerous contributions in: Klein DJ, Trinajstić N (eds) (1989) Valence-bond theory and chemical structure, Elsevier, Amsterdam

Scaling Properties of Topological Invariants

Jerzy Cioslowski*

Los Alamos National Laboratory, Theoretical Division, Group T-12, MS-J569, Los Alamos, New Mexico, USA

Table of Contents

---

* Present address: Department of Chemistry and the Supercomputer Computations Research Institute, Florida State University, Tallahassee, Florida 32306-3006, USA

Jerzy Cioslowski

The formalism of scaling relations between the reduced topological invariants is reviewed. First, the basic concepts of the chemical graph theory, the known approximate formulae for the total π-electron energy and the size extensivity of topological invariants are outlined. The scaling properties of the reduced topological invariants are introduced and the accuracy of various scaling functions is assessed. Finally, the stability index of benzenoid hydrocarbons and the lower and upper bounds to the reduced invariants are discussed.

1 Introduction

Hückel theory [1], the oldest quantum-mechanical approach to calculating the properties of organic molecules, has been outshined by the more rigorous semi-empirical and *ab initio* techniques for at least two decades. Even if the *ab initio* total energies of conjugated systems were found [2, 3] to parallel the Hückel π-electron energies, the Hückel method is certainly too crude to be of substantial value for *quantitative* considerations.

An elegant formalism that explicitly links the fields of Hückel theory and the molecular topology was proposed by Ruedenberg et al. [4] quite a long time ago and then vigorously pursued by, among others, Yugoslav researchers [5]. Certainly, the *chemical graph theory* shares all the deficiencies of the Hückel formalism, but it allows one to gain a deeper insight into the dependence of electronic properties on the molecular structure.

The ideas of chemical graph theory have a special meaning for benzenoid hydrocarbons (BHs). As the structures of BHs do not allow for the *cis/trans* isomers, there is a rigorous correspondence between the adjacency matrices of BHs and their properties. In other words, any property $P(\mathrm{G})$ of the benzenoid molecule G is a function of only its adjacency matrix $\mathbf{A}(\mathrm{G})$

$$P(\mathrm{G}) = P[\mathbf{A}(\mathrm{G})]\,. \tag{1}$$

In practice, it is much more convenient to use some quantities derived from $\mathbf{A}$ instead of the matrix itself. These quantities are usually called *topological invariants*. For example, the π-electron energy in the "variable β" calculations can be accurately approximated by topological invariants [6]

$$E_\pi(\omega) \approx E_\pi + (\omega/2)\,E_\pi^2/M\,, \tag{2}$$

where ω is a parameter. The meaning of other symbols will become clear to the reader in the next section.

In this review we discuss the scaling properties of topological invariants and the relations between them. First, we recall the basic concepts of the chemical graph theory. Then, we introduce reduced topological invariants and discuss the problem of size extensivity. Following a brief overview of the known approximate relations between topological invariants, we move to considerations of their scaling properties. Finally, we discuss some practical aspects of the present formalism.

2 Adjacency Matrix, Spectral Density Function and Topological Invariants

Much of the material we present in this paragraph is discussed exhaustively in two classical textbooks on the chemical graph theory [7, 8] and therefore we review it only very briefly. Let $\mathbf{A}(\mathrm{G})$ be the adjacency matrix pertinent to a molecular graph

G of some benzenoid hydrocarbons. We define the spectral density function corresponding to G as [9–13]

$$\Gamma(\mathrm{G}, t) = \sum_{i=1}^{N} \delta[t - x_i(\mathrm{G})], \tag{3}$$

where $\{x_i(\mathrm{G}), i = 1, N\}$ is the set of eigenvalues of $\mathbf{A}(\mathrm{G})$ and $\delta(x)$ is Dirac's delta function. As one should point out, there is a rigorous correspondence between G and $\mathbf{A}(\mathrm{G})$, but in general the spectral density $\Gamma(\mathrm{G})$ contains less information than G itself. There are so-called isospectral graphs that have the same function Γ [14]. At the present time, however, it is not clear whether the isospectral BHs do exist.

Knowledge of Γ allows us to calculate most of the topological invariants that are of interest for chemists. They include [7, 8]:

1. The moments of $\mathbf{A}(\mathrm{G})$

$$\mu_j(\mathrm{G}) = \sum_{i=1}^{N} x_i^j(\mathrm{G}) = \int_{-\infty}^{+\infty} \Gamma(\mathrm{G}, t)\, t^j\, dt. \tag{4}$$

In general case the following relations hold [7, 8]

$$\mu_0(\mathrm{G}) = N(\mathrm{G}) \quad \text{(the number of vertices in G)}, \tag{5}$$

and

$$(1/2)\,\mu_2(\mathrm{G}) = M(\mathrm{G}) \quad \text{(the number of edges in G)}. \tag{6}$$

BHs are alternant systems [7, 8]. Therefore

$$\mu_{2j-1}(\mathrm{G}) = 0 \quad \text{for} \quad j = 1, \ldots. \tag{7}$$

Eventually, for benzenoid systems [15, 16]

$$\mu_4(\mathrm{G}) = 18M(\mathrm{G}) - 12N(\mathrm{G}), \tag{8}$$

and [17]

$$\mu_6(\mathrm{G}) = 158M(\mathrm{G}) - 144N(\mathrm{G}) + 48 + 6n_b(\mathrm{G}), \tag{9}$$

where $n_b(\mathrm{G})$ is the number of bay regions or kinks in G [17].

2. The total π-electron energy

$$E_\pi(\mathrm{G}) = \sum_{i=1}^{N} |x_i(\mathrm{G})| = \int_{+\infty}^{-\infty} \Gamma(\mathrm{G}, t)\, |t|\, dt. \tag{10}$$

The above formula is valid for all neutral alternant systems.

3. The number of Kekulé structures

$$K(G) = \left(\prod_{i=1}^{N} |x_i(G)|\right)^{1/2} = \exp\left[(1/2) \int_{+\infty}^{-\infty} \Gamma(G, t) \ln |t| \, dt\right]. \tag{11}$$

The above formula is valid for, among others, benzenoid hydrocarbons.
4. The HOMO-LUMO separation

$$X(G) = 2 \min_{1 \leq i \leq N} |x_i(G)| . \tag{12}$$

Finally, we should make a remark that the descriptions of **A**(G) using $\{x_i(G), i = 1, ..., N\}$, Γ(G) or $\{\mu_i(G), i = 1, ... N\}$ are completely equivalent. In the following we will often skip the G designator and write, for example, N instead od N(G).

3 Reduced Topological Invariants

Let us introduce the following transformation [11]

$$\Gamma(G, t) \equiv (N^3/2M)^{1/2} \, \Omega[G, (N/2M)^{1/2} t] . \tag{13}$$

Eqs. (4)–(11) become

$$\int_{+\infty}^{-\infty} \Omega(G, t) \, dt = 1 , \tag{14}$$

$$\int_{+\infty}^{-\infty} \Omega(G, t) \, t^2 \, dt = 1 , \tag{15}$$

$$\int_{+\infty}^{-\infty} \Omega(G, t) \, t^j \, dt = (N/2M)^{j/2} (\mu_j/N) = \nu_j , \tag{16}$$

$$\int_{+\infty}^{-\infty} \Omega(G, t) \, dt = (2MN)^{1/2} E_\pi = e , \tag{17}$$

and

$$\int_{+\infty}^{-\infty} \Omega(G, t) \ln |t| \, dt = \ln [K^{2/N} (2M/N)^{-1/2}] = \ln k . \tag{18}$$

We also define

$$x = (2M/N)^{-1/2} (X/2) . \tag{19}$$

The new quantities, v_j, e, k and x, are the reduced (or normalized) topological invariants [10, 11]. e is also called the McClelland quotient [10], for it enters the classical McClelland formula [18]

$$E_\pi \leqq (2MN)^{1/2}, \quad \text{or} \quad E_\pi = e(2MN)^{1/2}. \tag{20}$$

It is not difficult to demonstrate that

$$0 < e \leqq 1, \quad 0 \leqq k \leqq 1, \quad \text{and} \quad 0 \leqq x \leqq 1. \tag{21}$$

The right-hand side equalities hold only for the molecular graph of the ethylene molecule.

4 Size Extensivity of Topological Invariants

Let $\{G_i, i = 1, \ldots\}$ be a homologous series (such as linear polyacenes, phenes, etc.) of BHs. It is obvious that the following limits exist

$$N_\infty = \lim_{i \to \infty} N(G_i)/i, \quad N_\infty > 0, \tag{22}$$

$$M_\infty = \lim_{i \to \infty} M(G_i)/i, \quad M_\infty > 0, \tag{23}$$

$$v_{j\infty} = \lim_{i \to \infty} v_j(G_i)/i, \quad v_{j\infty} > 0, \tag{24}$$

and

$$X_\infty = \lim_{i \to \infty} X(G_i), \quad X_\infty \geqq 0. \tag{25}$$

Also [19, 20]

$$\ln K_\infty = \lim_{i \to \infty} \ln K(G_i)/i, \quad \ln K_\infty \geqq 0, \tag{26}$$

N, M and $\ln K$ are *extensive* invariants, whereas X is *intensive*.

It is desirable that any (exact or approximate) relation between topological invariants preserves proper intensivity or extensivity. One finds out that all reduced invariants are intensive. We return to the problem of size extensivity in the next section.

5 Known Relations Between Topological Invariants

From the point of view of the organic chemist, the relations between E_π (which measures the thermodynamic stability of the molecule), X (which measures the energy of the first excited state relative to the ground state), N, M and K are of primary interest. Their importance stems from the fact that N, M and K can be quickly enumerated from the molecular graph. This makes a "pencil and paper" estimations of electronic properties of organic molecules possible.

The problem of approximate dependence of E_π upon N, M and K was addressed as early as in 1949 [21]. The equation proposed by Carter reads

$$E_\pi \approx AN + B \ln K + C, \tag{27}$$

where A, B and C are empirical parameters. Since then, more than twenty different formulae for E_π have been derived, most of them reviewed in [22]. There are also several upper bounds for E_π in terms of N, M, K (and μ_6) [12, 15, 16, 23, 24, 25]. The sharpest upper bound known reads [12]

$$E_\pi \leqq (NR + 2M)(P + 2R)^{-1/2}, \tag{28}$$

where

$$P = (91MN - 72N^2 + 24N + 3Nn_b - 18M^2)/(9MN - 6N^2 - 2M^2), \tag{29}$$

and

$$R = [(72MN - 72N^2 - 4M^2 + 48M + 6Nn_b)/(9MN - 6N^2 - 2M^2)]^{1/2}. \tag{30}$$

Since this upper bound yields the energies that are only 1–2% higher than the exact ones and because n_b does not contribute significantly to these values, this corroborates the empirical observation by Gutman et al. [22] that, in the case of BHs, "M and N determine not less than 99% of E_π". This also explains why various approximate formulae for E_π are rather accurate, even if their functional dependencies on K are quite different.

The bound (30) exhibits proper size extensivity. This is not true for other upper bounds [15, 16, 23, 24, 25]. Therefore, for large systems, they do not represent any improvement over the original McClelland estimate [18]. Similar problems are encountered with some approximate formulae for E_π [26]. Lower bounds for E_π are also known [12, 24].

Studies on the relations between X and other topological invariants are much less abundant in the chemical literature. There is only one report on an approximate formula that provides X with reasonable accuracy [10]

$$X \approx A(2M/N)^{1/2} + BK^{2/N}, \tag{31}$$

where A and B are empirical constants. In fact, validity of this approximate equation is one of the consequences of the scaling of topological invariants which we discuss below.

6 Scaling Properties of Topological Invariants

It is obvious that for molecular graphs with no more than N vertices, the knowledge of N independent topological invariants entirely determines their spectral density functions. In the general case, however, one needs an infinite number of invariants to fully specify the spectral density and, in turn, the values of E_π and X [27]. This can be formally expressed as

$$E_\pi = F_E(N, M, K, \mu_4, \ldots) \tag{32}$$

and

$$X = F_X(N, M, K, \mu_4, \ldots) \,. \tag{33}$$

From Eqs. (14)–(19) we learn that the same holds for the reduced invariants

$$e = F_e(k, \nu_4, \ldots) \tag{34}$$

and

$$x = F_x(k, \nu_4, \ldots) \,. \tag{35}$$

One should note that the entire dependence of E_π and X on N and M is absorbed here into the reduced quantities. Eqs. (34) and (35) are *exact* as long as F_e and F_x are functions of an infinite number of variables (reduced topological invariants). By limiting the number of variables, we introduce the element of approximation into Eqs. (34) and (35). Which and how many variables should be retained is a matter of choice. Of course, the simplest approximation is

$$e \approx \tilde{F}_e(k) \tag{36}$$

and

$$x \approx \tilde{F}_x(k) \,. \tag{37}$$

From (21) we expect that

$$\tilde{F}_e(1) = \tilde{F}_x(1) = 1 \,. \tag{38}$$

What we learn from Eqs. (36) and (37) is that both the reduced π-electron energy and the reduced HOMO-LUMO separation scale (approximately) according to a simple law. The analogy with the gas equation of state is appropriate here. One can define the reduced pressure, temperature and volume of the gas. Then, the reduced pressure is (approximately) a function of only the reduced volume and temperature, the nature of the gas under consideration bearing no significance. Similarly, for two different BHs with the same value of k, one can expect very similar values of both e and x.

Similar scaling relations apply to all reduced moments

$$\nu_j \approx \tilde{F}_j(k)\,, \qquad \tilde{F}_j(1) = 1\,, \qquad j = 4, 6, 8, \ldots \tag{39}$$

Two comments are necessary here. First, one can attain a better accuracy by using functions of more variables. The logical choice is to use the sixth reduced moment

$$\nu_6 = (N^2/8M^3)\,\mu_6 = (N/M)^3\,[(79M/4N) - 18 + (6/N) + (3n_b/4N)] \tag{40}$$

as the second variable (see Sect. 8).

Second, we should answer the question about scaling of other topological invariants. Research along this line has been pursued by Gutman [28]. The conclusion is that among size extensive invariants only ln K and the moments exhibit proper scaling. This means that only these quantities, together with the HOMO-LUMO separation, can be used as both the independent and dependent variables in the scaling relations.

In the next section of our review we present various approaches to approximating $\tilde{F}_e$ and $\tilde{F}_x$.

7 The Scaling Functions $\tilde{F}_e$ and $\tilde{F}_x$

In the previous section we demonstrated the existence of scaling relations between reduced topological invariants. However, our study cannot be completed without finding the explicit formulae for the scaling functions $\tilde{F}_e$ and $\tilde{F}_x$. This can be done in three distinct ways. We describe each of them separately.

7.1 Numerical Approximations

One can simply try to fit some algebraic functions to $\tilde{F}_e$ and $\tilde{F}_x$. A linear correlation is an obvious choice. It has been found (by the least square fit to the values of e and x for 18388 BHs) that [10, 13]

$$\tilde{F}_e(k) \approx 0.79848 + 0.13853k\,, \qquad \tilde{F}_e(1) = 0.93701\,,$$
$$\text{RMS error: } 9.9 \times 10^{-4} \tag{41}$$

and

$$\tilde{F}_x(k) \approx -2.58298 + 3.49979k\,, \qquad \tilde{F}_x(1) = 0.91681\,,$$
$$\text{RMS error: } 2.6 \times 10^{-2}\,. \tag{42}$$

As one learns from Eqs. (41) and (42), the linear approximation is not a very good one, especially for $\tilde{F}_e$. The quadratic fit fares slightly better for $\tilde{F}_e$

$$\tilde{F}_e(k) \approx 1.72534 - 2.24025k + 1.52480k^2\,, \qquad \tilde{F}_e(1) = 1.00989\,,$$
$$\text{RMS error: } 5.8 \times 10^{-4} \tag{43}$$

and brings almost no improvement for $\tilde{F}_x$ [10].

In the related study [29], Gutman et al. have pointed out that other nonlinear fits to $\tilde{F}_e$ are not much better than the quadratic one, but inclusion of the nonlinear terms is absolutely necessary to fulfill the condition (38).

The numerical fitting has its merits and disadvantages. Certainly, it is simple, but on the other hand it has empirical character and provides no insight into the mathematical nature of scaling relations.

7.2 Model Spectral Densities

Another approach to approximating the scaling functions is to construct some model spectral densities. In turn, Eqs. (14)–(19) yield the approximate $\tilde{F}_e$ and $\tilde{F}_x$. The choice of the Dirac delta function for Γ yields trivial relations

$$\tilde{F}_e(k) = \tilde{F}_x(k) = 1\,. \tag{44}$$

Therefore, the simplest, yet reasonable, choice is a sum of two delta functions of the same magnitude [30]. This results in

$$\tilde{F}_e(k) \approx [(1 + k^2)/2]^{1/2}\,, \qquad \text{RMS error: } 7.2 \times 10^{-3} \tag{45}$$

and

$$\tilde{F}_x(k) \approx [1 - (1 - k^4)^{1/2}]^{1/2}\,, \qquad \text{RMS error: } 2.8 \times 10^{-1}\,. \tag{46}$$

The approximation for $\tilde{F}_e$ has a moderate accuracy, whereas the result for $\tilde{F}_x$ is completely unsatisfactory.

A much better agreement is achieved when a saw-tooth model spectral density is used. The reader is referred to Ref. [11] for more details.

7.3 Template Molecules

The best results are attained within the template molecules approach. In order to arrive at approximations to the scaling functions we select a homologous series of molecules, $\{G_i\}$. For this series, the reduced topological invariants are functions of only the index i. We formally invert the functions $k(i)$ to obtain direct relations between the reduced invariants. One can say that the homologous series forms a template for the scaling relations.

We have found [13] that the linear polyenes constitute a convenient template. Let us introduce an auxiliary function

$$v(k) = (2k^2 - 1)/(4k^2 - 1)\,. \tag{47}$$

We have

$$\tilde{F}_e(k) \approx k^{-1}(2k^2 - 1)\,\{[\sin(\pi v(k)/2)]^{-1} - 1\}\,,$$
$$\text{RMS error: } 1.7\times 10^{-3} \tag{48}$$

and

$$\tilde{F}_x(k) \approx 2k^{-1}\sin(\pi v(k)/2)\,, \qquad \text{RMS error: } 4.7\times 10^{-1}\,. \tag{49}$$

Eq. (48) contains no empirical parameters, yet it has an impressive accuracy. The approximation for x is far worse. We believe that it could be vastly improved by using more sophisticated (such as the linear polyacenes) template. However, this would result in difficulties in inverting the function that describes the dependence of k on the index i.

8 Assessment of Accuracy of Scaling Relations

In the above considerations we dealt with the errors introduced by numerical or analytical approximations to the scaling functions. However, one could ask what is the best accuracy to be expected within the formalism of scaling relations. We attempt to answer this question by quoting the results of the following numerical experiment: We calculated various topological invariants for the sample of 18388 BHs. The molecules were divided into classes according to the values of selected invariants. For example, the entry (N, M) in Table 1 corresponds to BHs classified according to the values of N and M. We computed the maximal variations of both e and x within each class. Finally, the variations were averaged over the classes to yield their RMS (root mean square), STD (standard deviation) and MAX (maximal) values. There are several interesting observations to be made. First, the replacement of topological invariants by the reduced quantities does not increase the variations significantly compare the entries (N, M, K) and (k), (N, M, K, μ_6) and (k, ν_6). This proves the validity of scaling relations. Second, the variations in x are far larger than those in e. This means that the approximate formulae for x are necessarily less accurate than those for e. Third, the reduced fourth moment, ν_4, does not bear any substantial information; compare (ν_6) and (ν_4, ν_6), (k) and (k, ν_4) etc. The two-parameter scaling relations with k and ν_6 are bound to perform much better than the one-parameter scaling relations with k that we discussed in previous sections. Finally, there appears to be no room for a further improvement in approximating $\tilde{F}_e(k)$ and $\tilde{F}_x(k)$; compare the RMS errors of Eqs. (42) and (43) with the halves of variations for the (k)-entry in Table 1.

Table 1. Theoretical accuracy of various relations between topological invariants

Parameters	Variation of e within a class			Variation of x within a class		
	RMS	STD	MAX	RMS	STD	MAX
(N, M)	1.14×10^{-2}	9.73×10^{-3}	1.97×10^{-2}	2.15×10^{-1}	1.88×10^{-1}	3.13×10^{-1}
(N, M, μ_6)	4.18×10^{-3}	3.00×10^{-3}	9.51×10^{-3}	1.40×10^{-1}	1.08×10^{-1}	3.08×10^{-1}
(N, M, K)	1.23×10^{-3}	8.14×10^{-4}	6.58×10^{-3}	4.73×10^{-2}	3.63×10^{-2}	1.13×10^{-1}
(N, M, K, μ_6)	3.32×10^{-4}	2.20×10^{-4}	1.45×10^{-3}	2.47×10^{-2}	1.68×10^{-2}	8.32×10^{-2}
(ν_4)	1.21×10^{-2}	1.05×10^{-2}	1.97×10^{-2}	2.27×10^{-1}	2.00×10^{-1}	3.13×10^{-1}
(ν_6)	4.33×10^{-3}	3.13×10^{-3}	9.51×10^{-3}	1.44×10^{-1}	1.11×10^{-1}	3.10×10^{-1}
(ν_4, ν_6)	4.33×10^{-3}	3.13×10^{-3}	9.51×10^{-3}	1.44×10^{-1}	1.11×10^{-1}	3.10×10^{-1}
(k)	1.25×10^{-3}	8.29×10^{-4}	6.58×10^{-3}	4.77×10^{-2}	3.66×10^{-2}	1.24×10^{-1}
(k, ν_4)	1.24×10^{-3}	8.19×10^{-4}	6.58×10^{-3}	4.74×10^{-2}	3.63×10^{-2}	1.13×10^{-1}
(k, ν_6)	3.32×10^{-4}	2.20×10^{-4}	1.45×10^{-3}	2.47×10^{-2}	1.68×10^{-2}	8.32×10^{-2}
(k, ν_4, ν_6)	3.32×10^{-4}	2.20×10^{-4}	1.45×10^{-3}	2.47×10^{-2}	1.68×10^{-2}	8.32×10^{-2}

9 Stability Index (SI) of Benzenoid Hydrocarbons

The stability of BHs (and conjugated systems in general) is usually appraised through the topological resonance energy per electron, TRE(PE), [31]. Unfortunately, though we cannot deny that the concept of TRE has a certain elegance and simplicity, there are numerous examples of the stability predictions based on TRE that are contrary to the chemical sense. In particular, molecules with unpaired electrons (radicals) are predicted to be excessively stable within the TRE approach [32].

Eq. (41) is a convenient starting point for formulation of yet another measure of the stability of BHs [10]. It provides an approximation to E_π

$$E_\pi \approx 0.79848(2MN)^{1/2} + 0.13853NK^{2/N} \tag{50}$$

which enables one to partition the total π-electron energy into two terms. The first one is a "classical" energy due to formation of carbon-carbon bonds. The second term is an "extra" energy due to cyclic conjugation represented by K. This energy is proportional to N.

Table 2. Stability of some BHs

Molecule	TRE(PE)	SI
naphthalene	0.0389	1.246
anthracene	0.0339	1.219
phenanthrene	0.0390	1.258
pyrene	0.0374	1.251
perylene	0.0370	1.246
coronene	0.0395	1.283

In light of the above facts, we regard $K^{2/N}$ as a measure of the stability of BHs. We call it the stability index (SI). In most cases, the values of SI parallel those of TRE, but sometimes there are significant differences. In these cases SI is found to be the more reliable predictor [10]. In particular, SI = 0 for radical molecules and SI $\geqq$ 1 for the molecules with the singlet ground state. In Table 2 we compile SIs and TRE(PE)s of some benzenoid hydrocarbons.

10 Lower and Upper Bounds to Reduced Topological Invariants

In Table 3 we summarize the lower and upper bounds to k, e, x and SI. The lower bound for k can be derived from the fact that $K \geqq 1$ and $(2M/N)^{1/2} \leqq \sqrt{3}$. This gives $k \geqq 1/\sqrt{3}$. However, we conjecture that the smallest value of k corresponds to the linear ∞-acene and the largest one to the benzene molecule. We believe that it is also true for the McClelland quotient, e. A rigorous lower bound for e results from the inequality [33]

$$e \geqq 1 + \alpha \ln k; \qquad \alpha = (2 - \sqrt{3})/(1 - \ln \sqrt{3}) \approx 0.594526\,, \tag{51}$$

and the above lower bound for k. This yields $e \geqq [1 - (3 - \sqrt{3}) \ln \sqrt{3}]/(1 - \ln \sqrt{3})$. Finally, we note that the smallest value of h corresponds to the linear ∞-acene and we conjecture that the upper bound corresponds to the benzene molecule.

Table 3. Lower and upper bounds to some topological invariants

Invariant	Lower bound			Upper bound		
	RIG[a]	CON[a]	OBS[a]	RIG[a]	CON[a]	OBS[a]
k	0.577[b]	0.632[c]	0.713	1.000	0.891[d]	0.891
e	0.673[e]	0.887	0.895	1.000	0.943[f]	0.943
x	0.000	0.000[g]	0.016	1.000	0.707[h]	0.707
SI	1.000	1.000[g]	1.116	1.732[i]	?	1.294

a RIG: a rigorous bound, CON: a conjectured bound, OBS: the observed extremal values for the sample of 18388 BHs (see text).
b $3^{-1/2}$, c $(2/5)^{1/2}$, d $2^{-1/6}$, e see text, f $(8/9)^{1/2}$, g a rigorous bound, h $2^{-1/2}$, i $3^{1/2}$.

The rigorous lower bound to SI is equal 1 and this value is obtained for the linear ∞-acene. We also find that SI $\leqq \sqrt{3}$. However, this bound is too loose as revealed by the calculations on the sample of 18388 BHs. At the present, we do not know how to improve this result.

11 Conclusions

It is rewarding to learn that various topological invariants are interrelated through apparently simple scaling laws. The scaling relations enable one to derive approximate formulae for the invariants in a systematic way. Moreover, since the proper extensivity of the invariants is preserved, the resulting estimates have the same degree of accuracy for both large and small molecules.

With the dependence on N and M absorbed into the definitions of the reduced invariants, a single parameter k appears to control most of their variation observed in numerical calculations. If the accuracy provided by these one-parameter relations appears unsatisfactory, one can improve it vastly by including a second reduced invariant, namely ν_6.

It should be pointed out that the above discussion is not limited to benzenoid hydrocarbons. In fact, most of the equations are valid for all conjugated systems. We have already benefited from this fact in designing template molecules in Sect. 7.3 of this review.

Finally, we should come back to Eq. (2) in the introduction. We learn that the scaling relations are a *unique property of the Hückel Hamiltonian.* Even the simplest extension, such as introduction of the "variable β" approximation, invalidates the scaling. This is certainly not a good news from the point of view of the practical applications. Still, the author believes that the appealing simplicity of the Hückel Hamiltonian makes it worthwhile to investigate, even if just for the pure fun of mathematical adventure.

12 Acknowledgement

This work has been pursued under the auspices of the United States Department of Energy.

13 References

1. Hückel E (1931) Z. Physik 70: 204; (1932) 72: 310; (1932) 76: 628
2. Schaad LJ, Hess BA Jr (1972) J. Amer. Chem. Soc. 94: 3068
3. Ichikawa H, Ebisawa Y (1985) J Amer. Chem. Soc. 107: 1161
4. Ham NS, Ruedenberg K (1958) J. Chem. Phys. 29: 1215 and the references cited therein
5. see for example refs. [7] and [8]
6. Cioslowski J (1988) Intern. J. Quant. Chem. 34: 417
7. Gutman I, Polansky OE (1986) Mathematical concepts in organic chemistry, Springer, Berlin Heidelberg New York
8. Graovac A, Gutman I, Trinajstić N (1977) Topological approach to the chemistry of conjugated molecules, Springer, Berlin Heidelberg New York
9. Cioslowski J (1986) Match 20: 95
10. Cioslowski J (1987) Intern. J. Quant. Chem. 31: 581
11. Cioslowski J, Polansky OE (1988) Theor. Chim. Acta 74: 55
12. Cioslowski J (1988) Intern. J. Quant. Chem. 34: 217
13. Cioslowski J (in press) MATCH
14. Collatz L, Sinogowitz drop this (1957) Abh. Math. Sem. Univ. Hamburg 21: 64
15. Cioslowski J (1985) Z. Naturforsch. 48a: 1167

16. Gutman I, Türker L, Dias JR (1986) Match 19: 147
17. Dias JR (1985) Theor. Chim. Acta 68: 107
18. McClelland BJ (1971) J. Chem. Phys. 54: 640
19. Cioslowski J (1987) Intern. J. Quant. Chem. 31: 605
20. Graovac A, Babić D, Strunje M (1986) Chem. Phys. Lett. 123: 433
21. Carter PC (1949) Trans. Faraday Soc. 45: 597
22. Gutman I, Nedeljković Lj, Teodorović AV (1983) Bull. Soc. Chim. Beograd 48: 495
23. Türker L (1984) Match 16: 83
24. Gutman I, Teodorović AV, Nedeljković ILj (1984) Theor. Chim. Acta 65: 23
25. Cioslowski J, Gutman I (1986) Z. Naturforsch. 41a: 861
26. Hall GG (1981) Inst. Math. Appl. 17: 70
27. Gutman I (1974) Theor. Chim. Acta 35: 355
28. Gutman I (1987) Match 22: 269
29. Gutman I, Marković S, Marinković M (1987) Match 22: 277
30. Cioslowski J, Graovac A (1988) Croat. Chem. Acta 61: 797
31. Gutman I, Milun M, Trinajstić N (1977) J. Amer. Chem. Soc. 99: 1692
32. Gutman I (1981) Z. Naturforsch. 36a: 128
33. Cioslowski J (sorry, ou)

Molecular Topology and Chemical Reactivity of Polynuclear Benzenoid Hydrocarbons

Maximilian Zander

Rütgerswerke AG, D-4620 Castrop-Rauxel, FRG

Table of Contents

After a brief discussion of the notion of molecular topology and the analogy principle as related to topology/reactivity relationships more recent developments in the field of reactivity indices for polynuclear benzenoid hydrocarbons are reviewed. Reaction mechanisms and correlations of reactivity indices with rates of electrophilic substitution and Diels-Alder reactions, thermally induced polymerization, and biochemical transformations of benzenoid hydrocarbons are discussed.

1 Introduction

Relationships of molecular topology to chemical reactivity of polynuclear benzenoid hydrocarbons are of interest for both theoreticians and experimentalists. Reactivity indices, derived from topological approaches to the chemistry of benzenoid hydrocarbons, have proved useful in such varied fields as mechanistic organic chemistry, biochemistry, carbon science, and environmental science. It is not the aim of this review, however, to give a comprehensive account of reactivity indices of benzenoid hydrocarbons. Instead, the main emphasis is placed on the underlying basic principles of the relationship between topology and reactivity of benzenoid hydrocarbons.

2 Molecular Topology and its Significance for the Physical and Chemical Properties of Benzenoid Hydrocarbons

The basic concept of organic chemistry is that of Molecular Structure. It includes the idea that molecules can be regarded as isolated objects, i.e. as separable from their environment, that they "possess" a structure that determines their physical and chemical properties, and finally that this molecular structure can be adequately described by structural formulae. In terms of physics the notion of structure is directly related to the Born-Oppenheimer description of molecules. Although "Molecular Structure" makes no appearance in a quantum treatment of molecules starting from first principles [1, 2] the concept is clearly justified by its overwhelming success in organic chemistry.

Three characteristics of the molecular structure of a compound can be distinguished [3]:

1 The kind of atoms present in the molecule,
2 the pattern of bonds between the atoms, and
3 the geometry of the molecule, i.e. bond lengths and angles.

The second characteristic is usually termed "Molecular Topology" [4]. The topology of a molecule can be derived from its constitutional formula by abstracting from the nature of atoms and the type of bonds. By depicting each atom of the molecule by a small circle (vertex) and connecting these vertices by lines (edges) where in the constitutional formula a bond is indicated between the corresponding atoms, one arrives at the "complete molecular graph". By neglecting the hydrogen atoms this can be further reduced to the "skeleton graph". In the particular case of benzenoid hydrocarbons (or other fully conjugated systems) the skeleton graph is isomorphic with the "Hückel graph" which depicts the corresponding π-AO basis set of the system. As has been clearly shown by two different methods [5, 6] topological spaces can be defined on molecular graphs and hence the axiomatics and methods of mathematical topology can be applied to these graphs.

With some exceptions, i.e. non-planar systems, variances in geometry are small within the class of benzenoid (alternant) hydrocarbons and thus can be neglected. Hence the individual representatives of this class of compounds differ only with regard to their molecular topologies. Provided the influence of the non-topological structural characteristics (kind of atoms, geometry, additional electronic interactions that are not referred to in the constitutional formulae) on the physical and chemical properties of benzenoid hydrocarbons would be topology-invariant, then all changes in properties with structure would have to be regarded as being simply a result of changes in topology. In fact, the observation that all theoretical methods that take into account nearest-neighbour interactions only ("tight-binding approximations") are well suited to reproduce quantitatively many properties of benzenoid hydrocarbons in agreement with experiment corroborate the aforementioned assumption. These methods include HMO and VB theory as well as purely graph-theoretical approaches. The significance of topology for the properties of benzenoid hydrocarbons can be exceedingly clearly demonstrated by referring to Ruedenberg's Free Electron-Network model [7] (which is known to be mathematically isomorphic with the Hückel method). In the free electron model the π electrons of benzenoid hydrocarbons are treated as de Broglie electrons (in three-dimensional configuration space) "moving" along the walls of the "potential box". The relative energies of the electrons are obtained as a function of the shape of the potential box, i.e. as a function of molecular topology of the system.

The observation that the influence of the non-topological structural characteristics on the properties of benzenoid hydrocarbons is widely topology-invariant also means that the influence of these characteristics and of molecular topology must be regarded as independent from each other, i.e. as separable.

Of course, neglecting geometry in a theoretical treatment of benzenoid hydrocarbons is justified only because geometry can be regarded as approximately constant. It does not mean, however, that geometry does not play an important role with regard to the properties of benzenoid hydrocarbons. For example, it has been shown that it is only within the constraints of a symmetric hexagonal σ-framework that the π-electrons of benzene tend to delocalize [8].

While most of the π-electronic properties of benzenoid hydrocarbons depend strongly on molecular topology there are some properties that are almost exclusively determined by the number of carbon atoms (size) of the systems, i.e. values of those properties are nearly identical for isomeric hydrocarbons. An example is the total π-electron energy of a benzenoid hydrocarbon, the gross part ($> 95\%$) of which being determined by the size of the molecules [9]. In so-called "equi-topological series" of benzenoid hydrocarbons, however, both topology-dependent and independent properties are simply functions of molecular size. "Equi-topological series" of hydrocarbons are defined as those in which the respective subsequent compound is formally produced from the previous compound by addition of a certain fragment, with the same topological pattern always being used for the connection [10]. A well-known example is the acene series (anthracene, tetracene, pentacene ... etc) where all properties monotonously change with molecular size (Clar's Annellation Principle [11]). In equi-topological series also chemical reactivity is simply a function of the size of the systems. For example, the reaction rate constant of the endocyclic Diels-Alder addition of maleic anhydride (see sect. 5.2) to the different members of an equi-topological series increases to a good approximation linearly with molecular size of the systems [12].

3 The Analogy Principle and the Physical Meaning of Topology/Reactivity Relationships

Besides the concept of Molecular Structure the other fundamental concept in organic chemistry is the Analogy Principle. It was clearly formulated as early as the middle of the 19th century [13] and states (according to Hammett [14]) that "like substances react similarly and that similar changes in structure produce similar changes in reactivity". This holds also for "topology" instead of "structure" in cases where the molecular topology predominantly determines chemical properties.

The so-called "Quantitative Analogy Models" [15] constitute the quantitative version of the Analogy Principle. The particular types of quantitative analogy models referring to chemical reactivity are the well-known "Linear Free Energy Relationships" (LFER) where "free energy" stands for free energy of reaction or of activation depending on whether we are dealing with chemical equilibria or reaction rates.

LFER's for reaction rates rest on the Arrhenius equation

$$k = Ae^{-E^{\ddagger}/RT} \tag{1}$$

where $E^{\ddagger}$ is the activation energy and A is related to the entropy $\Delta S^{\ddagger}$ of reaction according to

$$A = \frac{kT}{h} e^{\Delta S^{\ddagger}/R} \tag{2}$$

(k = Boltzmann's constant, h = Planck's constant).

At constant temperature the ratio of rate constants k_1 and k_2 of two reactions having the same entropy of activation is given by

$$\log \frac{k_1}{k_2} = -\text{const.}\,(E_1^{\ddagger} - E_2^{\ddagger})\,. \tag{3}$$

When we compare Eq. (3) with the Hammett relation

$$\log \frac{k_1}{k_2} = \sigma \varrho \tag{4}$$

it becomes immediately clear that the product of "reactivity constant" σ and "reaction constant" ϱ is proportional to the difference of the activation energies of the two reactions being considered. The same holds e.g. for the Dewar equation (5) (and similar relations):

$$\log \frac{k_1}{k_2} = -\,\text{const.} \times \beta (N_1 - N_2) \tag{5}$$

where the resonance integral β corresponds to the reaction constant and the difference $(N_1 - N_2)$ of Dewar numbers to the reactivity constant.

In all these cases relative reaction rate constants (k_1/k_2) (k_2 being the rate constant of a reference reaction) are modeled by means of a product of two independent variables in which the influences of "external" and "structural" parameters respectively on activation energies are absorbed. In the case of type (5) relations the term (const. × β) contains the entire "physics" of the system while the "reactivity constant" (i.e. the Dewar number) is purely topological in nature. The separability of non-topological and topological influences on properties obviously also holds for the property "chemical reactivity". This is the reason for the surprising observation that the relative rate constants of entirely different types of reactions of benzenoid hydrocarbons, e.g. cycloaddition or electrophilic substitution reactions respectively can be calculated from one and the same topological reactivity constant (e.g. Dewar numbers) in good agreement with experiment.

For reactions that satisfactorily can be represented in terms of independent bond-breaking and bond-forming processes the Bell-Evans-Polanyi (BEP) Principle [16, 17] holds:

$$E^{\ddagger} = A + B\,\Delta H \tag{6}$$

where ΔH is the heat of reaction and A and B are constants. For similar reactions with identical entropy of reaction the ratio of equilibrium constants K is given by

$$\log \frac{K_1}{K_2} = -\frac{1}{RT}(\Delta H_1 - \Delta H_2)\,. \tag{7}$$

Thus by combining Eqs. (3), (6) and (7) we obtain:

$$\log \frac{k_1}{k_2} = B \log \frac{K_1}{K_2}\,. \tag{8}$$

In cases where $B = 1$ the Hammett relation also holds for equilibrium constants:

$$\log \frac{K_1}{K_2} = \log \frac{k_1}{k_2} = \sigma\varrho\,. \tag{9}$$

Finally, it should be emphasized that in the first place LFER's are nothing else but empirical models of similarity. Wold and Sjöström [15] have raised the interesting question whether each LFER is a special kind of "natural law", which can be derived for the particular application from underlying principles such as quantum mechanics, or whether the abundance of LFER rather indicates that they exhibit some general principle in the behavior of chemical systems.

4 Reactivity Indices for Benzenoid Hydrocarbons

Qualitative relationships between molecular topology and chemical reactivity of benzenoid aromatic hydrocarbons can readily be recognized by inspection of their molecular graphs. For example, the Hückel graphs of benzenoid hydrocarbons contain two types of vertices, viz. vertices of degree 2 and 3 (denoted here by $v_{(2)}$ and $v_{(3)}$) [18]. According to the neighbourhood relations the set of vertices $v_{(2)}$ can be further subdivided into three groups, vic. vertices $v_{(2)}$ adjacent to 2 vertices $v_{(2)}$ (e.g. the 2-position in anthracene (*1*)), vertices $v_{(2)}$ adjacent to 1 vertex $v_{(2)}$ and to another vertex $v_{(3)}$ (e.g. the 1-position in *1*), and vertices $v_{(2)}$ adjacent to 2 vertices $v_{(3)}$ (e.g. the 9-position in *1*). Generally the reactivity of vertices $v_{(2)}$ of benzenoid hydrocarbons increases in this order. Even in highly unsymmetric benzenoid hydrocarbons with many different reactive sites the most reactive site can in many cases thus be clearly identified by inspection of the graphs [19]. For example, the Hückel graph of benzo[a]pyrene (*2*) contains only one vertex $v_{(2)}$ of the type adjacent to 2 vertices $v_{(3)}$, and it is well established that this 5-position (see formula *2*) is the most reactive one of all positions present in benzo[a]pyrene.

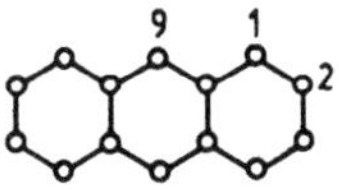

1

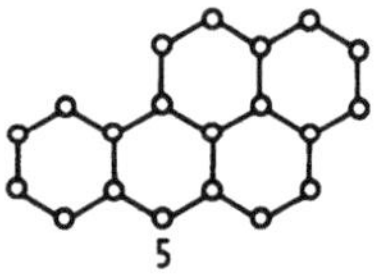

2

Simple relationships between molecular topology and reactivity can also be derived from the Clar formulae [20] of benzenoid hydrocarbons. For example, in a series of isomeric hydrocarbons the system with the largest number of inherent π-sextets is always the least reactive one. Due to the correspondence between number of π-sextets in Clar formulae and number of 120° angles [21] in the dualist graphs [22] of benzenoid hydrocarbons a similar relation holds for the dualist graphs.

Various reactivity indices have been derived for benzenoid hydrocarbons from the following purely topological approaches: the Hückel model (HMO), first-order perturbation theory (PMO), the free electron MO model (FEMO), and valence-bond structure resonance theory (VBSRT). Since many of the indices that have been known for a long time (index of free valence F_r, self-atom polarizability π_{rr}, superdelocalizability S_r, Brown's index Z, cation localization energy L_r^+, Dewar reactivity number N_r, Brown's para-localization energy L_p) have been described in detail by Streitwieser in his well-known volume [23] we will refer here only to some more recent developments.

Herndon's structure count ratio (SCR) [24] derived from VBSRT [25] is defined as

$$\mathrm{SCR} = \left(\frac{\mathrm{SC}_{\mathrm{P\,(or\,I)}}}{\mathrm{SC}_{\mathrm{R}}}\right) \tag{10}$$

where SC_R is the number of Kekulé structures[1] of the reacting hydrocarbon and $SC_{P\,(or\,I)}$ of the product (P) or the rate-controlling intermediate (I) respectively; ln SC values are proportional to resonance energies [25]. Accordingly lu SCR values correlate with rate constants. It depends on the particular type of reaction whether SC_P or SC_I has to be applied, e.g. SC_P in the case of Diels-Alder reactions and SC_I for electrophilic substitution reactions. The quickest way to obtain SC's is based on the correspondence of the sum of absolute values of the unnormalized non-bonding molecular orbital (NBMO) coefficients of odd alternant ions or radicals to the number of Kekulé structures for that ion or radical. SC's of even systems (e.g. SC_R) are obtained by deleting an orbital from the even system and then calculating the sum of absolute values of the unnormalized NBMO coefficients at points adjacent to the deleted orbital [26, 27]. The unnormalized NBMO coefficients are available by a "pencil-and-paper" method making use of the Longuet-Higgins zero-sum rule [28]. The method for obtaining SC_R and SC_I is illustrated in Scheme 1. For series of topologically related benzenoid hydrocarbons closed formulae for SCR's have been derived [29], and a method for calculating SCR's of non-alternant systems (fluoranthenes) has also been reported [30]. A computer program for calculating SC's according to Herndon's method [26, 27] of very large benzenoid systems (with up to 547 benzene rings) is available [31].

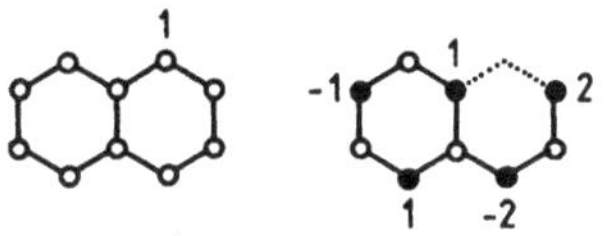

$SC_R = 3$
$SC_{J\,(1)} = 7$
$SCR_{(1)} = 2.33$

Scheme 1. Calculation of SCR for the 1-position of naphthalene. (Note, that in the case of endocyclic Diels-Alder adducts SC_p is given by the product of the SC's of the two isolated even-alternant systems occurring in the adduct, e.g. SC_p of the Diels-Alder adduct of anthracene = 4)

More recently a free electron version (PMO-F) of the first-order PMO method has been introduced [32, 33]. According to the branching conditions [34] of the FEMO model three types of resonance integral β' are used:

$\beta_1' = \beta$ for a bond linking 2 vertices v_2 (2 "nonjoints")
$\beta_2' = \sqrt{2/3}\,\beta$ for a bond linking a vertex v_3 (a "joint") with a vertex v_2
$\beta_3' = 2/3\,\beta$ for a bond linking 2 vertices v_3.

In order to obtain correct normalization, the contributions to the electron density have to be integrated along the three bonds converging to a "joint", whereas only two bonds converge to a "nonjoint". As a consequence the FE NBMO coefficients of joints and nonjoint are related by

$$c_{oj}^{FE} = \sqrt{3/2}\,c_{onj}^{FE} \tag{11}$$

1 The term "Kekulé structure" will refer to any valence-bond structure for unsaturated compounds in which single and double bonds alternate.

The calculus for obtaining reactivity numbers according to the PMO or PMO-F model respectively is outlined in Scheme 2.

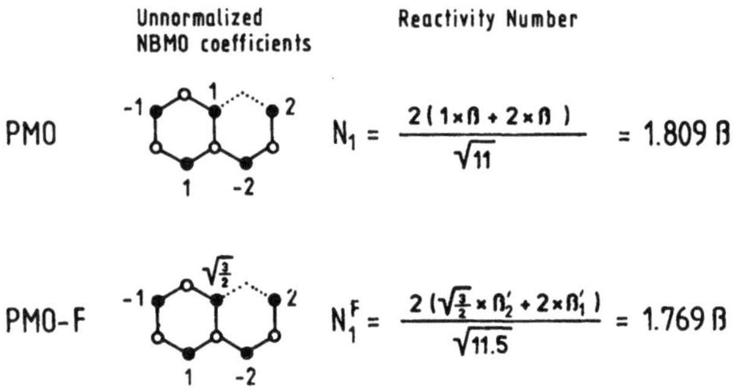

Scheme 2. Calculation of the reactivity number (localization energy) for the 1-position of naphthalene according to the PMO and PMO-F method. (The denominator follows from the normalization condition, i.e. the normalized NBMO coefficients are inversely proportional to the root of the sum of squares of the unnormalized coefficients)

Reference should also be made to a superdelocalizability index S_r^{FE} derived within the frame of the simple FEMO model [35]. Goodness of fit of correlations of S_r^{FE} values with relative rate constants for electrophilic aromatic substitution was found to be comparable with those based on CNDO/2 calculations.

In contrast to the aforementioned reactivity indices, which, in principle, can be regarded as being related to some model of reaction mechanism, the so-called "character orders" derived within Polansky's pars-orbital method [36] are explicitly measures of similarity [37] between different compounds. In this respect Polansky's method constitutes a consistent application of the analogy principle to quantum chemistry. The pars-orbital approach includes formally dividing a molecule into partial structures L and defining a "character order" whose value measures the analogy of the π-electron system of L with that of a reference compound. In the formulae *3b* to *d*, partial structures L are marked by thicker lines. Formula *3e* gives the benzoid and butadienoid character orders (multiplied by a factor of 1000) of the respective partial structures. The character orders are defined such that their value is greater the more pronounced is the analogy of the π-electron system of L with that of the reference compound – in the present case, benzene and butadiene, respectively. From the mathematical point of view, character orders are obtained by the projection of the wave function of the reference compound onto the wavefunction of the molecule considered. Character orders obtained in HMO approximation have been correlated with other HMO reactivity indices [38, 39]. On the basis of Polansky's pars-orbital method so-called pseudo-pars-orbitals have been defined enabling the calculation of localization energies that correlate with HMO cation localization energies L_r^+ [40]. Besides character orders other quantum chemical measures of similarity (based on the charge-bond order matrix) have also been derived [41] and tested in correlations with reaction rates [37].

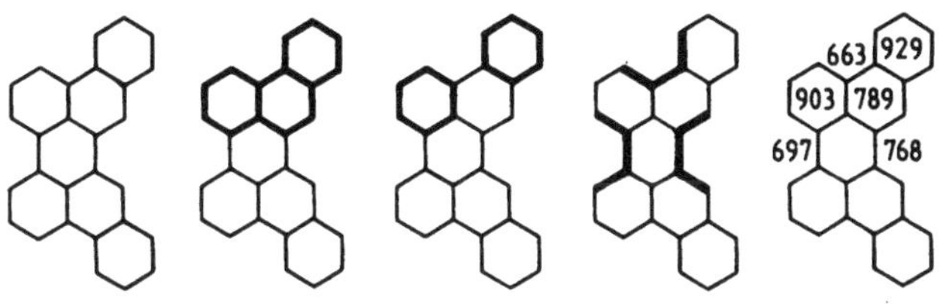

3a b c d e

As rate constants of reactions of benzenoid hydrocarbons can be related to differences in energy of reactants and products (or intermediates) Hess-Schaad resonance energy differences can also be used as reactivity indices. Hess-Schaad resonance energies are defined as the difference between HMO π energy and the additive contribution obtained by summing individual bond energies [42, 43]. Correlations of Hess-Schaad resonance energies with other HMO parameters have been discussed [44, 45].

There is increasing interest in the chemical behaviour of very large benzenoid hydrocarbons as they have been suspected to occur e.g. in interstellar matter [46]. Stabilities of homologous series of well-defined graphite-like layers with hexagonal symmetry containing up to 16000 carbon atoms have been studied [47]. Edge structure was found to be far more important than size as a determining factor for localization energies (calculated by the Dewar PMO method). In Fig. 1 edge structures of the series studied are depicted while the minimum, maximum and average values of Dewar reactivity numbers (PMO localization energies) for the different series are given in Table 1 [48].

The various types of reactivity indices can be classified in different ways, for example according to the quantum chemical method used for calculation e.g. purely topological methods (HMO, PMO, FEMO, VBSRT), SCF methods taking into account electron repulsion and other methods, or as related to the structure (energy) of reactant,

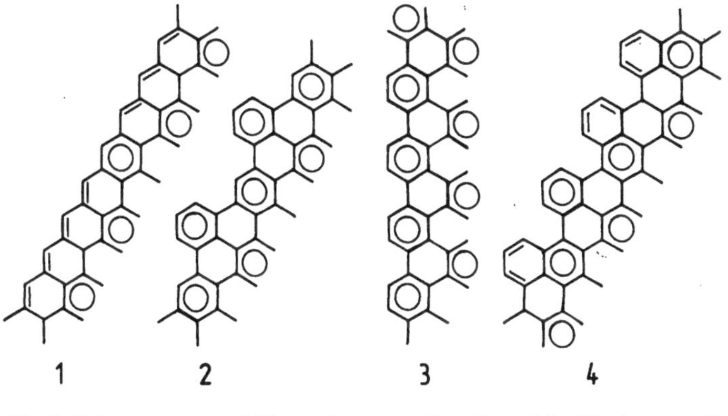

1 2 3 4

Fig. 1. Edge structure of 4 homologous series of graphite-like layers with hexagonal symmetry [47, 48]

Table 1. Dewar localization energies (β units) of different types of graphite-like layers with hexagonal symmetry [48]

Type[a]	C number	Localization energy Minimum	Maximum	Average
1	1014	0.03	2.27	0.69
2	1086	1.20	2.29	1.66
3	1014	1.23	2.31	1.66
4	978	0.26	2.23	1.19

a for edge structure see Fig. 1.

transition state or product respectively. A frequently used classification distinguishes between "models of isolated molecules", "localization models" and "transition state models".The models of isolated molecules are related to properties of the reactants. Transition state models requiring knowledge of the geometry of the transition state, can only rarely be applied.

A classification of reactivity indices according to the particular quantum chemical method used for calculation has the advantage that it does not refer to mostly unproven assumptions on the relation between physical meaning of the index and mechanism of reactions, although a pictorial interpretation of the physical meaning of the index may be of didactic and heuristic value.

5 Reaction Mechanisms and Correlations of Rates of Reaction with Reactivity Indices of Benzenoid Hydrocarbons – Some Selected Examples

5.1 Electrophilic Substitution

From the mechanistic point of view the prototype of electrophilic substitution of a hydrogen atom of a benzenoid aromatic hydrocarbon AH by an electrophile E^+ can be defined as a reaction that meets the following conditions: (a) reaction occurs via a σ-complex *C*, (b) formation of *C* is the rate-determining step, and (c) formation of product *P* is irreversible (Scheme 3). However, it is now well established that many reactions of benzenoid hydrocarbons which formally are electrophilic substitution reactions do not meet these conditions. It has been found, for example, that nitration of benzenoid hydrocarbons, e.g. anthracene or perylene, by nitrogen dioxide in dichloromethane (25 °C) occurs through an electron transfer mechanism, formation of the cation radical of the hydrocarbon being the rate-determining step [49]. Many alkylation and sulphonation reactions of benzenoid hydrocarbons are thermodynamically controlled, i.e. isomer distribution is not related to σ-complex stabilities [50]. However, most of the work that has been done on correlating reactivity indices of benzenoid hydrocarbons with rate constants of electrophilic substitution refers to the prototype mechanism.

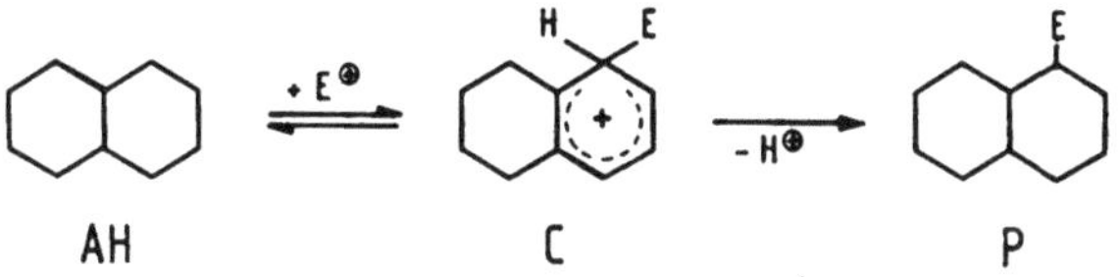

Scheme 3. Prototype mechanism of aromatic electrophilic substitution

Equilibrium or rate constants of protodetritiation of benzenoid hydrocarbons (Scheme 4) can be measured in well-defined experiments with high accuracy [51]. Referring to the Hammett-relation (4) (more correctly Hammett-Streitwieser relation) we may define

$$\varrho\,(\text{protodetritiation}) = 1 \cdot \tag{12}$$

Then

$$\log \frac{K_r}{K_o}\ (\text{or}\ \log \frac{k_r}{k_o})\ (\text{protodetritiation}) \tag{13}$$

where r denotes the particular position of a given hydrocarbon and K_o as well as k_o refer to a reference substance (usually benzene or naphthalene, 1-position) provides experimentally accessible reactivity constants σ_r^* [52]. As protodetritiation occurs according to the prototype mechanism of electrophilic substitution σ_r^* values can be regarded as measures of σ-complex stabilities or activation energies of σ-complex formation, respectively. Experimental gas-phase proton-affinity values [53, 54] constitute a related set of experimental values in which the measured proton affinity is assumed to refer to the most reactive molecular position. The value of ϱ in a given type of electrophilic substitution is a measure of the similarity of the transition state for the reaction to that of the σ-complex in protodetritiation [52].

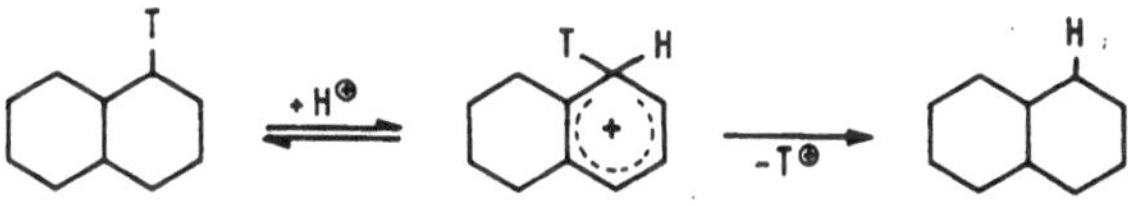

Scheme 4. Mechanism of protodetritiation

There are principally two different approaches of correlating experimental rate data of electrophilic substitution with reactivity indices: (1) Correlating the index with the rate data of a given reaction, e.g. bromination. For example, a satisfying correlation of Dewar reactivity numbers with the log of rate constants of the bromination of benzene, naphthalene (1- and 2-position), biphenyl (4-position), phenanthrene (9-position), and anthracene (9-position) has been observed [55]. In correlations of this type the reactivity index corresponds to the reactivity constant in the Hammett equation while the slope of the linear correlation corresponds to the reaction constant (see also Sect. 3); (2) correlating the index with experimental σ_r^* values.

Correlations of σ_r^* values with various types of quantum chemically calculated parameters have been studied by Streitwieser et al. [56]. A mediocre correlation was found for cation localization energies obtained with the simple HMO method (correlation coefficient $r = 0.893$, sample size: 23). A significantly improved correlation was observed with localization energies calculated by the ω-technique modification [57] of the HMO method which can be regarded as an approximate SCF method (correlation coefficient $r = 0.966$, sample size: 20). Optimum results were obtained with the CNDO/2 method (using standard geometry) in a version that gives proton affinities rather than localization energies (correlation coefficient $r = 0.979$, sample size: 27).

More recently, correlations of σ_r^* values with purely topological reactivity indices, vic. structure count ratio and Dewar reactivity number have been extensively studied by v. Szentpaly and Herndon [33, 58]. Rather satisfying correlations were obtained the correlation coefficients for obvious reasons (see Sect. 4) being nearly identical for both indices (0.959 and 0.960 respectively, sample size: 27). A significant improvement was achieved with Dewar reactivity numbers calculated according to the free electron version of the PMO treatment (correlation coefficient $r = 0.973$).

To cast some light on the relative importance of steric effects on the positional reactivities of benzenoid hydrocarbons, correlations of experimental σ_r^* values of phenanthrene (*4*), tetrahelicene (*5*), pentahelicene (*6*), and hexahelicene (*7*) with purely topological reactivity indices (Hückel cation localization energy, Dewar reactivity number and Herndon structure count ratio) have been studied [59].

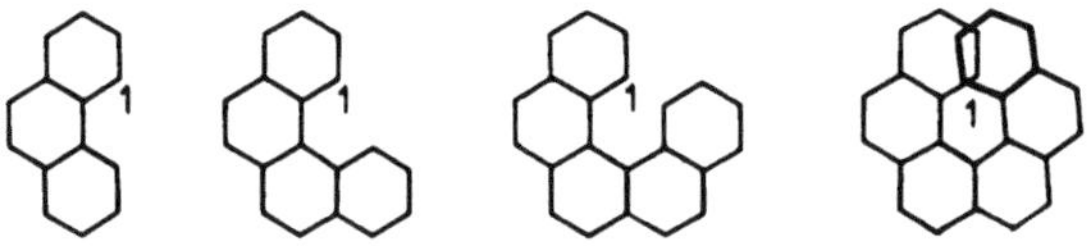

4 5 6 7

The observed correlations suggest that the regular increase in reactivity with ring number at a given site (except for the 1-position), seems to be due to the effect of differences in resonance energy and not to the increased distortion of the aromatic ring as previously suggested [60]. However, the results obtained also indicate that steric constraints exist being approximately constant for each particular type of reaction site.

5.2 Diels-Alder Reactions

The Diels-Alder reaction of benzenoid aromatic hydrocarbons with dienophiles was discovered more than fifty years ago [61, 62]. A classical example is the reaction of anthracene with maleic anhydride (Scheme 5). Although this type of reaction, termed "endogenic" or "endocyclic" Diels-Alder reaction, could be expected to be particularly well suited for correlating structure (topology) of benzenoid hydrocarbons with kinetic data, the problem has been systematically studied only very recently. Biermann and Schmidt in a series of publications [12, 29, 45, 63, 64] reported second-order rate constants (k_2), measured under standard conditions (1,2,4-trichlorobenzene, 91.5 ± 0.2 °C), for the endocyclic Diels-Alder reaction between maleic anhydride and 102 benzenoid hydrocarbons. Each rate constant was measured twice, the values usually agreeing within ± 3%. With few exceptions the positions of maleic anhydride addition, i.e. regioselectivity, were unambiguously concluded from the characteristic UV/visible absorption spectra of the adducts formed.

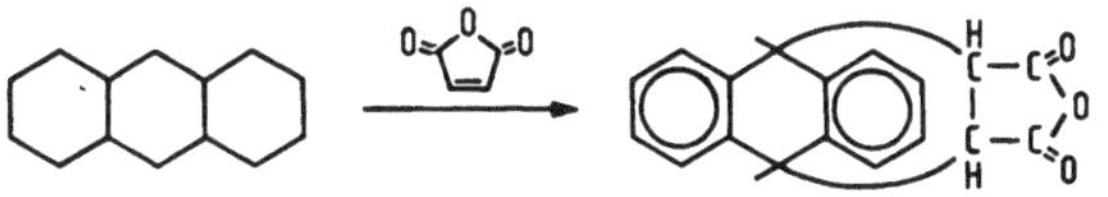

Scheme 5. Endocyclic Diels-Alder reaction of anthracene with maleic anhydride

A simple relation between regioselectivity of maleic anhydride addition to benzenoid hydrocarbons and Clar's π sextet formulae has already previously been observed [65]. In cases where the addition can lead to several isomeric adducts, the adduct whose Clar formula has the largest number of inherent π sextets is always the only one to be formed. Scheme 6 gives some examples. A necessary and, usually, also adequate condition for the endocyclic Diels-Alder reaction of benzenoid hydrocarbons with maleic anhydride is that at least one π sextet be "gained" by the addition, i.e. that the number of sextets in the product be at least one more than in the reactant.

The application of this principle has also led to the discovery of the first example of an endocyclic Diels-Alder reaction with a benzenoid hydrocarbon including dearomatization of the pyrene system: the hydrocarbon *8* reacts with maleic anhydride to form the Diels-Alder adduct *9* [66].

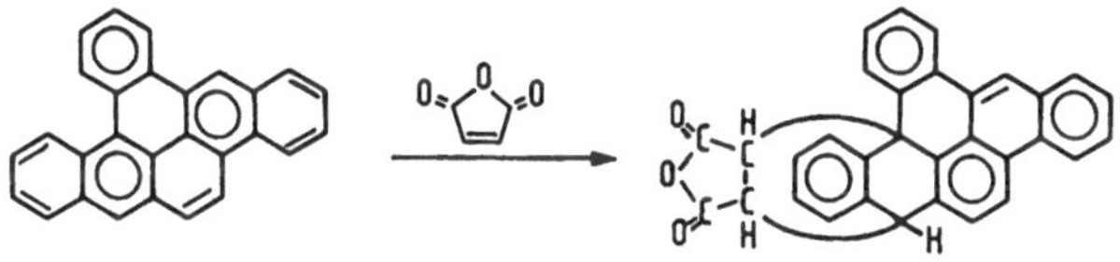

8 9

The relation observed between regioselectivity and number of Clar sextets of reactants and products is not unexpected for a reaction that is controlled by thermodynamics i.e. by the difference between free enthalpy of formation of reactants and products. Conversely, however, no conclusions regarding the mechanism of the reaction can be derived from the relation.

Experimental log k_2 values were correlated with Brown para-localization energies, Dewar reactivity numbers, Herndon structure count ratios, Hess-Schaad resonance energy differences, indices of free valence, and second-order perturbation stabilization energies. The latter are based on Fukui's frontier orbital theory [67] which classifies the Diels-Alder reaction of benzenoid hydrocarbons with maleic anhydride as mainly HOMO (aromatic hydrocarbon)-LUMO (maleic anhydride) controlled. However, the corresponding orbital interaction energy given by

$$E^{(2)} = \frac{(c_1c_3 + c_2c_4)^2}{\Delta E}\,\beta^2 \tag{14}$$

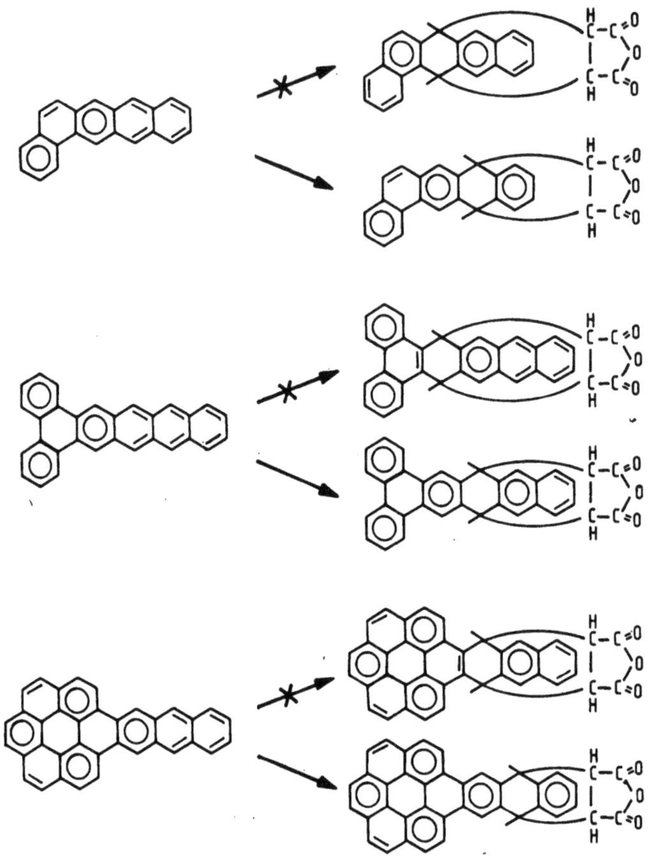

Scheme 6. Clar's π-sextet model and the endocyclic Diels-Alder reaction of benzenoid hydrocarbons

where c_1 and c_2 are the AO coefficients at the reacting centres of the hydrocarbon in its HOMO, c_3 and c_4 are the AO coefficients of the LUMO of maleic anhydride, ΔE is the difference in energies of the hydrocarbon-HOMO and maleic anhydride-LUMO and β the resonance integral, proved unsuitable to correlate experimentally observed rate behaviour [63]. Therefore an extended version of $E^{(2)}$ was used, $(\Sigma E^{(2)})$, obtained by summing over all orbital interactions where the individual terms are calculated according to Eq. (14).

As a measure for goodness of fit standard deviations of the different linear relations observed between log k_2 values and the theoretical figures are summarized in Table 2 [45, 63]. It is interesting to note that models considering only the starting hydrocarbon as well as those that are related to the difference in π-electron energies of products and reactants meet with comparable success. It is very unlikely in this case that the BEP-principle provides an explanation because there is theoretical evidence [68] that Diels-Alder reactions are typical Anti-BEP processes. The observation that reactivity indices derived from models presupposing an early transition state (e.g. second-order perturbation theory) or those presupposing a late transition state

Table 2. Standard deviations of the linear relations between reactivity indices and log of second-order rate constants (k_2) of the endocyclic Diels-Alder reaction of benzenoid hydrocarbons with maleic anhydride [45, 63, 64]

Theoretical method	standard deviation
Structure Count Ratio[a]	0.309
Second-order Perturbation[a]	0.320
Brown para-Localization Energy[a]	0.320
Hess-Schaad Resonance Energy[a]	0.290
Dewar Bis-Localization Energy[b]	0.250
Index of Free Valence[b]	0.325

a Sample size: 82. b Sample size: 46.

(e.g. structure count method) are comparably well suited to model experimental rate constants is simply due to the fact that all these different types of reactivity indices because of their common origin i.e. molecular topology, are interrelated [23].

Purely empirical approaches for correlating experimental k_2-values of the endocyclic Diels-Alder reaction of benzenoid hydrocarbons with maleic anhydride have also been described. A two-parameter expression based on invariants derived from the dualist graphs [22] of benzenoid hydrocarbons allows modelling of k_2-values in very good agreement with experimental figures (standard deviation of the linear correlation between calculated and experimental figures $r = 0.203$) [12]. While first ionization potentials (IP_1) of benzenoid hydrocarbons do not satisfactorily correlate with experimental k_2-values, a much better correlation has been observed with the difference $\Delta IP = IP_2 - IP_1$ and an explanation based on second-order perturbation theory has been given [63]. An empirical correlation between the triplet half-lifetimes $\tau_{1/2}$ (obtained from phosphorescence measurement) of the hydrocarbons and k_2-values has also been found: the product of both quantities appears to be roughly constant [63]. This relation probably arises from the facts that (a) the singlet-triplet splitting of the L_a state, the energy of which is related to the energy of the HOMO is nearly constant for hydrocarbons with more than two rings [69], i.e. triplet state energy decreases with decreasing HOMO-LUMO energy difference (of alternant systems) and (b) experimental triplet state lifetimes decrease with decreasing triplet state energies [70].

Benzenoid hydrocarbons, which like perylene (*10*) contain a peripheral cisoid C_4 arrangement (a "bay region" [71]), react with maleic anhydride in the presence of a suitable dehydrogenating agent to form fully aromatic dicarboxylic acid anhydrides *12* ("benzogenic Diels-Alder reaction") [72, 73]. Since in this process the rate-determining step is the "exocyclic" Diels-Alder reaction leading to *11*, the reactivity behaviour of the system is dominated by the π-electronic properties of the hydrocarbon centres at which this primary reaction occurs [74].

Correlations of experimental rate constants of benzogenic Diels-Alder reactions with Polansky (butadienoid) character orders [74] and Dewar reactivity numbers [75] have been observed. Herndon structure-count ratios, however, proved to be superior [76]. The standard deviation of the linear correlation between log k_2 and structure count ratio is 0.483, i.e. goodness of fit is much less compared to that of plots obtained

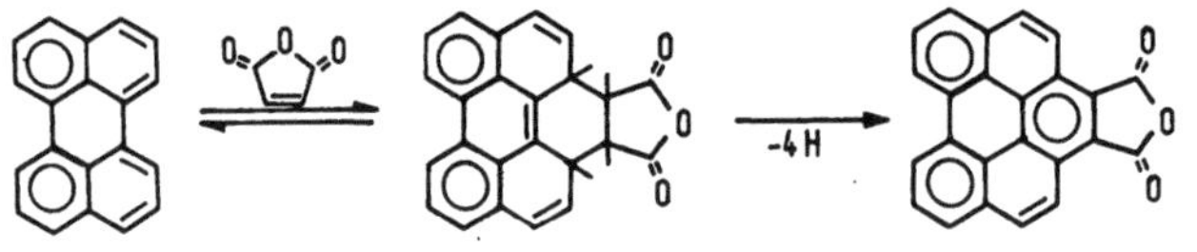

10 11 12

for the endocyclic Diels-Alder reaction (see Table 2). There is some experimental indication that the necessary prerequisite of reaction rate/structure relationships of the Hammett type, viz. constancy of activation entropy (see Sect. 3), is much better met for endocyclic than for exocyclic Diels-Alder-reactions of benzenoid hydrocarbons [76]. The regioselectivity of benzogenic Diels-Alder reactions can be explained and predicted on the basis of Clar's π-sextet formalism. In fact, this theory emerged from observations regarding the regioselectivity of benzogenic Diels-Alder reactions of benzenoid hydrocarbons [77].

5.3 Thermally Induced Polymerization

It is well established that liquid phase thermolysis of polynuclear benzenoid hydrocarbons at temperatures of about 400–500 °C occurs predominantly according to the general Scheme 7 [78]. Though polymerization through loss of hydrogen and intramolecular cyclization reactions are the main reaction pathways, intramolecular rearrangements [79], fragmentation reactions (due to hydrogen transfer and subsequent splitting of C–C single bonds initially formed) and alkylation reactions have also been observed [80].

Although the main features of the thermal chemistry of benzenoid hydrocarbons in liquid phase are well established many important mechanistic details need further

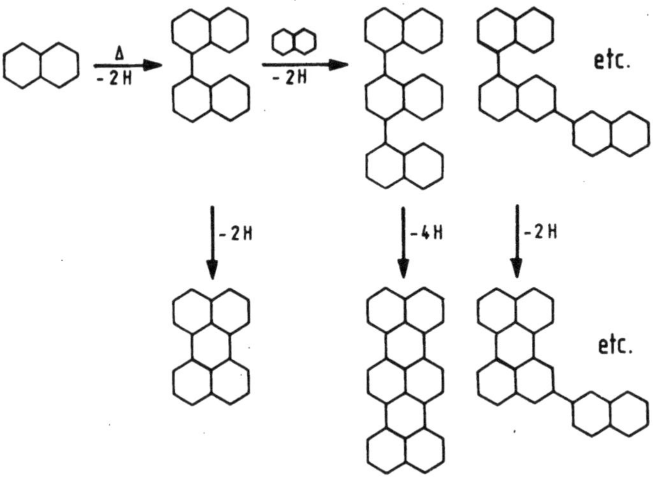

Scheme 7. Thermally induced polymerization of benzenoid hydrocarbons (simplified reaction scheme)

clarification. Different mechanisms of the initial step of thermally induced polymerization of benzenoid hydrocarbons have been discussed [81, 82]. According to recent findings by Stein et al. [83] the initial step in the pyrocondensation of anthracene is assumed to be the reversible formation of a biradical *13* via coupling of two ground molecules. The biradical then transforms to *14* by intramolecular hydrogen transfer. The preference for joining the 9-position of an anthracene molecule with the 2-position of another anthracene molecule is assumed to be a result of steric interference in the pathways leading to other 9-substituted isomers.

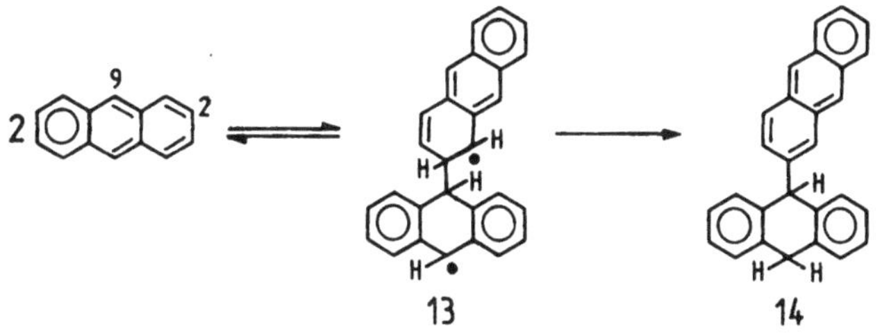

Lewis and Edstrom [84] have provided qualitative thermal reactivity data of various polynuclear benzenoid hydrocarbons. They classified the compounds as either thermally "reactive" or thermally "unreactive". The thermally "reactive" species possess sufficient reactivity in an atmospheric pressure system to undergo a condensation sequence in the liquid phase and yield a measurable amount of polymerized carbonaceous residue at 750 °C. The thermally "unreactive" species have sufficient stability so that such condensation reactions do not occur prior to complete volatilization. From 30 alternant (unsubstituted) systems studied 10 were found to be reactive.

Yokono et al. [85] have suggested that the results obtained by Lewis and Edstrom [84] can be understood in terms of the maximum value of the index of free valence as calculated by the HMO method. However, as Herndon [30] has shown, some discrepancies occur when the free valence approach is applied to the experimental findings. He found that the structure count ratio for the single position in each compound that would give rise to the most highly resonance stabilized radical is a reliable reactivity index to correlate and predict the qualitative aspects of the thermal behaviour of benzenoid hydrocarbons.

In a more recent study [86] carbonization rates of the hydrocarbons listed in Table 3 have been measured. Highly purified samples of each hydrocarbon were isothermally treated in sealed glass tubes at 430 °C, i.e. above the melting points of the compounds, for 4 h. The conversion rates were determined by UV/visible absorption spectroscopy and high-pressure liquid chromatography. The log of conversion rates observed (Table 3) were found to correlate linearly with the lowest Dewar reactivity number N_r (min) of each compound. For the seven cata-condensed systems studied the correlation coefficient of the linear correlation is $r = 0.9771$. Inclusion of the peri-condensed systems *8–10* lowers the correlation coefficient to $r = 0.9089$. However, both correlations were shown by statistical tests to be highly significant. Only benzo[a]pyrene (*11*) was recognized by statistical tests as an outlier.

Table 3 Minimum Dewar localization energy ($N_{r(min)}$) and thermally induced conversion rate (%) of benzenoid hydrocarbons (430 °C, 4 h)

Compound	$N_{r(min)}$	Conversion rate (%)
1 Tetracene	1.03	99
2 Benzo[a]anthracene	1.35	53
3 Chrysene	1.67	20
4 Triphenylene	2.12	9
5 Dibenzo[a, h]anthracene	1.51	50
6 Dibenzo[a, c]anthracene	1.50	36
7 Picene	1.69	24
8 Perylene	1.33	30
9 Benzo[ghi]perylene	1.55	27
10 Benzo[e]pyrene	1.63	40
11 Benzo[a]pyrene	1.15	23

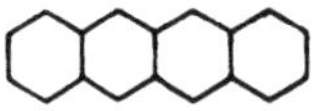
1

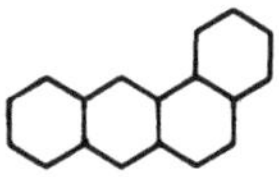
2

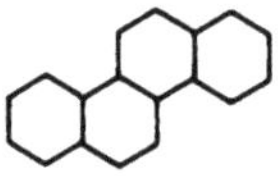
3

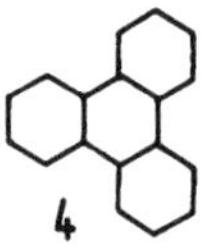
4

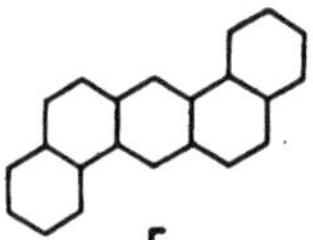
5

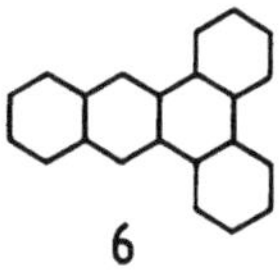
6

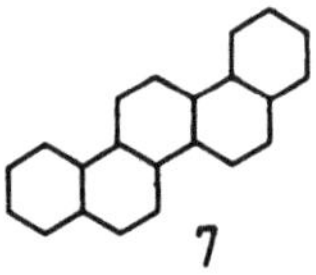
7

8

9

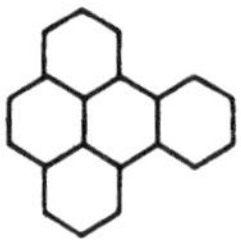
10

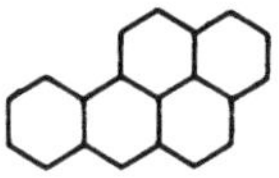
11

As the values of the minimum Dewar reactivity number can obviously be connected only with the initial step of the thermal chemistry of benzenoid hydrocarbons it was concluded that the initial step is rate-determining. However, it is also possible that some other factor, connected with the Dewar localization energies for topological reasons, determines the rates of reaction.

5.4 Biochemical Transformation

Certain benzenoid hydrocarbons are known to produce skin cancer in test animals [87]. Whether a hydrocarbon is carcinogenic or not depends strongly on its topology. For example, benzo[a]pyrene is a strong carcinogen while no carcinogenic (or mutagenic) effects have been observed with the isomeric benzo[e]pyrene. The unambiguous determination of any carcinogenity measure by animal tests, however, is hampered by the different response due to differences in age and nutrient state of animals, dose and route of administration of the carcinogens and other factors.

Nevertheless, experimental indices have been derived from animal tests that allow a rough quantitative differentiation of the carcinogenic potency of different hydrocarbons. The so-called Iball-Index [88] is defined as the percentage of papilloma-bearing mice (among those who survived beyond the shortest time of latent period) divided by the average length of the latent period in animals affected by cancer. Thus, the index is proportional to the fraction of subject animals that show a carcinogenic response divided by the mean latent period. The values of the Iball-Index range from 0 (e.g. triphenylene) to about 100 (e.g. dibenzo[a, i]pyrene, $I = 74$) and are thought to be reliable within a range of $\pm 10\%$ [89].

It is now widely accepted that the carcinogenic potency of benzenoid hydrocarbons is not a property of the hydrocarbons itself (although there is one contradicting theory [90]) but of certain metabolites formed under the influence of specific enzymes and reacting with biopolymers, particularely the DNA. For benzo[a]pyrene, as an example, the mechanism of the carcinogenic activity of benzenoid hydrocarbons according to our present knowledge is summarized in Scheme 8 [91]. With the participation of the enzyme cytochrome P 448 which is present in the endoplasmic reticulum of the cell benzo[a]pyrene is oxidized yielding the arene oxide *2*. During the next step the enzyme epoxide-hydratase [92] transforms *2* into the transdihydro diol *3*, which then again undergoes an epoxidation with the participation of cytochrome P 448 yielding the "ultimate carcinogen" *4*. The electrophile *4* reacts with nucleophilic bases of the DNA, guanine being the favoured base for the attack of the diol epoxide. It was found that from the two pairs of enantiomers of *4* only one isomer exhibits a high carcinogenic activity in experiments with mice while the other three have minor activity [93].

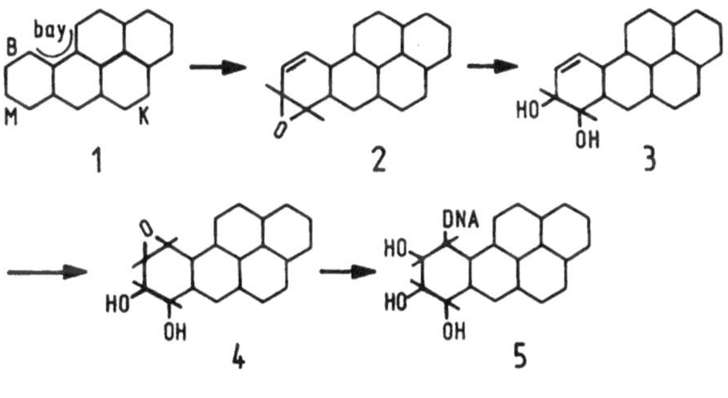

Scheme 8. Mechanism of carcinogenic activity of benzenoid hydrocarbons

Most of the more previous attempts to correlate carcinogenic potency with structural features of benzenoid hydrocarbons are related to the mechanism outlined above. The different quantitative models that have been developed can be distinguished with regard to the number of independent variables. One-variable theories (e.g. the so-called "bay-region theory" [94]) are normally inferior compared to two- [95] and three-variable theories which refer more explicitely to the different and partly competing metabolic reaction pathways. A particularely efficient model has been developed by v. Szentpaly [96]. In his so-called "MCS model" three important influences on carcinogenic potency are taken into account: M, the initial epoxidation of the M region (see formula *1* in Scheme 8) in competition with reactions on other centers of the hydrocarbon molecule; C, carbocation intermediate(s) in the reaction of the B region (formula *1*) diol epoxide with DNA and/or in the M region epoxide hydration; *S*, a size and solubility factor accounting for transport phenomena (the enzymatic epoxidation takes place in the microsomal endomembrane system, from where the metabolites have to reach the cell nucleus in order to react with the DNA).

Algebraic expressions for terms M and C were derived using Dewar's PMO method (for C in a version similar to the ω-technique [57] in order to calculate carbocation stabilization energies). The size factor *S* is simply a cubic function of the number of carbon atoms [97]. The three independent variables of the model were assumed to be linearly related to the experimental Iball indices (vide supra). By multilinear regression analysis (sample size = 26) an equation was derived for calculating Iball indices from the three theoretical parameters. The correlation coefficient for the linear relation between calculated and experimental Iball indices is $r = 0.961$.

v. Szentpaly [96] has correctly emphasized that caution is required in applying MO reactivity indices to enzymatic reactions. However, he thought, that as a specific reaction at a given molecular region in a group of related compounds is investigated, the effects of differing reactivities should predominate over differences in binding to the enzyme. In fact, his successful attempt in modelling experimental Iball indices supports this view.

From the efficiency of the MCS model, which is purely topological in nature, it may be concluded that even very complex chemical behaviour of benzenoid hydrocarbons is dominated by molecular topology.

6 Concluding Remark

The topological approach to polynuclear benzenoid hydrocarbons is very successful and provides elegant descriptions of structure/property relationships. The benzenoid hydrocarbons, however, constitute a highly homogenous class of compounds exhibiting only small variances with regard to chemical bonding i.e. type of C–C bonds, bond lengths and angles. Therefore "Molecular Topology" and "Molecular Structure" of the systems are redundant in this case and hence their significance for the chemical and physical properties of the compounds is indistinguishable for simple logical reasons. However, a more recently discovered topological phenomenon [3, 4, 98–100] is not restricted to a particular class of structurally related compounds but applies to structurally very different compounds exhibiting a wide variance with regard to kind of atoms and bonding type, charge distribution and geometry (bond lengths and angles). This clearly indicates that molecular topology provides the frame for molecular physics and chemistry [4].

7 References

1. Woolley RG (1978) J. Am. Chem. Soc. 100: 1073
2. Claverie P, Diner S (1980) Israel J. Chem. 19: 54
3. Zander M, Polansky OE (1984) Naturwiss. 71: 623
4. Gutman I, Polansky OE (1986) Mathematical concepts in organic chemistry, Springer, Berlin Heidelberg New York

5. Merrifield RE, Simmons HE (1983) In: King RB (ed) Chemical applications of topology and graph theory, Elsevier, Amsterdam; pp. 1–16; (1980) Theor. Chim. Acta 55: 55
6. Polansky OE (1986) Z. Naturforsch. 41a: 560
7. Ruedenberg K, Scherr CW (1953) J. Chem. Phys. 21: 1565
8. Hiberty PC, Shaik, SS, Lefour J-M, Ohanessian G (1985) J. Org. Chem. 50: 4657
9. Graovac A, Gutman I, Trinajstic N (1977) Topological approach to the chemistry of conjugated molecules, Springer, Berlin Heidelberg New York (Lecture Notes in Chemistry No. 4)
10. Zander M (1986) Match 19: 171
11. Clar E (1964) Polycyclic hydrocarbons, Academic, New York vol 1, pp 41–69
12. Balaban AT, Biermann D, Schmidt W (1985) Noveau J. Chim. 9: 443
13. Kekulé A (1861) Lehrbuch der Organischen Chemie, Band 1, Verlag von Ferdinand Enke, Erlangen, pp 124–132
14. Hammett LP (1940) Physical organic chemistry, McGraw-Hill, New York p 348
15. Wold S, Sjöström M (1978) In: Chapman NB, Shorter J (eds) Correlation analysis in chemistry, Plenum, New York, chapt 1
16. Dewar MJS, Dougherty RC (1975) The PMO theory of organic chemistry, Plenum, New York
17. Klumpp GW (1978) Reaktivität in der Organischen Chemie, Thieme, Stuttgart vol 2, pp 367–369
18. Terminology according to Harary F (1974) Graphentheorie, R. Oldenbourg, Munich
19. Zander M (1985) Z. Naturforsch. 40a: 636
20. Clar E (1972) The aromatic sextet, Wiley, London
21. Zander M (1982) Naturwissenschaften 69: 436
22. Balaban AT, Harary F (1968) Tetrahedron 24: 2505
23. Streitwieser AJr (1961) Molecular orbital theory for organic chemists, Wiley, New York
24. Herndon WC (1975) J. Org. Chem. 40: 3583
25. Herndon WC (1980) Israel J. Chem. 20: 270
26. Herndon WC (1973) Tetrahedron 29: 3
27. Herndon WC (1974) J. Chem. Educ. 51: 10
28. Longuet-Higgins HC (1950) J. Chem. Phys. 18: 265, 275, 283
29. Biermann D, Schmidt W (1980) Israel J. Chem. 20: 312
30. Herndon WC (1982) Tetrahedron 38: 1389
31. Brown RL (1983) J. Comput. Chem. 4: 556
32. v Szentpaly L (1981) J. Photochem. 17: 112
33. v Szentpaly L, Herndon WC (1988) In: Ebert LB (ed) Polynuclear aromatic compounds, American Chemical Society, Washington DC, Chapt 17 (Advances in Chemistry Series 217)
34. Kuhn H (1948) Helv. Chim. Acta 31: 1441; (1949) Helv. Chim. Acta 32: 2247
35. v Szentpaly L (1981) Chem. Phys. Lett. 77: 352
36. Polansky OE, Derflinger G (1967) Int. J. Quantum Chem. 1: 379
37. Mehlhorn A, Fratev F, Polansky OE, Monev V (1984) Match 15:3
38. Sofer H, Derflinger G, Polansky OE (1968) Mh. Chem. 99: 1879; 1895
39. Zander M (1979) Z. Naturforsch. 34a: 521
40. Golebiewski A (1974) Acta Phys. Polonica A46: 719
41. Fratev F, Polansky OE, Mehlhorn A, Monev V (1979) J. Mol. Struct. 56: 245
42. Hess BA Jr, Schaad LJ (1971) J. Am. Chem. Soc. 93: 305
43. Schaad LJ, Hess BA Jr (1974) J. Chem. Educ. 51: 640
44. Herndon WC (1981) J. Org. Chem. 46: 2119
45. Hess BA Jr, Schaad LJ, Herndon WC, Biermann D, Schmidt W (1981) Tetrahedron 37: 2983
46. Léger A, d'Hendecourt L, Boccara N (eds) (1987) Polycyclic aromatic hydrocarbons and astrophysics, D. Reidel, Dordrecht
47. Stein SE, Brown RL (1985) Carbon 23: 105
48. Stein SE, Brown RL (1987) Mol. Struct. Energ. 2: 37
49. Pryor WA, Gleicher GJ, Cosgrove JP, Church DF (1984) J. Org. Chem. 49: 5189
50. Marsh J (1977) Advanced Organic Chemistry, McGraw Hill, New York
51. Baker R, Eaborn C, Taylor R: J. Chem. Soc. Perkin II 1972: 97

52. Krygowski TM (1972) Tetrahedron 28: 4981
53. Aue DH, Bowers MT (1979) In: Bowers MT (ed) Gas-phase ion chemistry, Academic, New York Chapt 9
54. Meot-Ner (Mautner) M (1980) J. Phys. Chem. 84: 2716
55. Altschuler L, Berliner E (1966) J. Am. Chem. Soc. 88: 5837
56. Streitwieser A Jr, Mowery PC, Jesaitis RG, Lewis A (1970) J. Am. Chem. Soc. 92: 6529
57. Streitwieser A Jr, (1960) J. Am. Chem. Soc. 82: 4123
58. v Szentpaly L, Herndon WC (1984) Croatica Chemica Acta 57: 1621
59. Shawali AS, Hassaneen HM, Párkányi C, Herndon WC (1983) J. Org. Chem. 48: 4800
60. Archer WJ, Shafig YE, Taylor RJ (1981) J. Chem. Soc. Perkin Trans. 2: 675
61. Clar E (1931) Ber. Dtsch. Chem. Ges. 64: 1682
62. Diels O, Alder K (1931) Justus Liebigs Ann. Chem. 486: 191
63. Biermann D, Schmidt W (1980) J. Am. Chem. Soc. 102: 3163
64. Biermann D, Schmidt W (1980) J. Am. Chem. Soc. 102: 3173
65. Franck HG, Zander M (1966) Chem. Ber. 99: 1272
66. Blümer G-P, Gundermann K-D, Zander M (1976) Chem. Ber. 109: 1991
67. Fukui K (1970) Fortschr. Chem. Forsch. 15: 1; Houk KN (1975) Acc. Chem. Res. 8: 361
68. Dewar MJS (1984) J. Am. Chem. Soc. 106: 209
69. Zander M (1965) Angew. Chem. Intern. Ed. Engl. 4: 930
70. Siebrand W (1966) J. Chem. Phys. 44: 4055
71. Bartle KD, Jones DW (1972) Adv. Org. Chem. 8: 317
72. Clar E (1932) Ber. Dtsch. Chem. Ges. 65: 846
73. Clar E, Zander M: J. Chem. Soc. 1957: 4616; Zander M (1960) Angew. Chem. 72: 513
74. Zander M (1969) Justus Liebigs Ann. Chem. 723: 27
75. Zander M (1978) Z. Naturforsch. 33a: 1395
76. Biermann D (1981) Thesis, Ludwig-Maximilians-University, Munich
77. Clar E, Zander M: J. Chem. Soc. 1958: 1861
78. Lewis IC, Singer LS (1988) In Ebert LB (ed) Polynuclear Aromatic Compounds, American Chemical Society, Washington DC, Chapt 16 (Advances in Chemistry Series 217)
79. Lewis IC (1982) Carbon 20: 519
80. Zander M, Haase J, Dreeskamp H (1982) Erdöl und Kohle · Erdgas · Petrochem. 35: 65
81. Lewis IC (1980) Carbon 18: 191
82. Stein SE (1981) Carbon 19: 421
83. Stein SE, Griffith LL, Billmers R, Chen RH (1987) J. Org. Chem. 52: 1582
84. Lewis IC, Edstrom T (1963) J. Org. Chem. 28: 2050
85. Yokono T, Miyazawa K, Sanada Y, Marsh H (1979) Fuel 58: 692
86. Zander M (1986) Fuel 65: 1019
87. IARC Monographs (International Agency for Research on Cancer), vol 3, Lyon 1973
88. Iball I (1939) Am. J. Cancer 35: 188
89. Herndon WC: Private Communication cited in ref. [96]
90. Popp FA (1976) In: Deutsch F (ed) Molecular basis of malignancy, Georg Thieme, Stuttgart, pp 47–55
91. For a review see Zander M (1980) In: Hutzinger O (ed) Anthropogenic compounds, Springer, Berlin Heidelberg New York pp 109–131 (Handbook of environmental chemistry vol 3, Part A)
92. Oesch F (1973) Xenobiotica 3: 305
93. Buening MK (1978) Proc. Nat. Acad. Sci. USA 75: 5358
94. Jerina DM et al. (1976) In: de Serres FJ et al. (eds) In: In vitro metabolic activation in mutagenesis testing, Elsevier/North Holland, Amsterdam, pp 159–177
95. Qianhuan D (1980) Scientia Sinica 23: 453
96. v Szentpaly L (1984) J. Am. Chem. Soc. 106: 6021
97. Herndon WC (1974) Int. J. Quantum Chem., Quantum Biol. Symp. 1: 123
98. Polansky OE, Zander M (1982) J. Mol Struct. 84: 361
99. Motoc I, Polansky OE (1984) Z. Naturforsch. 39a: 1053
100. Polansky OE, Mark G, Zander M (1987) Der topologische Effekt an Molekülorbitalen (TEMO), Schriftenreihe des Max-Planck-Instituts für Strahlenchemie No. 31, Mülheim an der Ruhr (FRG)

A Periodic Table for Benzenoid Hydrocarbons

Jerry Ray Dias

Department of Chemistry, University of Missouri, Kansas City, Missouri 64110, U.S.A.

Table of Contents

The formulas for all benzenoid hydrocarbons are found in Table PAH6 which has x,y-coordinates of (d_s, N_{Ic}). Recursive construction of this table was accomplished using an aufbau principle. This table complies to a sextet rule analogous to the octet rule for the periodic table of elements, and d_s is analogous to the outer-shell electronic configuration and N_{Ic} is analogous to the principle quantum number associated with the periodic table of elements. The aufbau principle and the excised internal structure and strictly pericondensed concepts were evolved in conjunction with the formula periodic table for benzenoid hydrocarbons and represent new ideas formulated by this author. These fundamentals led to the first examples of enumeration of the strictly peri-condensed, non-Kekulean, and total resonant sextet benzenoid groups and has lead to the identification of many topological properties characteristic of benzenoids.

1 Introduction

The uniform consecutive change in physicochemical properties of a homologous series of organic compounds forms the basis of the approach by which chemists study different classes of chemical compounds. The first order structural factor in a homologous series is principally connectivity which governs the eigenvalues associated with the corresponding molecular graphs and consequently the overall molecular energies. A homologous series of compounds generated by the consecutive annelation of some specific fragment to a specific parent molecule organizes the set of compounds into a partial order with hierarchial relationships. This constructive (aufbau) process leads to a periodic table of compounds [1]. If each successive annelation can be performed in several different ways, then each coordinate of the periodic table will represent a set of isomers. In this latter case to select out a particular homologous series, one needs to prescribe a set of restrictions for successive annelation. Herein, we describe a formula periodic table for benzenoid hydrocarbons composed exclusively of fused hexagonal rings (PAH6) [2]. This table unifies all benzenoid hydrocarbons and polyhexes into a systematic framework and in itself has a mathematical structure [3]. Since the chemical formula is an invariant, two new graphical invariants of d_s and N_{Ic} (*cf.* Table 1 for glossary of terms) were defined and related to the chemical formula with the aid of the relationship of $d_s + N_{Ic} = r - 2$. Recursive construction of a Formula Periodic Table for Benzenoid Hydrocarbons (Table PAH6) was accomplished where d_s is analogous to the outer-shell electronic configuration and N_{Ic} is analogous to the principal quantum number. Table PAH6 establishes a partially ordered set (poset) of isomer groups having *x,y*-coordinates of (d_s, N_{Ic}). Isomers are thus analogous to isotopes, and this table complies with a sextet rule analogous to the octet rule for the Periodic Table of Elements. The structural concepts of strictly

Table 1. Glossary of terms

d_s	– net tree disconnections of internal graph edges (positive values) or connections (negative values – called negative disconnection)
E_π	– total $p\pi$ energy
N_c	– total number of carbon atoms in a PAH
N_H	– total number of hydrogen atoms in a PAH
N_{Ic}	– number of internal carbon atoms in a PAH having a degree of 3
N_{Pc}	– number of peripheral carbon atoms in a PAH having a degree of 3
PAH6	– polycyclic aromatic hydrocarbon containing exclusively fused hexagonal rings; also referred to as benzenoid and polyhex.
$\|P\| = p = N_c$	– total number of graph points
$P(G; X)$	– characteristic polynominal of a molecular graph
$\|Q\| = q$	– number of graph edges (lines or C–C bonds)
q_I	– number of internal graph edges
q_p	– number of peripheral graph edges
r	– number of rings
τ	– number of rings obtained upon deletion of all internal third degree vertices from a PAH6 σ-bond graph.
Table PAH6	– Formula Periodic Table for PAH6s
$X = (\varepsilon - \alpha)/\beta$	= graph eigenvalue

peri-condensed and excised internal structure were important in the simplified enumeration of the former type of benzenoid hydrocarbons (PAH6s).

Benzenoid hydrocarbons are ubiquitous pyrolytic chemical pollutants and have highly ordered arrangements of hexagonal rings analogous to a mosaic constructed from a single tile shape. The repeating motif of benzenoids is the hexagonal ring of carbon atoms. According to well established theorems of polyhexagonal systems, only a small list of rotational symmetries is possible for benzenoids. They can have twofold, threefold, or sixfold axes of rotational symmetry and no other possibilities.

In deciphering the information content of benzenoid molecular graphs, our preliminary work has focused on combinatorial derivations for rapidly determining the associated characteristic polynomials by topological inspection of their graphical invariants [4–7]. The equations derived revealed that the coefficients in the characteristic polynomial of a molecular graph carry topological information. Coupled with embedding of smaller alternant (bipartite) substructures onto larger ones for identifying the presence of common eigenvalues, these equations have lead to a powerful method for rapidly determining the eigenvalues of small molecular graphs that can be used to correlate chemical properties. Table PAH6 sorts polyhex graphs into isomer groups and identifies their group properties and molecular graph eigenvalues examine and discern the molecular graph properties within each isomer group.

Since computer enumeration methods are still hampered by the explosive growth in the number of polyhex isomers having more than eleven hexagonal rings, the most practical approach is the enumeration of select benzenoid groups of interest to chemists [8].

2 Concepts and Definitions

2.1 Benzenoid/Polyhex Graphs

A polyhex graph PH = (V, E) with a finite vertex set, $V(|V| = N_c)$, and a finite edge set, $E\ (|E| = q)$, is composed of k hexagons $H_i(1 \leqq i \leqq k)$. Each H_i is a subgraph of PH where $V(H_i)$ and $E(H_i)$ denote the vertex set and edge set, respectively, of H_i. The collection of hexagons defining a polyhex PH is denoted by

$$P = \{H_1, H_2, \ldots, H_k\}$$

where $PH = H_1 \cup H_2 \ldots \cup H_k$. A spanning subgraph of PH having components (cycles) with all vertices of degree-2 is called a 2-factor subgraph. A single component 2-factor subgraph of PH is called a Hamiltonian circuit C_m ($m = 4n + 2; n = 1, 2, 3 \ldots$) where $m = N_c$ and $N_c - 2$ is divisible by four [3, 9]. A multicomponent 2-factor subgraph of a PH has an odd number of cycles C_m if $N_c \equiv 2 \pmod 4$ and an even number if $N_c \equiv 0 \pmod 4$ where $\Sigma m = N_c$. A sextet 2-factorable structure is a polyhex structure having 2-factor subgraphs composed of exclusively hexagonal components (cycles) where $N_c \equiv 0 \pmod 6$. A benzenoid hydrocarbon molecular structure (PAH6) is isomorphic to a polyhex structure. Benzenoid structures having Hamiltonian 2-factors are called Hamiltonian benzenoid hydrocarbons.

A 1-factor of a graph G is a spanning subgraph of G consisting of only K_2 components. A 1-factor subgraph of a polyhex structure is isomorphic to a Kekulé Graph where K, the Kekulé structure count, is the number of distinct 1-factor subgraphs associated with PH. The number of components in a 1-factor of PH is odd for $N_c \equiv 2 \pmod 4$ and even for $N_c \equiv 0 \pmod 4$. If a graph is 2-factorable, then it is 1-factorable; conversely, if a PH graph is not 1-factorable, then it is not 2-factorable [in the language of logic: $2\text{-}F(\mathrm{PH}) \Rightarrow 1\text{-}F(\mathrm{PH}) \Leftrightarrow \neg 1\text{-}F(\mathrm{PH}) \Rightarrow \neg 2\text{-}F(\mathrm{PH})$]. Each Hamiltonian 2-factor corresponds to two 1-factors, and, in general, each 2-factor subgraph of n components corresponds to 2^n distinct 1-factor subgraphs. Throughout this work the carbon and hydrogen atoms and the $p\pi$-bonds in all molecular graphs will be omitted and only the C–C σ-bond skeleton will be shown. The various types of vertices referred to herein are methylene $=C{<}^{H}_{H}$ shown as a primary (degree one) graph vertex, methine $=C{<}^{H}$ shown as a secondary (degree two) graph vertex, and carbon $=C{<}$ shown as a tertiary (degree three) graph vertex.

2.2 Dualist Graph

A dualist or inner dual graph is to be distinguished from a dual graph. The dualist graph is the terminology used by Balaban [10], and the inner dual graph is used by Trinajstić; Hall uses bual (*b*astard d*ual*) [11, 12]. A dualist graph is defined as a graph produced by placing points at the center of each hexagon of a polyhex and joining them by edges passing perpendicularly through a side shared by two adjoining hexagons. A dual graph has additionally a point located in the outer infinite region with edges passing through the periphery of the polyhex. The dual graph of a dual graph regenerates the original graph. The dualist graph of a polyacene with r rings is a linear sequence of r-1 line segments with r points, and the dualist graph of benzo[*a*]acenes is a linear sequence of r-2 line segments terminated by a line segment having a 120° angle to the linear sequence. The dualist graph of a hexagonal lattice is a triangular lattice. Pyrene and its dualist graph are shown below where the latter has N_{Ic} triangles and r vertices (points).

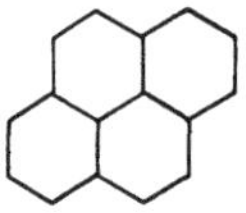

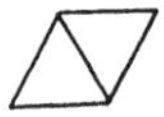

2.3 Excised Internal Structure

An excised internal structure or insular structure is a structure obtained by excising out the set of connected internal vertices usually associated with a strictly peri-condensed benzenoid hydrocarbon [3, 13]. A strictly peri-condensed benzenoid PAH6 has all its internal third degree vertices mutually connected, has no cata-condensed appendages ($n_4 = 0$), and has a formula found on or toward the extreme left-hand diagonal formula boundary found in Table PAH6. The excised internal structure of pyrene ($C_{16}H_{10}$) is ethene and of coronene ($C_{24}H_{12}$) is benzene as shown below by the dotted lines.

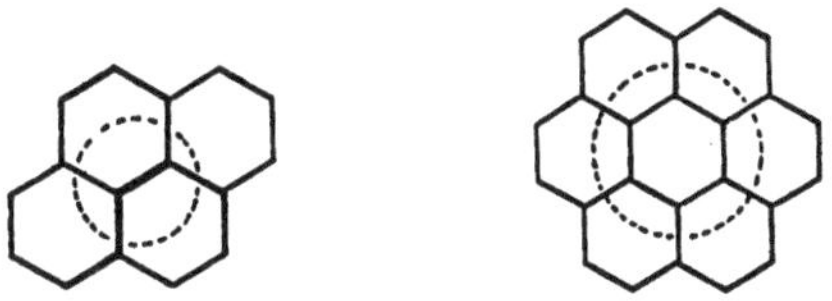

Pyrene/Ethene Coronene/Benzene

A dualist graph of a homogeneous polycyclic structure is generally defined as a graph with points (vertices) located at the center of each equilateral polygon and edge connecting these points through adjacent polygons sharing a common side. The dualist graph of a polyhex graph with r rings and N_{IC} internal vertices is a triangular lattice with r vertices and N_{Ic} triangular rings, and the dualist graph of this latter dualist graph is the excised internal structure of the original polyhex with N_{Ic} vertices. These dualist graph relationships are illustrated as follows with coronene. In coronene the vertices for the incipient dualist graph are shown in each hexagonal ring, and in its dualist graph the vertices for the incipient excised internal are shown in each trigonal ring.

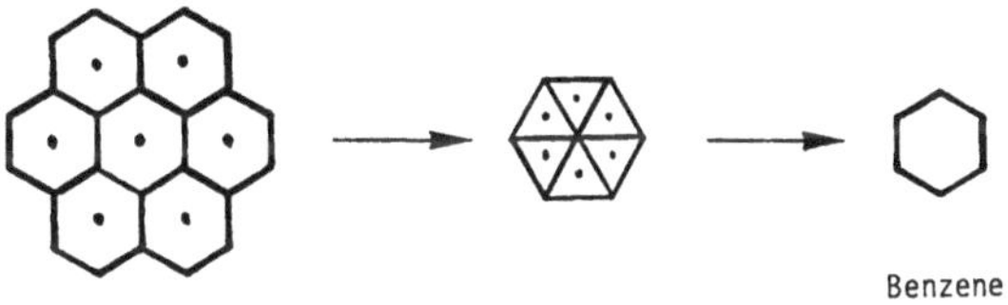

Coronene Dualist graph Benzene Excised internal structure

The excised internal structure was called insular orbitals by Schmidt and co-workers [14]. In gas-phase photoelectron spectroscopy (PE), cata-condensed hydrocarbons differ from peri-condensed ones in that the latter have additional PE bands originating from the orbitals of the insular structures. For example, all coronenes in which an inner benzene nucleus is surrounded by an outer $4n + 2$ annulene perimeter show a double band at about 8.6 and 9.0 eV, in addition to the more or less regularly spaced bands arising from the orbitals of the peripheral ring; similarly, most pyrenes show a spectral band at about 9.1 eV which arises from the ethene insular. A Hückel correlation diagram showing the evolution of the occupied π-levels of a cata-condensed hydrocarbon from those of the corresponding annulene can be constructed by allowing the resonance integrals β for bridging interaction to increase from 0 to 1. Similarly, a Hückel correlation diagram for the evolution of the occupied π-levels of a pericondensed hydrocarbon can be constructed by allowing the inter-fragment resonance integrals between the annulene perimeter and insular structure to increase from 0 to 1.

If a benzenoid excised internal structure is 1-factorable, 2-factorable, strictly peri-condensed, has one or more bay regions, and/or has one or more selective lineations, then the corresponding larger daughter PAH6 structure formed by circumscribing a perimeter of carbon atoms around the excised internal structure and incrementing it with six hydrogens will also have these attributes [3, 4].

2.4 Strictly Peri-Condensed

Strictly peri-condensed benzenoid hydrocarbons have all their internal edges mutually connected ($d_s \geqq 0$) and have formulas found at the extreme left-hand diagonal edge of Table PAH6 [2, 3]. Strictly peri-condensed benzenoids have no cata-condensed appendages (i.e., $n_4 = 0$) and may contain a maximum number of internal carbon vertices. Formulas at the extreme left-hand diagonal edge of Table PAH6 will correspond to benzenoid hydrocarbons having only strictly peri-condensed isomers, and formulas just adjacent to these will correspond to benzenoid hydrocarbons having both strictly peri-condensed and non-strictly peri-condensed isomers. For example, $C_{28}H_{14}$ has 8 nonradical benzenoid isomers, seven that are strictly peri-condensed and benzo[*a*]coronene which is not. Strictly peri-condensed benzenoid hydrocarbons have no adjacent bay regions, i.e., their perimeters have no coves (two proximate bay regions) or fjords (three proximate bay regions). For a given number of carbon atoms, strictly peri-condensed benzenoid hydrocarbons are among the more stable benzenoids, have a maximum number of internal carbon atoms, and possess a minimum number of bay regions [15].

2.5 Aufbau Principle

Recursive construction of all PAH6 isomers of a specified formula from two sets of PAH6 isomers having formulas immediately above it and to the left of it in Table PAH6 was a key concept espoused in my seminal paper [1] of 1982. For the benzenoid hydrocarbons having formulas on the upper and diagonal edges of Table PAH6 only one set of precursor isomers exist leading to a simplified enumeration process for cata-condensed and strictly peri-condensed benzenoids. For example, all the cata-condensed benzenoids have no precursor formulas above them but do have formulas at their left. Thus, all cata-condensed benzenoids are ultimately generated from naphthalene by successive annelation of C_4H_2 units in all combinatorial ways. The conversion of naphthalene to anthracene and phenanthrene in Scheme I is illustrated. Other aufbau or building-up processes are also illustrated in Scheme I. Of particular importance is the attachment of C_2 units to bay regions to give PAH6 structures possessing two additional internal third degree vertices as shown in the conversion of phenanthrene to pyrene. This latter aufbau process can be used to generate all the strictly peri-condensed benzenoid structures having formulas along the left-hand sloping diagonal edge of Table PAH6. For example, attachment of a C_2 unit to the bay region of benzo[*ghi*]perylene ($C_{22}H_{12}$) generates coronene ($C_{24}H_{12}$), and the attachment of a C_2 units to the bay region of dibenzo[*bc,ef*]coronene ($C_{30}H_{14}$); gives ovalene ($C_{32}H_{14}$) dibenzo[*bc,ef*]coronene is generated from coronene by attachement of two C_3H units per Scheme I.

The generation of diradical PAH6 structures from nonradical precursors can only be achieved through the appropriate attachment of C_3H units. For example, diradical

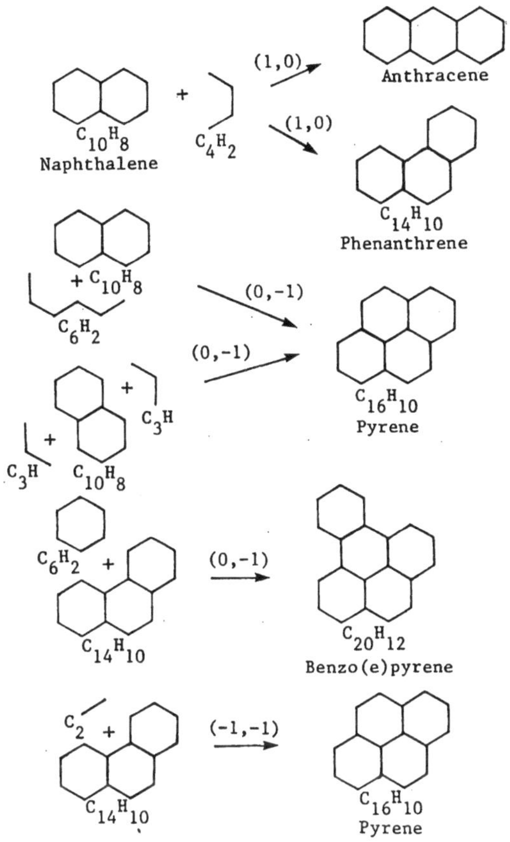

Scheme I. Recursive aufbau construction of benzenoid hydrocarbons

triangulene ($C_{22}H_{12}$) is generated from pyrene ($C_{16}H_{10}$) and diradical dibenzo[*de,hi*]-naphthacene from naphthacene by appropriate attachment of two C_3H units.

3 Formula Periodic Table for Benzenoids

3.1 Structure of Table PAH6

Given the disconnection set $DS = \{\ldots, -2, -1, 0, 1, 2, \ldots\}$ and the internal vertex set $IV = \{0, 2, 4, \ldots\}$, the product $DS \times IV$ is the collection of ordered pairs (d_s, N_{Ic}). The relation R from set DS to IV is given by Table PAH6 and is the subset of R = Table PAH6 $\subset DS \times IV$. We say d_s is related to N_{Ic} (by the relation R) if $(d_s, N_{Ic}) \in$ (Table PAH6). Thus

$$\text{Table PAH6} \subset \{(d_s, N_{Ic}); d_s + N_{Ic} = r - 2, r \geqq 2, N_{Ic} \geqq 0\}$$

and is a hierarchial structure which is the focus of this work.

Table PAH6. Formula periodic table for benzenoid polycyclic aromatic hydrocarbons (PAH6)

$d_s = -9$	$d_s = -8$	$d_s = -7$	$d_s = -6$	$d_s = -5$	$d_s = -4$	$d_s = -3$	$d_s = -2$	$d_s = -1$	$d_s = 0$	$d_s = 1$	$d_s = 2$	$d_s = 3$	N_{Ic}
	Forbidden region						Most reactive		$C_{10}H_8$	$C_{14}H_{10}$	$C_{18}H_{12}$	$C_{22}H_{14}$	... 0
									$C_{16}H_{10}$	$C_{20}H_{12}$	$C_{24}H_{14}$	$C_{28}H_{16}$	... 2
					$N_c/N_H < 2$				$C_{22}H_{12}$	$C_{26}H_{14}$	$C_{30}H_{16}$	$C_{34}H_{18}$	... 4
		$C_{00}H_{00}$						$C_{24}H_{12}$	$C_{28}H_{14}$	$C_{32}H_{16}$	$C_{36}H_{18}$	$C_{40}H_{20}$	... 6
								$C_{30}H_{14}$	$C_{34}H_{16}$	$C_{38}H_{18}$	$C_{42}H_{20}$	$C_{46}H_{22}$	... 8
				$2 < \frac{N_c}{N_H} < 3$			$C_{32}H_{14}$	$C_{36}H_{16}$	$C_{40}H_{18}$	$C_{44}H_{20}$	$C_{48}H_{22}$	$C_{52}H_{24}$	... 10
							$C_{38}H_{16}$	$C_{42}H_{18}$	$C_{46}H_{20}$	$C_{50}H_{22}$	$C_{54}H_{24}$	$C_{58}H_{26}$	... 12
						$C_{40}H_{16}$	$C_{44}H_{18}$	$C_{48}H_{20}$	$C_{52}H_{22}$	$C_{56}H_{24}$	$C_{60}H_{26}$	$C_{64}H_{28}$	... 14
$\frac{N_c}{N_H} > 3$					$C_{42}H_{16}$	$C_{46}H_{18}$	$C_{50}H_{20}$	$C_{54}H_{22}$	$C_{58}H_{24}$	$C_{62}H_{26}$	$C_{66}H_{28}$	...	
					$C_{48}H_{18}$	$C_{52}H_{20}$	$C_{56}H_{22}$	$C_{60}H_{24}$	$C_{64}H_{26}$	$C_{68}H_{28}$	...		
				$C_{50}H_{18}$	$C_{54}H_{20}$	$C_{58}H_{22}$	...	...	...	...			
			$C_{52}H_{18}$	$C_{56}H_{20}$	$C_{60}H_{22}$	...							
Least Reactive		$C_{54}H_{18}$	$C_{58}H_{20}$	$C_{62}H_{22}$	...								
		$C_{60}H_{20}$	$C_{64}H_{22}$	...			Intermediate reactive						
	$C_{62}H_{20}$	$C_{66}H_{22}$	...										
$C_{64}H_{20}$	$C_{68}H_{22}$	...											
$C_{70}H_{22}$	...												
...													

In Table PAH6 every compound in a particular column has the same d_s value, and every compound in the same row has the same N_{Ic} value [2]. Member compounds in the same column with the same d_s value belong to the same column series ($N_c = 3N_H - 14 - 2d_s$, where d_s is specified), and member compounds in the same row with the same N_{Ic} value belong to the same row series ($N_c = 2N_H - 6 + N_{Ic}$, N_{Ic} is specified). For example, all PAH6 isomers of the formulas $C_{16}H_{10}$ and $C_{20}H_{12}$ have the same value of $N_{Ic} = 2$ and similarly, all PAH6 isomers of the formulas $C_{20}H_{12}$ and $C_{26}H_{14}$ have the same value of $d_s = 1$. Table PAH6 extends infinitely in three directions–horizontally to the right, vertically to the bottom, and in a slanting direction to the left. The ratio of N_c/N_H for any column approaches 3 as the number of rings increases. Every formula in the column beginning with $C_{54}H_{18}$ belongs to the $N_c = 3N_H$ ($d_s = -7$) column series, and thus $N_c/N_H = 3$ for every member in this column. Columns to the left of the $N_c = 3N_H$ series have N_c/N_H ratios that approach 3 from larger values and columns to the right from smaller values. Similarly, the ratio of N_c/N_H for any row approaches 2 as the number of rings increases. Every formula in the row beginning with $C_{24}H_{12}$ belongs to the $N_c = 2N_H$ ($N_{Ic} = 6$) row series, and thus $N_c/N_H = 2$ for every member in this series. Rows above the $N_c = 2N_H$ series have N_c/N_H ratios that approach 2 from smaller values and the rows below from larger values.

There are four quadrants to Table PAH6, which are determined by the range of values in the ratio of N_c/N_H for the possible formulas of PAHs. The intersection of these quadrants occurs at the nonexistent formula of $C_{00}H_{00}$. The horizontal line is defined by the linear array of formulas having $N_c/N_H = 2$, and the vertical line is defined by the linear array of formulas having $N_c/N_H = 3$; these lines intersect at a hypothetical formula that should have a N_c/N_H ratio that is simultaneously equal to both 2 and 3 which is impossible ($N_c/N_H = 0/0$ for $C_{00}H_{00}$ and is undefined). In the upper left-hand corner quadrant is the forbidden region where no reasonable formula can exist and the boundary diagonal line passes through the nonexistent $C_{xy}H_{00}$ formula array which is parallel to the diagonal linear array of formulas all having the same number of hydrogens (e.g. $C_{xy}H_{22}$). The lower left-hand quadrant contains formulas where $N_c/N_H > 3$, the lower right-hand quandrant has $2 < N_c/N_H < 3$, and the upper right-hand quadrant has $N_c/N_H < 2$.

In the rows of the Formula Periodic Table for Benzenoid PAH6s (Table PAH6), the number of formula carbons (N_c) increases from left to right according to the even residue classes of congruent modulo 4. For example, in the $N_{Ic} = 0$ and $N_{Ic} = 2$ row series, the number of formula carbons follow $N_c \equiv 2 \pmod 4$ and $N_c \equiv 0 \pmod 4$, respectively; these two respective relationships are each applicable for every other row. In the columns of the Formula Periodic Table for Benzenoid PAH6s, the number of formula carbons increases from top to bottom according to the even residue classes of congruent modulo 6. In the $d_s = 0$, $d_s = 1$, and $d_s = 2$ column series, the number of formula carbons are given by $N_c \equiv 4 \pmod 6$, $N_c \equiv 2 \pmod 6$, and $N_c \equiv 0 \pmod 6$, respectively; this sequence of congruent modulo 6 relationships successively repeats for all the columns in Table PAH6. The basis for Table PAH6 is that all PAH structures having a common CH formula must comply with $d_s + N_{Ic} = r - 2$ where d_s is the net number of disconnections (or connections) of internal edges and N_{Ic} is the number of internal third degree carbon vertices. For example, anthracene/phenanthrene have two internal edges that are disconnected ($d_s = 1$) and no internal third degree vertices ($N_{Ic} = 0$) and anthanthrene/benzo[*ghi*]perylene have all their internal edges connected ($d_s = 0$) and four internal third degree vertices ($N_{Ic} = 4$). $N_{Ic} + d_s = r - 2$ is valid for all three connected nets. Not only is this relationship applicable to polyhex graphs, but it can be easily verified for the Schlegel planar representations of the cube, dodecahedron, and truncated icosahedron.

Two vectors (2-dimensional) are perpendicular if their dot (scalar) product is zero: $\bar{v} \cdot \bar{t} = v_1 t_1 + v_2 t_2 = 0$. Thus if a vector $\bar{v} = (v_1, v_2)$, then its perpendicular is $\bar{v}_p = (-v_2, v_1)$ since $\bar{v} \cdot \bar{v}_p = -v_1 v_2 + v_2 v_1 = 0$. Two vectors are parallel if $\bar{t} \cdot \bar{v}_p = -t_1 v_2 + t_2 v_1 = 0$. Also, parallel vectors have proportional ordered pairs. All ordered pairs associated with the PAH6 formulas in the $d_s = -7$ column series of Table PAH6 correspond to a set of parallel vectors represented by (54,18) $\cong$ (3,1); similarly all ordered pairs associated with the $N_{Ic} = 6$ row series correspond to a set of parallel vectors represented by (24,12) $\cong$ (2,1). The dot product of $(-1,3)$ with any member of a set of vectors associated with some particular column of formulas in Table PAH6 will have the same value, and the dot product of $(-1,2)$ with any member of vectors belonging to a given row in Table 1 will have the same value. In other words, $(-1,3) \cdot (N_c, N_H) = 14 + 2d_s$ and $(-1,2) \cdot (N_c, N_H) = 6 - N_{Ic}$ which are equations given above for Table PAH6.

Only isomeric benzenoids in Table PAH6 have the same number of σ-bond edges (q) even though formulas with shift coordinates (Δd_s, $\Delta N_{Ic}/2$) of (3,2) have corresponding PAH structures with the same number of carbons and formulas with shift coordinates of (2,1) have structures with the same number of rings. Since each coordinate location (d_s, N_{Ic}) in Table PAH6 corresponds to a unique q value, the locants in Table PAH6 represent a countable infinite set as there exists a one-to-one correspondence between them and the set of all natural numbers.

3.2 Isomer Enumeration

Our prior work presented two concepts for enumeration of benzenoid PAH6s. The first concept was that all benzenoid PAH6 isomers of a particular formula could be recursively enumerated from the benzenoid isomers having the corresponding formulas immediately above it and to the left of it in the Formula Periodic Table for Benzenoid PAH6s (Table PAH6) [1]. This is illustrated in Scheme I for the enumeration of anthracene/phenanthrene ($C_{14}H_{10}$) and pyrene ($C_{16}H_{10}$) from naphthalene ($C_{10}H_8$); Figures 1 and 2 summarize this algorithmic process. Attachment of a C_4H_2 unit to naphthalene (N) gives anthracene (A) and phenanthrene (P), and continuing this process leads to all the possible catacondensed isomers (Fig. 1) [2]. From Fig. 1, it is evident that only acenes beget acenes and the benzo[*a*]acenes derive from either acenes or lower benzo[*a*]acenes. Benzo[*a*]anthracene (B[*a*]A), chrysene (C), benzo[*c*]phenanthrene (B[*c*]P), and triphenylene (T) are obtained by attachment of C_4H_2 units to the four different sides of phenanthrene (P). There are 19 different attachments of C_4H_2 units onto the 5 isomers of $C_{18}H_{12}$ leading to 12 distinct $C_{22}H_{14}$ isomers, and there are 70 different attachements of C_4H_2 units onto the 12 isomers of $C_{22}H_{14}$ leading to 37 distinct $C_{26}H_{16}$ isomers and so forth. The ordering hierarchy used in Fig. 1 is the magnitude of the total p_π energy. Acenes have the lowest E_π

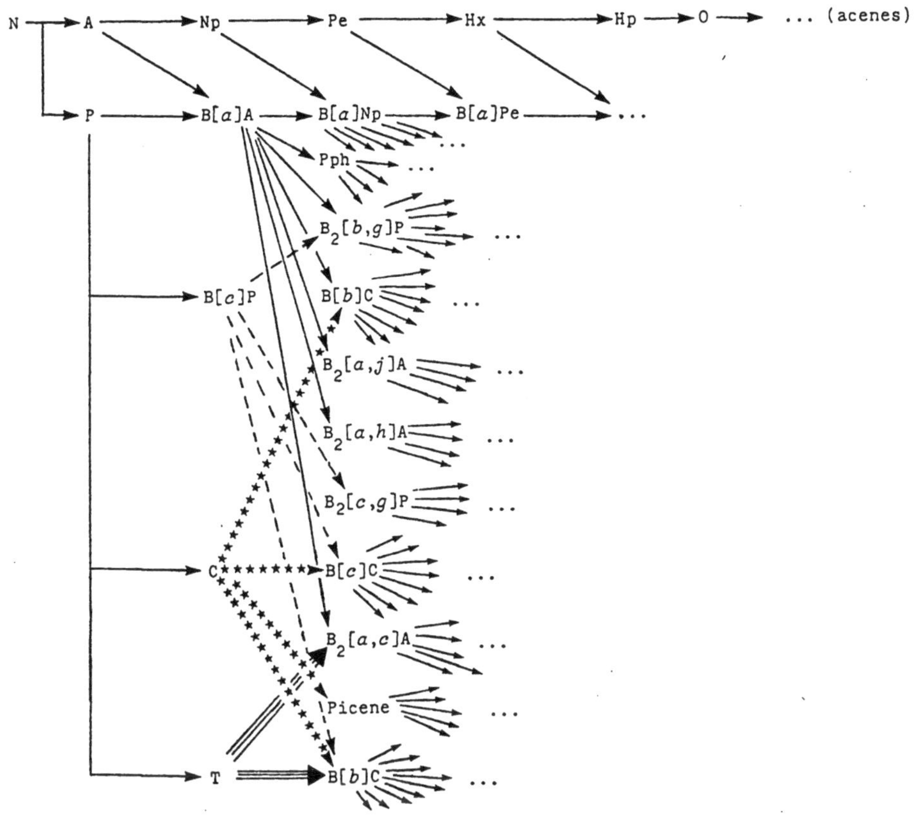

Rank order is increasing E_π from top to bottom and left to right.

Fig. 1. Scheme for recursive enumeration of the cata-condensed benzenoids

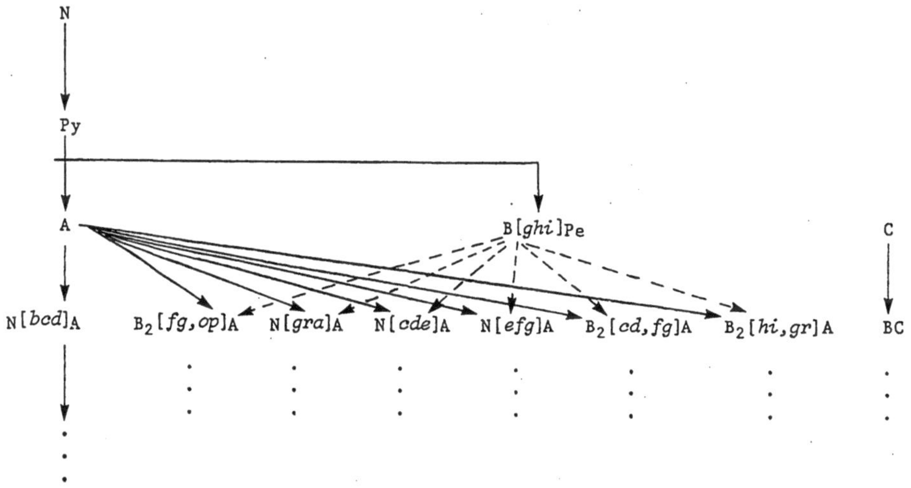

Rank order is increasing E_π from top to bottom and left to right.

Fig. 2. Scheme for recursive enumeration of some peri-condensed benzenoids

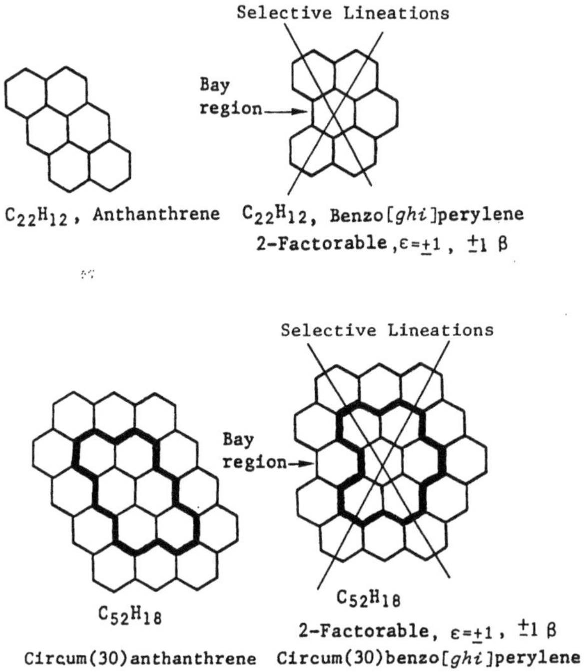

Fig. 3. The corresponding parent/daughter isomers of $C_{22}H_{12}/C_{52}H_{18}$, respectively

energy and the smallest HOMO magnitude and benzo[*a*]acenes are next. The E_π of chrysene has a magnitude larger than benzo[*a*]anthracene but smaller than benzo[*c*]-phenanthrene, and finally triphenylene has the largest E_π value. Figure 2 shows a similar pattern where pyrene is derived from naphthalene by attachment of either two C_3H units or one C_6H_2 unit. Anthanthrene (A) and benzo[*ghi*]perylene (B[*ghi*]Pe) derive from pyrene. The magnitude of the HOMO and E_π of anthanthrene is smaller than that of benzo[*ghi*]perylene. Again all less stable and symmetrical "rhomboid" pericondensed isomers can be only derived from their lower homologs (N → Py → A → N[*bcd*]A → ...) and have the smallest HOMO values. Note that benzo[*a*]coronene (BC) derives only from coronene by the attachment of a C_4H_2 unit (Fig. 2).The second concept was that strictly peri-condensed PAH6s without catacondensed appendages have formulas along the left-hand diagonal edge of Table PAH6 and their isomers can be enumerated by enumerating the number of isomers associated with their excised internal structures [13]. Thus the number of nonradical isomers of $C_{22}H_{12}$ is two, anthanthrene and benzo[*ghi*]perylene, and the corresponding number of circumscribing isomers ($C_{52}H_{18}$) is two (Fig. 3). Since circulene and helicene structures are not possible for strictly pericondensed PAH6s, the ambiguity associated with other benzenoid isomer enumeration studies is absent in the enumeration of strictly peri-condensed PAH6s. The one-isomer series derived from naphthalene (Fig. 4), pyrene, and coronene are strictly peri-condensed PAH6s. Via this important concept the number of nonradical benzenoid isomers for $C_{38}H_{16}$, $C_{40}H_{16}$, $C_{42}H_{16}$,

$C_{48}H_{18}$, $C_{50}H_{18}$, $C_{52}H_{18}$, $C_{60}H_{20}$, $C_{62}H_{20}$, and $C_{64}H_{20}$ were determined for the first time to be 10, 3, 1, 22, 7, 2, 32, 12, and 3, respectively [3]. From Figures 3 and 4 it is evident that the number of bay regions and selective lineations present in a benzenoid excised internal structure is transmitted to the daughter strictly peri-condensed benzenoid structures derived by circumscribing a perimeter of carbon atoms around the smaller parent benzenoid excised internal structure. Table 2 summarizes the formulas for some strictly peri-condensed isomer series.

By merging our prior aufbau concept with the Formula Periodic Table for Benzenoid Hydrocarbons (Table PAH6), the enumeration process itself will have a

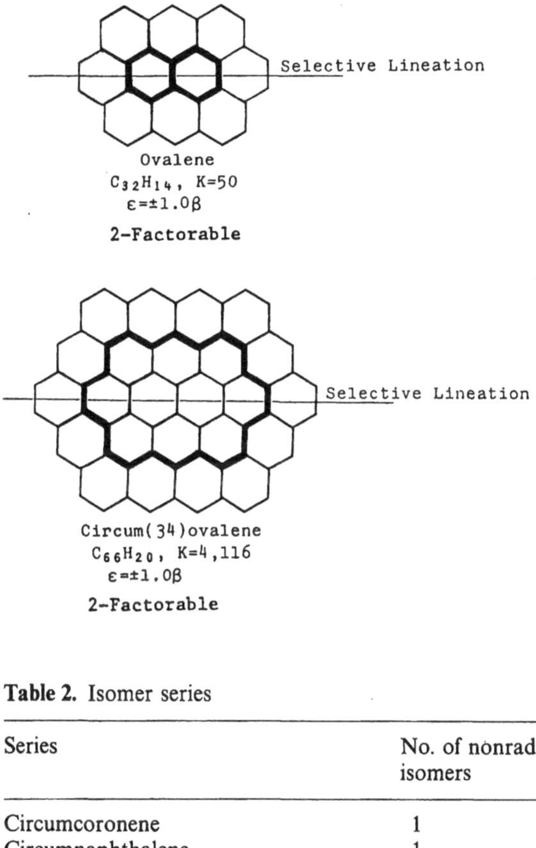

Fig. 4. One-isomer series of circumnaphthalene

Table 2. Isomer series

Series	No. of nonradical isomers	N_c[a]	N_H[a]
Circumcoronene	1	$6n^2$	$6n$
Circumnaphthalene	1	$6n^2 + 4n$	$6n + 2$
Circumpyrene	1	$6n^2 + 8n + 2$	$6n + 4$
Circumanthanthrene	2	$6n^2 + 12n + 4$	$6n + 6$
– circumbenzo[*ghi*]perylene			
Circumnaphtho[*abc*]coronene	3	$6n^2 + 16n + 8$	$6n + 8$
– circumdibenzol[*bc, ef*]coronene			
– circumdibenzo[*bc, kl*]coronene			
$C_{40}H_{16}$, ...	3	$6n^2 + 20n + 14$	$6n + 10$
$C_{50}H_{18}$, ...	7	$6n^2 + 24n + 14$	$6n + 12$
$C_{76}H_{22}$, ...	12	$6n^2 + 32n + 38$	$6n + 16$
$C_{106}H_{26}$, ...	38	$6n^2 + 40n + 60$	$6n + 20$

a $n = 1, 2, 3, \ldots$

structure based on the thermodynamics of the benzenoid structures being progressively generated. The premise espoused here is that an enumeration process following the thermodynamics of the structures being generated should be the most efficient process possible.

A mapping of a set A into a set B is a subset of $A \times B$ in which each element of A occurs once and only once as the first component in the elements of the product subset. Figures 1 and 2 are examples of mapping from one isomer set to another. In any mapping α of A into B, the set A is called the *domain* and the set B is call the *co-domain* of α. If the mapping is *onto*, B is called the *range* of α; otherwise, the range of α is the proper subset of B consisting of the images of all elements of A. The mapping of the set $\mathbb{C}_4$ (four cata-condensed rings) of the 5 $C_{18}H_{12}$ cata-condensed benzenoid isomers onto the set $\mathbb{C}_5$ (five catacondensed rings) of the 12 $C_{24}H_{14}$ isomers is a subset of the product set $\mathbb{C}_4 \times \mathbb{C}_5$ of $\mathbb{C}_4$ and $\mathbb{C}_5$. The mapping of set $\mathbb{C}_4$ onto set $\mathbb{C}_5$ uses arrows to connect associated elements in Fig. 1 and consists of 19 correspondences which is a subset the maximum number of $|\mathbb{C}_4 \times \mathbb{C}_5| = 5 \times 12 = 60$ possible correspondences. Since mappings that are single valued at the left with only one arrow directed away from each left-hand element are also functions, the mappings in Figures 1 and 2 are not functions. A mapping of set A into a set B is called a *one-to-one* mapping of A into B if the images of distinct elements of A are distinct elements of B; if, in addition, every element of B is an image, the mapping is called a *one-to-one* of A *onto* B. The mappings in Figures 1 and 2 are not one-to-one; for example, many of the distinct elements of $\mathbb{C}_4$ have a common image element in $\mathbb{C}_5$.

The successive mapping from one isomer set to another in Figures 1 and 2 reveals the interrationships and partial ordering between and within each corresponding locant in Table PAH6. In Fig. 1, only one path of arrows or one-to-one correspondence exists for the acene (N → A → Np → Pe → ...), zigzag (P → C → Picene → ...), and helicene (P → B[*c*]P → B_2[*c*, *g*]P → ...) series. Since B_2[*a*, *j*]A and B_2[*a*, *h*]A both exclusively derive from B[*a*]A whereas B_2[*c*, *g*]P derives exclusively from B[*c*]P, the former two isomers show a closer kinship to each other when their E_π and ε_{HOMO} are compared.

Starting with naphthalene, two principally different series of polycyclic aromatic hydrocarbons are obtained depending on the direction of annelation. Linear annelation gives the acene series, and angular annelation gives the aphene series. An acene has only one sextet shared among all its rings and an aphene has two. If the number of sextets remain the same in an annelation series, then regular shifts toward the red (i.e., toward longer wavelengths) of the α, β, p absorption bands are observed as the number of rings increase [16]. The maximum bathochromic shifts is observed in the acene series; this is made evident by the comparison of progressive color change in going from colorless anthracene to green hexacene and from colorless phenanthrene to yellow hexaphene. Other illustrative annelation series with a fixed number of sextets include perylene (yellow) to bisanthrene (dark blue), triphenylene (colorless) to coronene (yellow), dibenzo[*fg*, *op*]naphthacene (colorless) to ovalene (orange), and pyrene (colorless) to anthanthrene (yellow) [16]. The annelation method of Clar for correlating UV absorption spectral data is a mapping process where the number of bay regions and resonant sextets were generally held constant within an annelated series of benzenoid compounds.

3.3 Structure/Energy Correlations of Table PAH6

All PAH6 formulas having the same number of hydrogens are found in a linear diagonal array of formulas (Table PAH6) and have the same perimeter length ($q_p = 2N_H - 6 = \text{constant}$ for $N_H = \text{constant}$) [9, 13]. The formulas in the $N_c = 2N_H - 6$ row series have no internal third degree carbon vertices ($N_{Ic} = 0$ and $N_c = q_p$). Thus the two isomers of $C_{14}H_{10}$ both have a perimeter length of 14 carbon atoms, and pyrene ($C_{16}H_{10}$) has two internal third degree carbon vertices ($N_{Ic} = 2$) with a perimeter length of 14 carbon atoms giving a total of 16 carbon atoms in its formula. Similarly, all five PAH6 isomers of $C_{18}H_{12}$ have a perimeter length of 18 carbon atoms, and the three isomers of $C_{20}H_{12}$ have $N_{Ic} = 2$, the two PAH6 isomers of $C_{22}H_{12}$ have $N_{Ic} = 4$, and coronene ($C_{24}H_{12}$) has $N_{Ic} = 6$ all with a perimeter length of 18 carbon atoms. In hexagonal graphs the maximum number of internal third degree carbon vertices that can be contained in a 10 carbon atom peripheral cycle is zero (naphthalene), in a 14 carbon atom peripheral cycle it is two (pyrene), and in an 18 carbon peripheral atom cycle it is six (coronene). Polycircumnaphthalene, polycircumpyrene, and polycircumcoronene are three series of formulas appearing on the left-hand diagonal boundary of Table PAH6. Each formula of these series has only one corresponding PAH6 structure. These terminal formulas have PAH6 structures containing the maximum number of internal third degree carbon vertices that each corresponding set of peripheral $q_p = 4\tau + 2$ ($\tau = 2, 3, 4 \ldots$) carbon atoms can contain.

The 1-factor and 2-factor subgraphs of PAH6 structures having formulas belonging to the alternate rows of $N_c = 2N_H - 6$, $N_c = 2N_H - 2, \ldots$ [i.e., $N_c \equiv 2 \pmod 4$] of Table PAH6 have an odd number of components and an even number of components if they have formulas belonging to the other set of alternate rows [i.e., $N_c \equiv 0 \pmod 4$] [3, 9]. A graph G is Hamiltonian if it has a single spanning cycle. Thus PAH6 structures containing Hamiltonian subgraphs can only have corresponding formulas found in the $N_c \equiv 2 \pmod 4$ row series of Table PAH6 [9]. Note that if the strictly peri-condensed benzenoid parent excised internal structure has one or more bay regions, is 2-factorable, and/or has one or more selective lineations, then the corresponding larger daughter PAH6 structure also has one or more bay regions, is 2-factorable, and/or has one or more selective lineations. Thus in Fig. 3 benzo[*ghi*]perylene has a bay region, is 2-factorable, and has two eigenvalues of $\varepsilon = \pm 1\beta$ and so does its corresponding $C_{52}H_{18}$ PAH6 structure. On the same basis, all the PAH6 structures having formulas belonging to the one-isomer series (for example Fig. 4) have no bay regions and possess eigenvalues of $\varepsilon = \pm 1\beta$.

According to the sextet rule all benzenoid PAH6 structures having 2-factor subgraphs composed exclusively of hexagons are predicted to be more stable and have associated chemistry which tend to preserve this structural identity [17, 18]. Triphenylene has a 2-factor composed of three hexagons and is the most stable $C_{18}H_{12}$ isomer. Only the $N_c \equiv 0 \pmod 6$ column series in Table PAH6 can have PAH6 isomers possessing 2-factor subgraphs composed exclusively of hexagon components; the number of hexagon components will be odd for the $N_c \equiv 2 \pmod 4$ row series and even for the $N_c \equiv 0 \pmod 4$ row series [2, 3, 9]. The recent work of Stein and Brown has shown that total resonant sextet benzenoids which are also strictly peri-condensed represent the most stable benzenoid hydrocarbons possible [18].

Threefold symmetry can only be exhibited by nonradical benzenoid molecular graphs with $N_{Ic} \equiv 0 \pmod 6$ and $N_c \equiv 0 \pmod 6$ [19]. Triphenylene and related starphenes have $N_{Ic} = 0$, and coronene-related threefold graphs have $N_{Ic} = 6$. The strictly peri-condensed $C_{48}H_{18}$ benzenoid circum(30)triphenylene, has $N_{Ic} = 18$. Total resonant sextet benzenoids *14* ($C_{42}H_{18}$) and *53* ($C_{54}H_{24}$) have threefold symmetry and $N_{Ic} = 12$, and a threefold symmetrical $C_{72}H_{24}$ total resonant sextet benzenoid having $N_{Ic} = 30$ has been portrayed in a previous paper by the author. The triangular-shaped $C_{60}H_{24}$ total resonant sextet benzenoid has $N_{Ic} = 18$. Thus, like total resonant sextets, nonradical threefold symmetrical benzenoids are only found in the $N_c \equiv 0 \pmod 6$ column series of the Formula Periodic Table for Benzenoids.

If the eigenvalues of a smaller graph are contained among the eigenvalue spectrum of a larger graph, then the smaller graph is said to be subspectral to the larger one [11]. Previously, it was shown that threefold molecular graphs possessed doubly degenerate eigenvalue subsets. Whenever a molecular graph possesses degenerate eigenvalue subsets, a (zero) node vertex can be obtained at any selected molecular graph position by taking an appropriate linear combination of these degenerate eigenvalues subsets [7]. Thus, deletion of any atomic vertex or placement of any heteroatom or polyene substitutent at any position on a molecular graph having doubly degenerate eigenvalue subsets will lead to successor molecular graphs still retaining one of these eigenvalue subsets. This principle was used to present a compact compendium of select eigenvalues for well over 2000 molecules [7].

Connected isospectral graphs are isomeric molecular graphs having exactly the same eigenvalues. A graph may contain a pair of nonequivalent vertices which have the property that removal of each vertex in turn produces either two identical or isospectral graphs. Vertices having this property may be called isospectral points because isospectral graphs can be generated from them by attaching two different fragments in a reciprocal fashion to such vertices. Isospectral points possess equal magnitude eigenvectors for all nondegenerate eigenvalues. It has been shown that molecular graphs with threefold (or higher) symmetry can be used to generate isospectral graphs.

It is now obvious that strictly peri-condensed and essentially strain-free total resonant sextet benzenoids have similar features in regard to their molecular graph perimeter topology and stability per carbon. Benzenoid hydrocarbons with threefold symmetry possess special properties that result in them possessing doubly degenerate eigenvalue subsets, parenthood to isospectral derivatives, and relatively higher π-electron stability. The Formula Periodic Table for Benzenoid Hydrocarbons (Table PAH6) sorts both total resonant sextet and threefold benzenoids into the same column series of $N_c \equiv 0 \pmod 6$ [19].

The perimeter topology of essentially strain-free total resonant sextet or strictly peri-condensed benzenoids does not permit them to undergo pyrolytic ring closure concomitant with H_2 loss. This is because these systems are devoid of two adjacent bay regions (cove) or three bay regions (fjord). Benzo[*c*]phenanthrene possesses a cove which can undergo closure to a pentagonal ring, and dibenzo[*c*, *g*]phenanthrene (pentahelicene) possesses a fjord which can undergo closure to a hexagonal ring to give benzo[*ghi*]perylene. On the average, strictly peri-condensed, total resonant sextet, and threefold benzenoid hydrocarbons are among the more stable isomers [19].

The necessary and sufficient collective graphical requirements for a PAH6 benzenoid to be cata-condensed are $d_s \geqq 0$, $n_3 = 0$, $n_4 \geqq 2$, and $N_{\mathrm{Ic}} = 0$. Similarly, strictly peri-condensed benzenoid PAH6s have $d_s \leqq 0$, $n_4 = 0$, and usually $N_{\mathrm{Ic}}(\max)$ for a given perimeter $q_p = 4\tau + 2$ ($\tau = 2, 3, \ldots$).

If two polyhexes have the same number of edges q, then they must be isomers. Note that the number of edges is odd for every other column in the Formula Periodic Table for Benzenoid Hydrocarbons (PAH6s). Thus benzenoid hydrocarbons having formulas belonging to the $d_s = \ldots, -2, 0, 2, 4, \ldots$ column series have q = odd number.

Expansion of the Hückel orbital (HMO) secular determinant for a PAH graph gives the characteristic polynomial $P(\mathrm{G}; X) = \det|X\mathbf{I} - \mathbf{A}|$ where $\mathbf{I}$ is the identity matrix and $\mathbf{A}$ is the adjacency matrix for the corresponding graph [11]. The characteristic polynomial of a N carbon atom system has the following form

$$P(\mathrm{G}; X) = \sum_{n=0}^{N} a_n X^{N-n} = 0 = X^N - qX^{N-2} - 2r_3X^{N-3} + a_4X^{N-4}$$
$$- [2r_5 - 2(q - 3 - \alpha_3)\, r_3]\, X^{N-5} + a_6X^{N-6} + \ldots$$

where a_n are coefficients that can alternatively be obtained by the graphical Sachs' method. The factors of this equation give the eigenvalues for the corresponding PAH graph. By definition $a_0 = 1$ but the other coefficients convey graphical and topological information. For even carbon PAH6s $a_{\mathrm{odd}} = 0$. The number of edges (σ-bonds) in a PAH graph gives the coefficient $a_2 = -q$. For PAH6 graphs $a_N = \pm K^2$ where K is the number of Kekulé structures (1-factors); for the PAH6s in Table PAH6, the negative sign applies ($a_N = -K^2$) for the $N_c \equiv 2 \pmod 4$ row series and the positive sign ($a_N = K^2$) applies for the $N_c \equiv 0 \pmod 4$ row series.

Previously the author derived the following general equations for the fourth and sixth coefficients of the characteristic polynomial

$$a_4 = (1/2)(q^2 - 9q + 6N_c) - 2r_4 - d_1 - d_4 - 3d_5 - 6d_6 - \ldots \tag{1}$$

$$a_6 = -(1/6)(q^3 - 27q^2 + 116q) - N_c(3q - 16) - e(3{,}3) - 2r_6 + (q - 6)\,e(2{,}1) + (q - 5)\,e(3{,}1) + 2(q - 4 - \alpha_4)\,r_4 + r_3 \tag{2}$$

where q = No. of C–C σ-bonds, N_c = No. of carbon vertices, r_n = No. of rings of size n, d_i = No. of carbon vertices of degree-i, $e(i, j)$ = No. of edges (C–C σ-bonds) with end points of degree-i and degree-j, and α_4 = No. of attachments or branches on the tetragonal ring [4, 5]. Equation (1) is valid for all graphs and Eq. (2) is valid for all graphs having vertices of degree-1 to -3 and having no more than one trigonal or one tetragonal ring. With Eq. (1) and Eq. (2) the characteristic polynomial of small molecules can be quickly generated by inspection. The derivation of Eq. (2) utilized the periodicity associated with Table PAH6.

Our work has already identified a number of eigenvalue correlations [20]. The presence and minimum degeneracy of eigenvalues of $\varepsilon = 0$ and $\pm 1.0\beta$ can be rapidly determined by graph theoretical methods. The following two rules are germane.

Eigenvalue of Zero

Rule 1. Whenever the excised internal structure has an eigenvalue of zero, then the nonbisanthrene-like strictly peri-condensed benzenoid structure also has an eigenvalue of zero.

Phenylenyl monoradical ($C_{13}H_9^{\cdot}$) is a strictly peri-condensed PAH6 with the methyl radical as an excised internal structure and both have $\varepsilon = 0$. The diradical $C_{22}H_{12}$ isomer, triangulene, is a strictly peri-condensed benzenoid hydrocarbon and has trimethylenemethane diradical as an excised internal structure; both triangulene and trimethylenemethane diradicals have two eigenvalues of $\varepsilon = 0$.

Eigenvalue of One

Rule 2. Whenever a benzenoid structure can have a succession of edges bisected with a straight line drawn from one side of the molecule to the other with the terminal rings being symmetrically convex relative to the line, then those rings intersected by the line can be embedded with a perpendicular succession of ethene substructures and the benzenoid structure as a whole will have at least one eigenvalue pair of plus and minus one. This straight line will be called a selective lineation.

If a parent benzenoid excised internal structure has one or more selective lineations, then the larger daughter benzenoid structure will also have an identical number of selective lineations. For each distinct selective lineation present in an alternant hydrocarbon there will be a corresponding eigenvalue pair of $\varepsilon = \pm 1.0\beta$. Thus, coronene has three selective lineations and is triply degenerate in $\varepsilon = \pm 1.0\beta$, and perylene has four selective lineations and is quadruply degenerate in $\varepsilon = \pm 1.0\beta$.

Through a vector addition analog method, we previously showed that total $p\pi$-energy (E_π^L) of a large benzenoid hydrocarbon can be estimated from the known E_π values of smaller ones [15]. Thus when the number of carbon (N_c^L) and hydrogen (N_H^L) atoms of a large benzenoid hydrocarbon is related by $(N_c^L, N_H^L) = (N_c, N_H) + (N_c', N_H') = (N_c + N_c', N_H + N_H')$ to the number of carbon and hydrogen atoms of smaller benzenoid hydrocarbons, then $E_\pi^L = E_\pi + E_\pi'$. Since both $N_c^L = N_c + N_c'$ and $q^L = q + q'$ are overall conserved quantities, the level accuracy of the vector predicted E_π^L is better than the approximations due to McClelland ($E_\pi \leqq \sqrt{2qN_c}$) or Hall $\left(E_\pi = q + \frac{1}{3}N_c\right)$ because of the input of known E_π and E_π' values for the smaller benzenoids. For example, in the vector addition of $(10,8) + (22,12) = (38,22)$ naphthalene ($E_\pi = 13.68\beta$) plus a $C_{22}H_{12}$ isomer ($E_\pi' = 40.08 - 40.10\beta$) would go to a $C_{38}H_{22}$ PAH6 isomer ($E_\pi^L = 53.26 - 53.83\beta$) giving a better E_π^L estimate (53.8β) than $E_\pi^L = \sqrt{2 \cdot 46 \cdot 38}\beta = 59.13\beta$ or $E_\pi^L = \left(46 + \frac{1}{3} \cdot 38\right)\beta = 58.67\beta$ due to McClelland or Hall, respectively [21, 22]. Since no vector sum corresponds to a strictly peri-condensed benzenoid having a formula on the extreme edge of Table PAH6, these benzenoid species can not have their E_π estimated by this method. Incidentally, it should be noted that vector addition between any diagonally located pair of formulas in Table PAH6 equals the vector sum of the formulas located on the opposing diagonal. For example, $(10,8) + (20,12) = (16,10) + (14,10) = (30,20)$ and correspondingly $E_\pi(C_{10}H_8) + E_\pi(C_{20}H_{12}) = E_\pi(C_{16}H_{10}) + E_\pi(C_{14}H_{10}) = 13.6832\beta + 28.2220\beta = 25.5055\beta + 19.4483\beta = 41.9052\beta = 41.9538\beta$ for naphthalene, benzo[*a*]pyrene, pyrene, and phenanthrene, respectively, where the net number of bay regions and internal third degree carbon vertices equate on both sides of the equation. Table 3 gives E_π data for total resonant sextet isomers that can be used by this method along with data from other standard sources to

Table 3. Minimally strained total resonant sextet isomers[a]

Formula	No. of Isomers	K	E_π, β
$C_{18}H_{12}$	1	9	25.27
$C_{24}H_{14}$	1	20	34.16
$C_{30}H_{18}$	1	40	42.55
$C_{30}H_{16}$	1	45	43.07
$C_{36}H_{20}$	1	89	51.44
$C_{36}H_{18}$	3	104	51.98
$C_{42}H_{24}$	1	178	59.82
$C_{42}H_{22}$	2	198	60.33
$C_{42}H_{20}$	3	227	60.87
$C_{42}H_{18}$	1	250	61.37
$C_{48}H_{26}$	2	396	68.71
$C_{48}H_{24}$	4	449	
$C_{48}H_{22}$	8	520	69.78
$C_{48}H_{20}$	1	575	
$C_{54}H_{30}$	3	793	77.09
$C_{54}H_{28}$	5	889	
$C_{54}H_{26}$	11	1009	
$C_{54}H_{24}$	12	1149	
$C_{54}H_{22}$	4	1320	79.20
$C_{60}H_{32}$	4	1762	
$C_{60}H_{30}$	13	1960	86.49
$C_{60}H_{28}$	26	2270	
$C_{60}H_{26}$	27	2550	
$C_{60}H_{24}$	9	3100	
$C_{60}H_{22}$	1	3250	88.61

a The K and corresponding E_π values given are the median or representative ones for sets with more than one isomer.

estimate E_π for larger benzenoids with greater reliability than any other currently known method [3, 19].

The sequence of absolute values for the coefficients of the acyclic and characteristic polynomials are unimodal (i.e., increase uniformly at first and then decrease in magnitude). The largest magnitude coefficient of $|a_{2t}^{ac}|$ appears at $t = (N_c/2 + 1)/2 + 1$ for the $N_c \equiv 2 \pmod 4$ row series and at $t = (N_c/2 + 2)/2 + 1$ for the $N_c \equiv 0 \pmod 4$ row series [20]. The acyclic and characteristic polynomials of PAH6s has an even number of coefficients, $N_c/2 + 1 =$ even No., for the $N_c \equiv 2 \pmod 4$ row series and an odd number of coefficients, $N_c/2 + 1 =$ odd No., for the $N_c \equiv 0 \pmod 4$ row series. Let $e(i,j)$ be the number of edges with end-points of degree-i and degree-j. For 3(2)-polyhexagonal graphs, the number of internal edges q_I is given by $e(3,3)_I = 5q - 6N_c + 6$ and is fixed for any specific set of polyhex isomers [4, 23]. The values of 3(2,2), $e(2,3)$, and $e(3,3)_p$ found on the boundary of a polyhex are variable for different isomers and are governed by the relationships of $e(2,2) - e(3,3)_p = 6$ and $2e(3,3)_p = 6$ and $2e(3,3)_p + e(2,3) = 2N_{pc} = 2N_H - 12$ where $e(2,3)$ must be an even number and $e(3,3) = e(3,3)_I + e(3,3)_P$. For all acenes and one-isomer strictly peri-condensed PAH6s, $e(3,3)_p = 0$ since $n_0 = e(2,2) - 6$.

Every PAH6 isomer set beyond pyrene ($C_{16}H_{10}$) in the $N_{Ic} = 2$ row series of Table PAH6 contains isomers of $K = 0$ and $K = 9$. In the $N_{Ic} = 4$ row series every isomer set beyond $C_{34}H_{18}$ has isomers of $K = 0$ and $K = 9$. In the $N_{Ic} = 6$ row series every isomer set beyond $C_{40}H_{20}$ has isomers with $K = 0$ and $K = 9$.

4 Concluding Remarks

Benzenoid hydrocarbons are ubiquitous pyrolytic chemical pollutants, and the above approach has systematized this class of compounds into a unified framework for the first time. Two new graphical invariants of d_s and N_{Ic} were defined and related to the chemical formula with the aid of the relationship of $d_s + N_{Ic} = r - 2$. Recursive construction of a Formula Periodic Table for Benzenoid Hydrocarbons was accomplished where d_s is analogous to outer-shell electron configuration and N_{Ic} is analogous to outer-shell or principal quantum number. This formula periodic table establishes a partially ordered set (poset) of isomer groups having x,y-coordinates of (d_s, N_{Ic}). Isomers are thus analogous to isotopes, and this table complies with a sextet rule analogous to the octet rule for the Periodic Table of Elements. Our structural concepts of strictly peri-condensed and excised internal structure were important in the simplified enumeration of the former type of benzenoid hydrocarbons.

Our work related to combinatorics deals with identifying and associating graphical invariants of molecular graphs with the arrangement and selection of mathematical elements belonging to sets and configurations. We have been principally concerned with enumeration of select molecular graphs with specific fixed invariants and with deciphering and correlating their information content. In deciphering the information content of benzenoid molecular graphs, my research has focused on combinatorial derivations for rapidly determining the associated characteristic polynomials by topological inspection of their graphical invariants. The equations derived showed that the coefficients in the characteristic polynomial of a molecular graph carry topological information. Coupled with embedding of smaller alternant (bipartite) substructures onto larger ones for identifying the presence of common eigenvalues, these equations have lead to a powerful method for rapidly determining eigenvalues of molecular graphs that can be used to correlate chemical properties. Thus, the Formula Periodic Table for Benzenoid Hydrocarbons sorts polyhex graphs into isomer groups and identifies their group properties, and molecular graph eigenvalues examines and discerns the individual molecular graph properties within each isomer group.

The merging of mathematical graph theory with chemical theory is the formalization of what most chemists do in a more or less intuitive mode. Chemists currently use graphical images to embody chemical information in compact form which can be transformed into algebraic sets. Chemical graph theory provides simple descriptive interpretations of complicated quantum mechanical calculations and is, thereby, in-itself-by-itself an important discipline of study.

5 References

1. Dias JR (1982) J. Chem. Inf. Comput. Sci. 22: 15
2. Dias JR (1985) Acc. Chem. Res. 18: 241
3. Dias JR (1987, 1988) Handbook of polycyclic hydrocarbons, Parts A and B, Elsevier, Amsterdam

4. Dias JR (1985) Theoret. Chim. Acta 68: 107
5. Dias JR (1987) J. Chem. Educ. 64: 213
6. Dias JR (1987) Can. J. Chem. 65: 734
7. Dias JR (1988) J. Mol. Struct. (Theochem) 165: 125
8. Balaban AT, Brunvoll J, Cioslowski J, Cyvin BN, Cyvin SJ, Gutman I, He WC, He WJ, Knop JV, Kovačević M, Müller WR, Szymanski K, Tošić R, Trinajstić N (1987) Z. Naturforsch. 42a: 863
9. Dias JR (1985) Nouv. J. Chim. 9: 125
10. Balaban AT, Harary F (1968) Tetrahedron 24: 2505
11. Trinajstić N (1983) Chemical graph theory, CRC Press, Boca Raton, FL
12. Hall GG (1988) Theoret. Chim. Acta 73: 425
13. Dias JR (1984) J. Chem. Inf. Comput. Sci. 24: 124; (1984) Can. J. Chem. 62: 2914
14. Clar E, Robertson J, Schlögl R, Schmidt W (1981) J. Am. Chem. Soc. 103: 1320
15. Dias JR (1986) J. Mol. Struct. (Theochem) 137: 9
16. Clar E (1964) Polycyclic hydrocarbons, vols 1, 2, Wiley, New York
17. Clar E (1972) The aromatic sextet, Wiley, New York
18. Dias JR (1987) Thermochim. Acta 122: 313; Stein SG, Brown RL (1985) Carbon 23: 105
19. Dias JR (1989) J. Mol. Struct. (Theochem) 185: 57
20. Dias JR (1987) J. Mol. Struct. (Theochem) 149: 213
21. McClelland B (1974) J. Chem. Soc., Faraday Trans. 2 70: 1453
22. Hall GG (1987) Inst. Math. App. 17: 70; (1986) Theoret. Chim. Acta 70: 323
23. Dias JR (1983) Match 14: 83

Calculating the Numbers of Perfect Matchings and of Spanning Trees, Pauling's Orders, the Characteristic Polynomial, and the Eigenvectors of a Benzenoid System

Peter John and Horst Sachs

Technische Hochschule Ilmenau, Postfach 327, DDR-6300 Ilmenau,
German Democratic Republic

Table of Contents

A survey of relevant papers is given, and five simple and simply handleable algorithms of low complexity based on results contained in these papers are described (without proofs):

Algorithms **A** and **C** enable the number of Kekulé structures and Pauling's bond orders to be determined and the characteristic polynomial and the eigenspaces (eigenvalues, eigenvectors) to be calculated for all benzenoid systems.

Algorithms **B** and **D** enable the same to be done, in a more efficient way, for those benzenoid systems whose dualist graph is a tree (representing catacondensed benzenoid hydrocarbons).

Algorithm **E** enables, in a particularly efficient way, the number of spanning trees of any benzenoid system to be determined.

All of these algorithms are variants of a simple summation procedure following the edges in a cycle-free directed graph.

1 Introduction

1.1 Definitions

For the concept of a graph see any textbook. All graphs considered in this paper are finite and have no loops and no multiple edges.

A *hexagonal cell* (briefly: a *cell*) is a closed plane domain bounded by a regular hexagon of unit side length.

A *benzenoid system* (BS) is a finite 2-connected plane graph in which the closed hull of every finite region is a cell.

A *catacondensed benzenoid system* (CBS) is a BS in which no vertex belongs to more than two cells. Two cells of a BS are *adjacent* if they have an edge in common. The *dualist graph* D = D(B) of a benzenoid system B has as its vertices the centers of the cells of B where two vertices of D are connected by a straight line segment if and only if the corresponding cells are adjacent in B.

A CBS is a BS whose dualist graph is a tree and conversely. A *generalized benzenoid system* (GBS) is a connected subgraph of a BS in which the length of the boundary of any region is $4s + 2$ ($s = 1, 2, \ldots$).

A *matching* of a graph G is a set of pairwise disjoint edges of G.

A *perfect matching* or *1-factor* of a graph G is a matching of G that covers all vertices. If, in particular, G is a BS then a 1-factor of G is usually called a *Kekulé structure* (KS).

A *Kekuléan benzenoid system* (KBS) is a BS which has at least one KS.

1.2 Chemical Background

The structural formula of a benzenoid hydrocarbon (BH) consists of a benzenoid system B spanned by the carbon atoms and some hanging edges such that each carbon atom which belongs to only one cell (to more than one cells) of B is connected

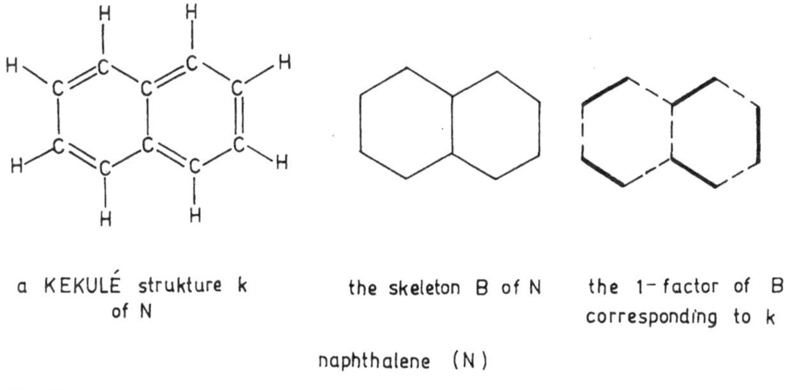

Fig. 1

to exactly one hydrogen atom (or to no hydrogen atom, respectively) (Fig. 1); let us call B the *skeleton* of the BH.

According to the Kekulé model of a BH, the edges of B represent single or double bonds, and since the valence of each carbon atom is four, those edges representing double bonds form a KS of B (Fig. 1). Therefore, the skeleton of a BH is necessarily a KBS.

1.3 Notation

Let B be a KBS and let k be a KS of B. The edges of B that belong or do not belong to k are referred to as the red or blue edges, respectively.

The number of KSs of B is denoted by $K = K(B)$; the number of KSs of B which contain or do not contain a given edge e is denoted by $r(B, e)$ or $b(B, e)$, respectively; thus $r(B, e) + b(B, e) = K$ for every edge e of B. If all KSs are considered equally likely, the number $p(B, e) = r(B, e)/K$ is the probability of finding e in a KS. $p(B, e)$ equals the probability of finding a double bond between the pair of carbon atoms represented by e; $p(B, e)$ is called *Pauling's bond order* [1].

Note that for every vertex v of B,

$$\sum p(B, e) = 1\,, \quad \text{equivalently}\,, \quad \sum r(B, e) = K(B) \tag{1}$$

where the sum is taken over the two or three edges e which are incident to v.

2 The Literature on Benzenoid Hydrocarbon Calculations

2.1 Kekulé Structures and Pauling's Bond Orders

Let $K = K(B)$ denote the number of Kekulé structures contained in a benzenoid system B.

M. Gordon and W. H. T. Davison [2] (1952) were the first to develop a simple algorithm which allows K to be determined for any CBS. Another basic paper is that of M. J. S. Dewar and H. C. Longuet-Higgins [3] (1952) who found the formula

$$K^2(B) = |\det A(B)| \tag{2}$$

where B is any BS and $A(B)$ denotes its adjacency matrix.

This formula can easily be deduced from a theory due to P. W. Kasteleyn [4] (1961) which allows the number of 1-factors of any planar graph G with an even number of vertices to be expressed as the value of the Pfaffian $|\mathrm{Pf}\, S| = \sqrt{\det S}$ of some skew-symmetric matrix S connected with G. Elementary proofs of Eq. (2) (not using Kasteleyn's formula) for plane graphs in which every face F is a $(4k + 2)$-gon (where k depends on F) were also given by D. Cvetković, I. Gutman and N. Trinajstić [5] (1972) and H. Sachs [6] (1986).

For some special classes of BSs, explicit formulae for K were given by T. F. Yen [7] (1971). Using the coefficients of the non-bonding molecular orbitals (NBMO-s), W. C. Herndon [8] (1973) found an interesting method for calculating K which, however, is not generally applicable. D. Cvetković and I. Gutman [9] (1974) developed a procedure for determining K for any unbranched CBS which is based on simple matrix concatenation. M. Randić [10] (1976) was the first to explicitly formulate the simple and, in a way, most natural "decomposition principle" which however, practically fails if BSs of some extent are concerned.

In this connection, also a paper of O. E. Polansky and I. Gutman [11] (1980) on so-called "all-benzenoid aromatic hydrocarbons" and a paper by B. Džonova-Jerman-Blažič and N. Trinajstić [12] (1982) should be mentioned.

During the last decade, the number of papers concerning the determination of K for special classes of BSs has rapidly increased; we must content ourselves with referring the reader to the following recent books: N. Trinajstić [13] (1983), I. Gutman and O. E. Polansky [14] (1986), and S. J. Cyvin and I. Gutman [15] (1988).

The Pauling bond order (PBO) concept was introduced by L. Pauling, L. O. Brockway, and J. Y. Beach [1] (1935). The determination of the PBOs requires counting Kekulé structures in GBSs. N. S. Ham [16] (1958) proved that, for a KBS, Pauling's bond order p_{ij} for neighbouring carbon atoms i, j equals the bond order defined by K. Ruedenberg [17] (1954) which results from LCAO-MO considerations; further, he showed that p_{ij} is equal to the entry in position (i, j) of the inverse A^{-1} of the adjacency matrix A of the system (note that, by Eq. (2), $|\det A| = K^2 > 0$).

Algorithms **A** and **B** given in this paper (see also Refs. [18, 19, 20, 21, 22]) both allow K and the PBOs to be determined, **A** (which is derived from a much more general algorithm, see Ref. [19]) is applicable to any GBS with as many peaks as valleys (see Sect. 3.1) whereas **B** is restricted to CBSs. **A** is based on the Dewar and Longuet-Higgins formula Eq. (2) [22] but allows the order of the determinant to be drastically reduced; **B** is closely related to the Gordon-Davison algorithm [21]. **A** and **B** generalize and/or simplify all the main approaches found in the literature.

2.2 Eigenvalues and Eigenvectors

E. Heilbronner [23] (1953) invented the "composition principle"; he was the first to utilize the symmetries of a hydrocarbon in order to simplify the calculation of its characteristic polynomial by means of some folding operations [24] (1954).

Independently, L. M. Lihtenbaum [25] (1956) and L. Collatz and U. Sinogowitz [26] (1957) elaborated the fundamentals of a general theory of graph spectra. Also independently, M. Milić [27], H. Sachs [28], and L. Spialter [29] (1964) established a formula expressing the coefficients of the characteristic polynomial of a graph in terms of its cyclic structure.

Much information about the early approaches and results is contained in the renowned reference book "Dictionary of Electron Calculations" by C. A. Coulson and A. Streitwieser, Jr. [30] (1965). Since that time, hundreds of papers have appeared from which we mention only H. Hosoya [31] (1972), R. B. Mallion, A. J. Schwenk and N. Trinajstić [32] (1975), J.-I. Aihara [33] (1976), and K. Balasubramanian [34] (1982). Infinite periodic graphs were first considered by L. Collatz [35] (1978). For more details, the reader is referred to the monographs "Spectra of Graphs" by D. M. Cvetković, M. Doob and H. Sachs [36] (1980) and "Recent Results in the Theory of Graph Spectra" by D. M. Cvetković, M. Doob, I. Gutman and A. Torgašev [37] (1988).

Algorithms **C** and **D** given in this paper (see also Refs. [18, 19]) are based on a general graph-theoretic procedure for calculating determinants and solving systems of linear equations, see K. Al-Khnaifes [19] (1988); they allow the eigenvalues and eigenspaces of any BS simultaneously to be calculated.

2.3 Spanning Trees

In his classic paper on electric networks, G. Kirchhoff [38] (1847) implicitly established the celebrated "Matrix-Tree-Theorem" which, in modern terminology, expresses the complexity (i.e., the number of spanning trees) of any finite graph G as the determinant of a matrix which can easily be obtained from the adjacency matrix of G. Simple proofs were given by R. L. Brooks, C. A. B. Smith, A. H. Stone and W. T. Tutte [39] (1940), H. Trent [40] (1954), and H. Hutschenreuther [41] (1967) (for relations between the complexity and the spectrum of a graph see Ref. [36] pp 38, 39, 49, 50).

For a benzenoid system B, the calculation of its complexity can be remarkably simplified by considering the dualist graph of B instead of B; this observation is due to I. Gutman, R. B. Mallion and J. W. Essam [42] (1983).

Algorithm **E** given in this paper (see also Refs. [18, 19]) is based on the Gutman-Mallion-Essam result; it allows a further considerable reduction of the order of the determinant to be calculated.

3 Algorithms for Calculating the Number of Kekulé Structures and Pauling's Bond Orders in Kekuléan Benzenoid Systems

3.1 The General Case

Let $B = (V, E)$ be a KBS on n vertices $v_1, v_2, \ldots, v_n$. Assume B to be drawn in the plane in such a way that some of its edges are vertical: clearly this can be done in six ways. Let $\tilde{B}$ be such a drawing. Suppose that the vertices of $\tilde{B}$ are coloured white or black such that

(i) the top vertices of the cells are white,

(ii) every edge connects a white vertex with a black one.

It is easy to see that such a colouring exists and that it is unique.

A *peak* x (*valley* y) of $\tilde{B}$ is a vertex whose neighbours lie all below x (above y). Necessarily, all peaks are white and all valleys are black (Fig. 2).

Let the peaks and valleys be labelled $1, 2, \ldots, \tilde{p}$ and $1, 2, \ldots, \tilde{v}$, respectively, and let the number of white or black vertices of B be n_w or n_b, respectively. It is well known that the existence of a KS in B implies $n_w = n_b = n/2$ as well as $\tilde{p} = \tilde{v}$ [46]. For our purposes it is convenient to choose the drawing $\tilde{B}$ such that $\tilde{p}$ is minimum.

Let p_{ik} denote the number of monotonic paths in $\tilde{B}$ each issuing from peak k and terminating in valley i. The matrix $P = (p_{ik})$ $(i = 1, 2, \ldots, \tilde{v}; k = 1, 2, \ldots, \tilde{p}1)$ which is called the path matrix of B can very easily be calculated (see Theorem 2 below).

Suppose $\tilde{p} = \tilde{v}$; recall that K denotes the number of Kekulé structures contained in B.

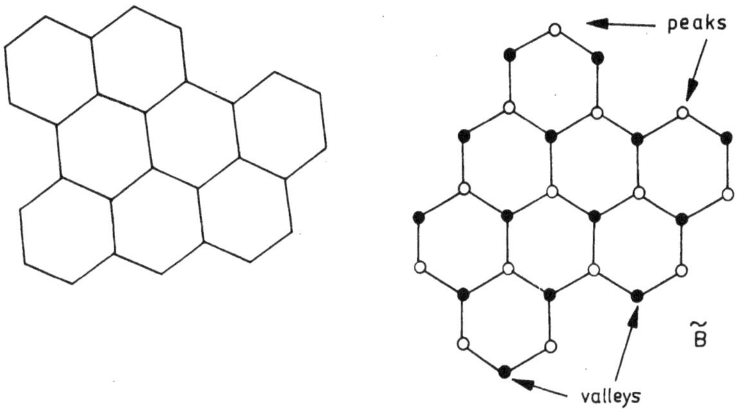

Fig. 2

Theorem 1 [19, 20, 22].

$$K = |\det \mathbf{P}| . \qquad (3)$$

We shall now explain how to calculate **P**. Fix some peak x^0 and a vertex v of $\tilde{B}$ which is not a peak and suppose that $\{v'\}$ or $\{v', v''\}$ is the set of neighbours of v which lie above v (Fig. 3.1); further assume that the numbers of monotonic paths issuing from x^0 and terminating in v' or v'' – a' and a'', say – have already been calculated. Then, clearly, the number a of monotonic paths issuing from x^0 and terminating in v equals a' or $a' + a''$, respectively (Fig. 3.2).

This observation can be used to calculate the numbers p_{ik}: see Fig. 4.1 (4.2) where x^0 is identified with the first (second) peak.

For the HS of Fig. 4 Theorem 1 now yields

$$K = \text{abs} \begin{vmatrix} 10 & 1 \\ 4 & 3 \end{vmatrix} = 26 .$$

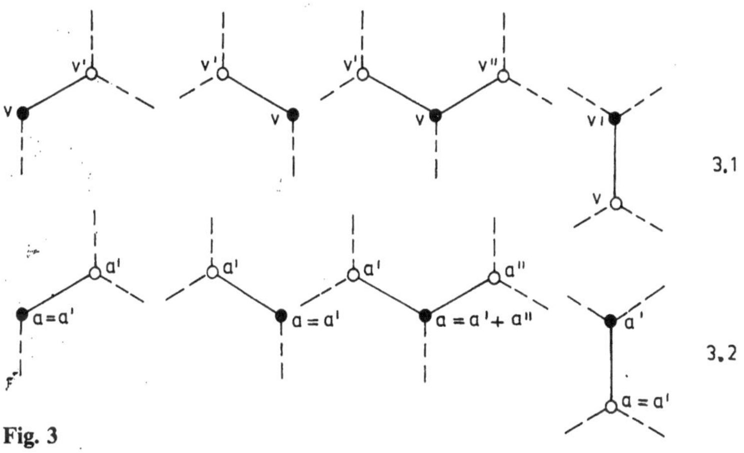

Fig. 3

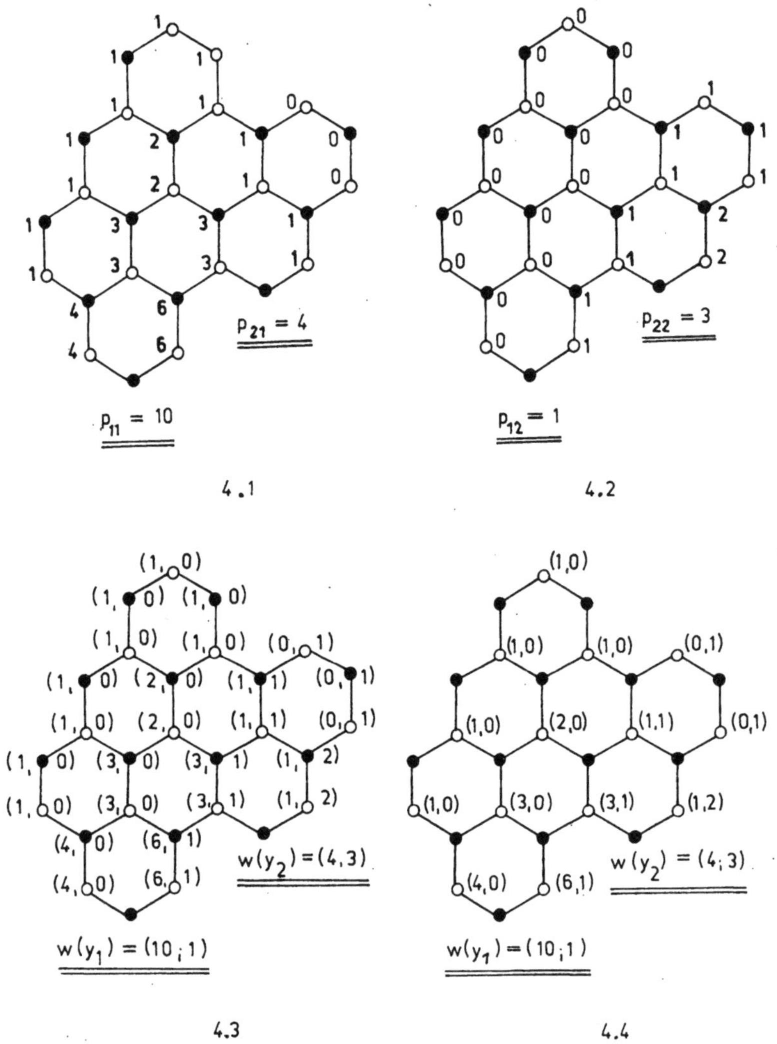

Fig. 4

Applying the method outlined above simultaneously to all the peaks, we obtain the following algorithm.

Algorithm A

Assign to every vertex v of $\tilde{B}$ a vector $\mathbf{w}(v) = (w_1(v), w_2(v), \ldots, w_{\bar{p}}(v))$ according to the following rules: Start with the peaks. Let x_k denote the kth peak and put

$$w_i(x_k) = \delta_{ik} = \begin{cases} 0 & \text{if} \quad i \neq k \\ 1 & \text{if} \quad i = k, \end{cases}$$

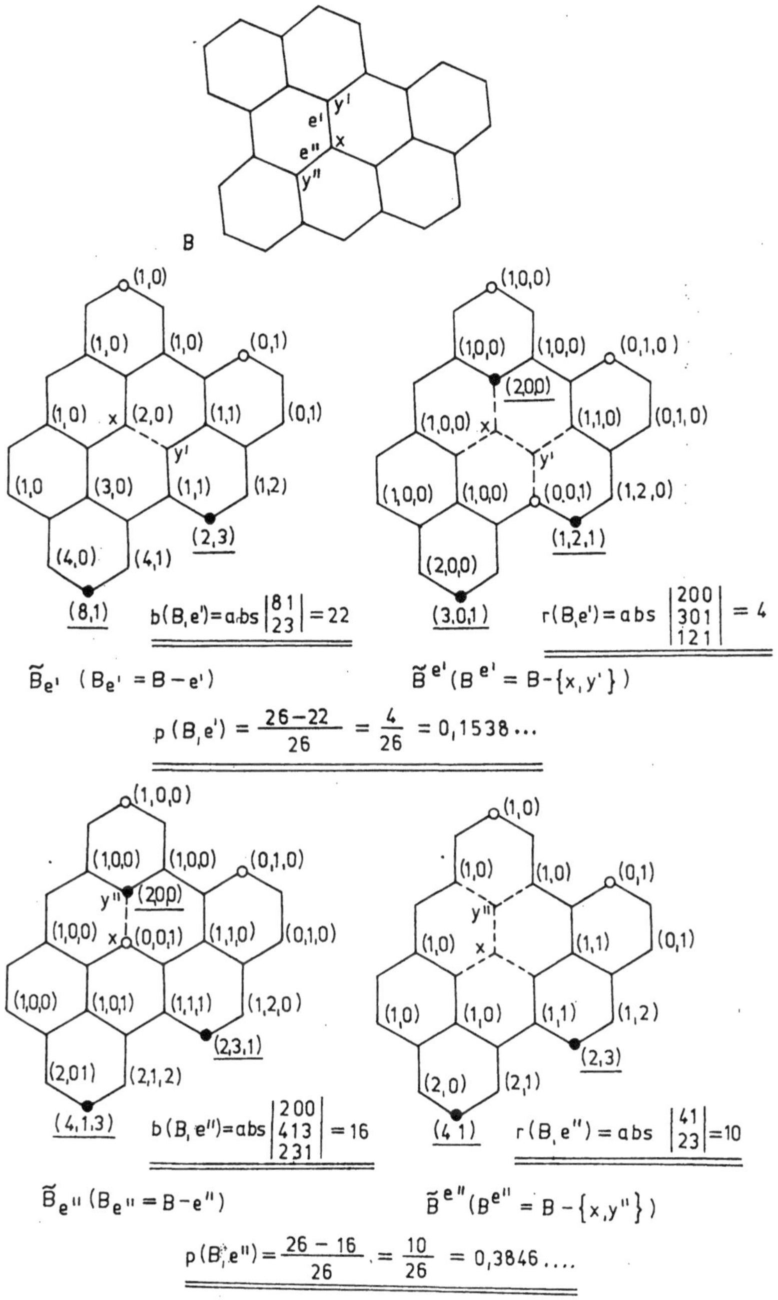
B
e'
e''
x
y'
y''
(8,1)
b(B,e')=abs |8 1 / 2 3| = 22
$\tilde{B}_{e'}$ ($B_{e'}$ = B − e')
(3,0,1)
r(B,e')=abs |2 0 0 / 3 0 1 / 1 2 1| = 4
$\tilde{B}^{e'}$ ($B^{e'}$ = B − {x, y'})
p(B,e') = (26−22)/26 = 4/26 = 0,1538...
(4,1,3)
b(B,e'')=abs |2 0 0 / 4 1 3 / 2 3 1| = 16
$\tilde{B}_{e''}$ ($B_{e''}$ = B − e'')
(4 1)
r(B,e'')=abs |4 1 / 2 3| = 10
$\tilde{B}^{e''}$ ($B^{e''}$ = B − {x, y''})
p(B,e'') = (26 − 16)/26 = 10/26 = 0,3846....

Fig. 5

i.e.

$$\mathbf{w}(x_k) = \overset{1\,\dots\dots\quad k\,\dots\dots\quad \tilde{p}}{(0, \dots, 0, 1, 0, \dots, 0)}\,.$$

Running through $\tilde{B}$ from top to bottom, assign to every vertex v which is not a peak the sum of the vectors assigned to the (one or two) vertices above v and adjacent to v (see Figs. 4.3, 4.4 where Fig. 4.4 simplifies Fig. 4.3).

Clearly, $w_k(v)$ equals the number of monotonic paths issuing from peak x_k and terminating in v. In particular, if y_i denotes the ith valley, then $w_k(y_i) = p_{ik}$ implying

Theorem 2 [19, 20, 22].

$$\mathbf{P} = \begin{bmatrix} \mathbf{w}(y_1) \\ \mathbf{w}(y_2) \\ \vdots \\ \mathbf{w}(y_{\tilde{p}}) \end{bmatrix} = (w_k(y_i))\,. \tag{4}$$

It is important to note that Theorems 1 and 2 also apply to generalized hexagonal systems.

Now we turn to the calculation of Pauling's bond orders. Let B be the skeleton of some BH, let $\tilde{B}$ be a drawing of B as described above, and let $e = (x, y)$ be an edge of B. Put $B_e = B - e$, $B^e = B - \{x, y\}$ (see Fig 5, where $e = e'$ or $e = e''$). Note that B_e and B^e are generalized benzenoid systems. Clearly,

$$b(B, e) = K(B_e)\,, \qquad r(B, e) = K(B^e)\,.$$

Thus $b(B, e)$ and $r(B, e)$ can be calculated by applying Theorem 1 and 2 to B_e and B^e, respectively, and so can Pauling's bond order;

$$p(B, e) = r(B, e)/K(B) = (K(B) - b(B, e))/K(B) \tag{5}$$

Depending on whether e is a vertical or an oblique edge of $\tilde{B}$ it is convenient to use the second or the third term, respectively, of (5) for calculating $p(B, e)$ (see the examples treated in Fig. 5). There are many ways of calculating $p(B, e)$ for *all* edges of B in a more or less efficient manner, e.g.: Calculate $r(B, e)$ for the h vertical edges e' which bound some hexagon from the right side (Fig. 6.1, the bold edges), as explained above (h = number of hexagons). Calculate $p(B, e'')$ for the remaining edges e'' making use of (1). This is always possible and straightforward since the edges e'' form a tree (Fig. 6.2, the bold edges).

3.2 Catacondensed Benzenoid Systems

Let B be a CBS with at least two cells. An edge of B is called *internal* if it is the intersection of two cells, and *external* otherwise. Cell c is an *end cell (bifurcation cell)*

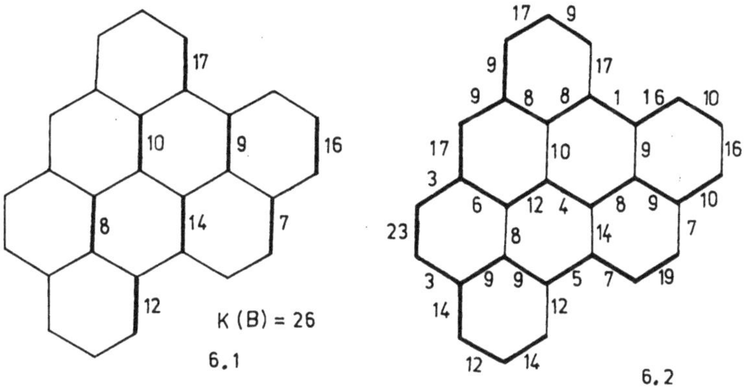

The numbers r(B, e)

Fig. 6

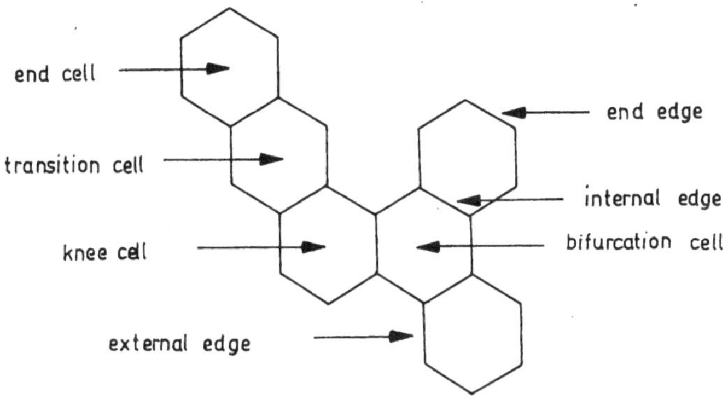

Fig. 7

of B if and only if the boundary of c has exactly one (three) internal edges. An *end edge* of B is an edge of an end cell c which lies opposite to the internal edge of c. Cell c is a *transition (knee) cell* of B if and only if the boundary of c has exactly two internal edges which lie (do not lie) in opposite position to each other (Fig. 7.).

Note that every CBS has a KS; to see this, colour all internal edges blue and the external edges alternatingly red and blue. Let B be a CBS and e an internal edge of B. To *cut B along the edge e* means to create a pair of smaller CBSs, say B' and B'', whose union is B and whose intersection is e with its endvertices, as indicated in Fig. 8 where the external edges e', e'' correspond to e.

Gluing is the inverse of cutting. Let B', B'' be CBSs with external edges e', e'', respectively, whose end vertices have valency 2 (as in Fig. 8). To *glue B' and B'' together along the edges e', e''* means to create a larger CBS, say B, by identifying e' with e'' to e, as indicated in Fig. 8. It is not difficult to prove the following.

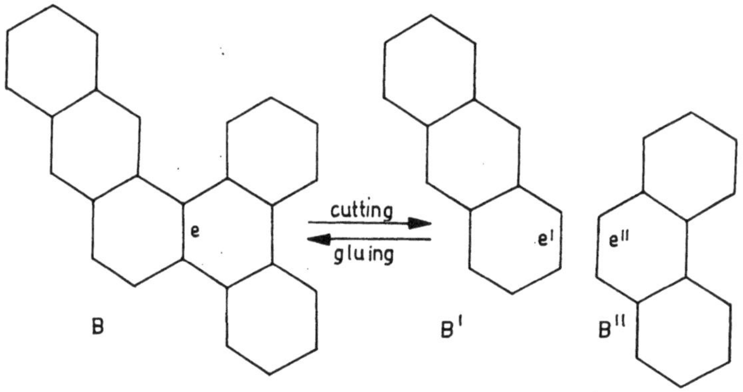

Fig. 8

Proposition 1. *Choose any pair of Kekulé structures k′ of B′ and k″ of B″ such that not both edges e′ and e″ are blue. Glue B′, B″ together along e′, e″, colour e red if both of e′, e″ are red and blue otherwise, and retain the colours of all other edges. Then the resulting colouring defines a Kekulé structure of B, and all Kekulé structures of B can be obtained this way.*

Proposition 1 immediately implies the following simple reduction formulae.

$$\begin{aligned} r(B, e) &= r(e) = r(e') \cdot r(e'') \\ b(B, e) &= b(e) = r(e') \cdot b(e'') + b(e') \cdot r(e'') \end{aligned} \tag{6}$$

(see Fig. 8) where $r(a)$, $b(a)$ refer to the respective CBS from which the edge a is taken.

Algorithm B

Preliminaries:

Consider a cell c and an internal edge e of B lying on the boundary of c. By saying that "e is marked π internally (externally) with respect to c" we mean that the mark π is written on that "bank" of e which belongs (does not belong) to c (see Fig. 9). Let R by the dualist graph of B; select an end vertex M of R and direct all edges of R towards M: speaking intuitively R is thus turned into a "river" $\bar{R}$ with its tributaries, sources and mouth M (Fig. 10). The end cells corresponding to the sources and the mouth are called the source cells and the mouth cell c_M, respectively. The end edge of c_M is called the *mouth edge* of B and denoted by e_M.

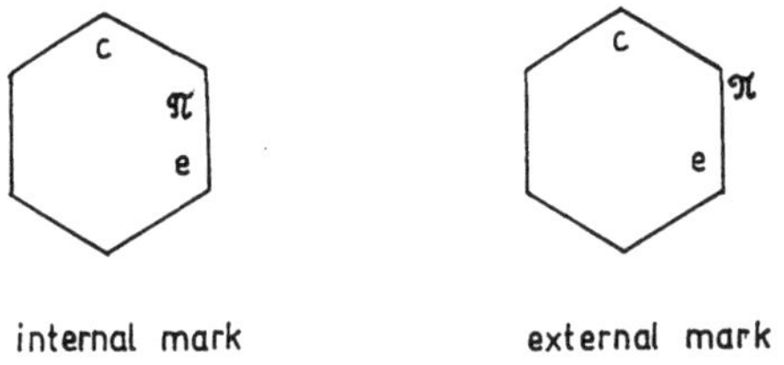

Fig. 9

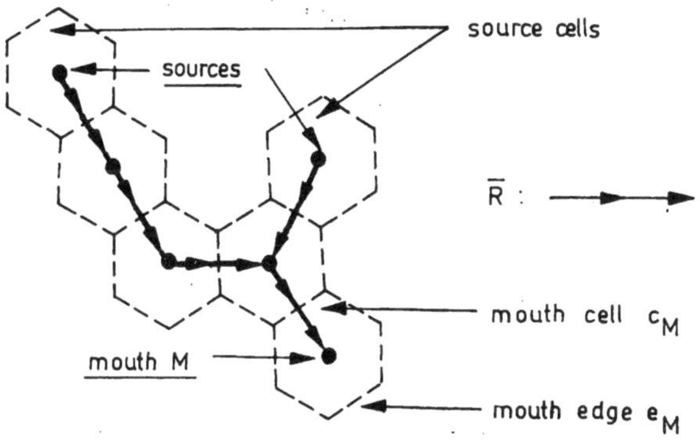

Fig. 10

Starting from the sources and following the river, we shall reconstruct B from its cells by glueing cells or branches stepwise together and simultaneously calculate numbers $\bar{r}(e)$, $\bar{b}(e)$ for the interior edges and the mouth edge e_M which are identical with the r- and b-values of the "actual" mouth edges, i.e. of the mouth edges of the parts of B so far constructed; thus in particular,

$$\begin{aligned} \bar{r}(e_M) &= r(e_M) = r(B, e_M) \\ \bar{b}(e_M) &= b(e_M) = b(B, e_M) \\ \bar{r}(e_M) + \bar{b}(e_M) &= K(B)\,. \end{aligned} \tag{7}$$

Description of the algorithm

1) Let c_0 be a source cell and e_0 its unique internal edge: mark e_0 internally $\lambda_0 = (\bar{r}(e_0); \bar{b}(e_0)) = (1; 1)$; cf. Fig. 11.

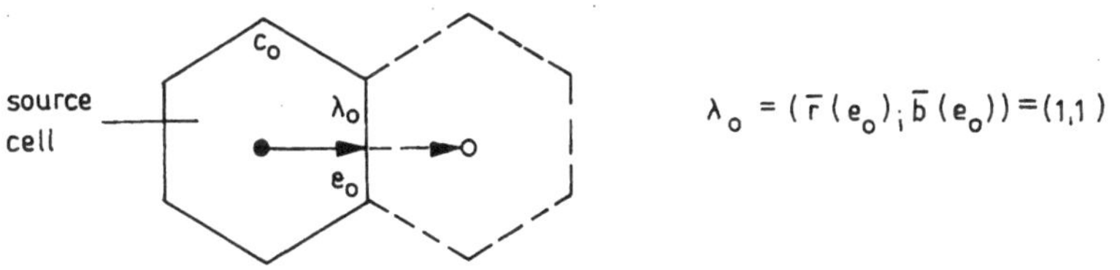

Fig. 11

2) Let c be a transit cell and e, f its pair of internal edges. If e is marked externally $\lambda = (\bar{r}(e); \bar{b}(e))$, mark f internally $\mu = (\bar{r}(f); \bar{b}(f)) = (\bar{r}(e); \bar{r}(e) + \bar{b}(e))$; cf. Fig. 12.
3) Let c' be a knee cell and e', f' its pair of internal edges. If e' is marked externally $\lambda' = (\bar{r}(e'); \bar{b}(e'))$, mark f' internally $\mu' = (\bar{r}(f'); \bar{b}(f')) = (\bar{r}(e') + \bar{b}(e'); \bar{r}(e'))$; cf. Fig. 13.

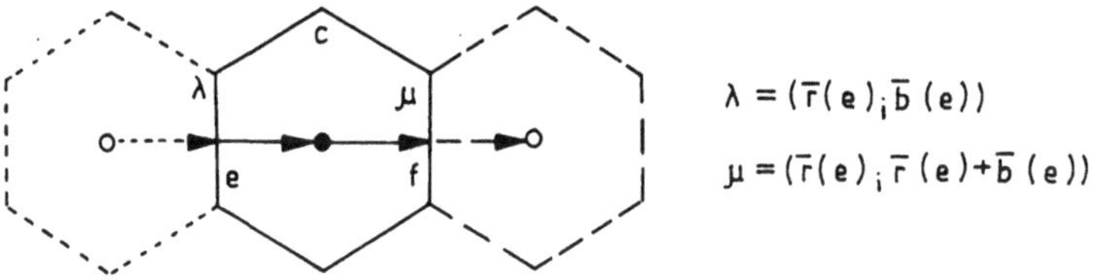

Fig. 12

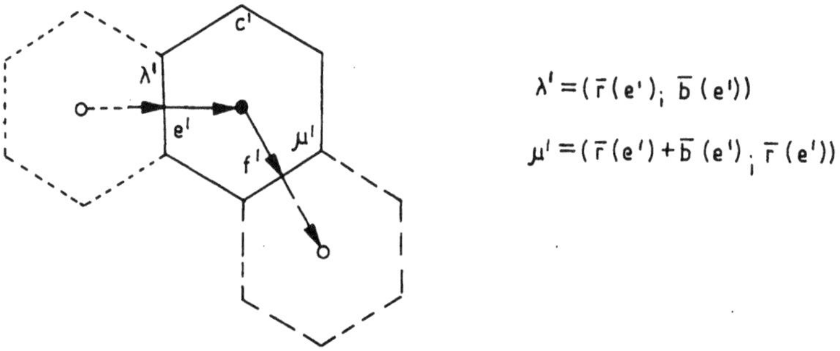

Fig. 13

4) Let c^* be a bifurcation cell and e^*, f^*, g^* its triple of internal edges. If e^* and f^* are marked externally $\lambda^* = (\bar{r}(e^*); \bar{b}(e^*))$ and $\mu^* = (\bar{r}(f^*); \bar{b}(f^*))$, respectively, mark g^* internally $\nu^* = (\bar{r}(g^*); \bar{b}(g^*)) = ([\bar{r}(e^*) + \bar{b}(e^*)] \cdot [\bar{r}(f^*) + \bar{b}(f^*)]; \bar{r}(e^*) \cdot \cdot \bar{r}(f^*))$; cf. Fig. 14.
5) If the unique internal edge e^0 of c_M is marked externally $\lambda^0 = (\bar{r}(e^0); \bar{b}(e^0))$, mark e_M internally $\lambda_M = (\bar{r}(e_M); \bar{b}(e_M)) = (\bar{r}(e^0); \bar{r}(e^0) + \bar{b}(e^0))$; cf. Fig. 15.
Clearly, after step 5 all internal edges are marked on exactly one side (the river $\bar{R}$ crosses each

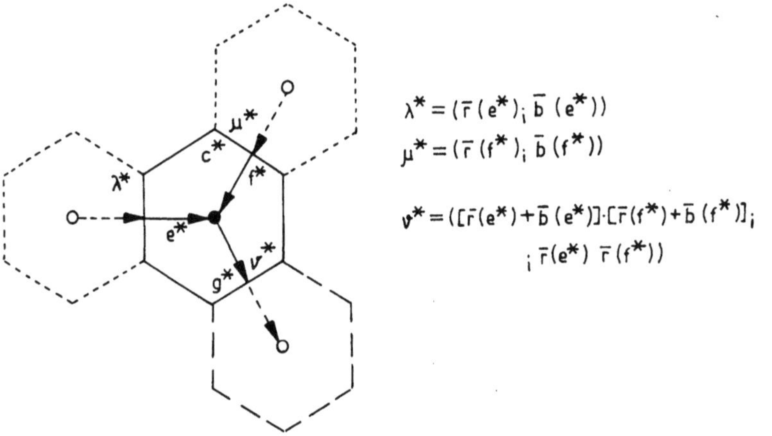

Fig. 14

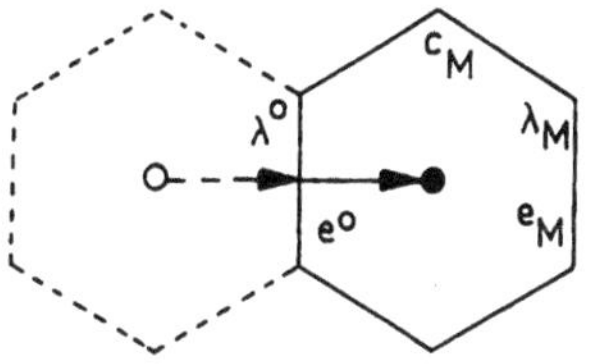

$$\lambda^o = (\bar{r}(e^o); \bar{b}(e^o))$$
$$\lambda_M = (\bar{r}(e^o); \bar{r}(e^o) + \bar{b}(e^o))$$

Fig. 15

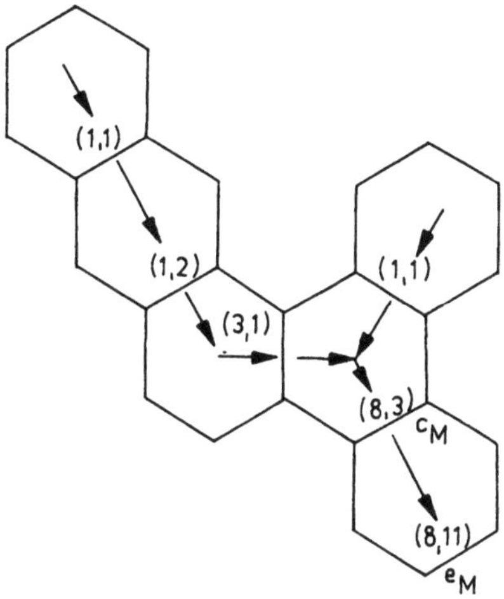

$$K(B) = r(B, e_M) + b(B, e_M) = 8 + 11 = 19$$

Fig. 16

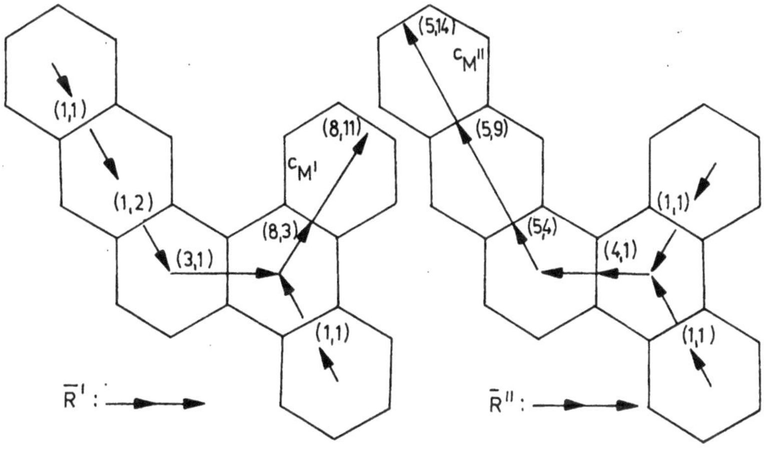

Fig. 17

internal edge from the marked side to the unmarked side) and equations (7) hold. For the example of Fig. 16, we obtain $r(B, e_M) = 8$, $b(B, e_M) = 11$ and $K(B) = 19$.

We can now select another end vertex of R as a new mouth M' and repeat the procedure with a new river $\bar{R}'$ and a new mouth cell c_M', and so on (Fig. 17). It turns out that for the second run we can use part of the information obtained in the first run; and so on.

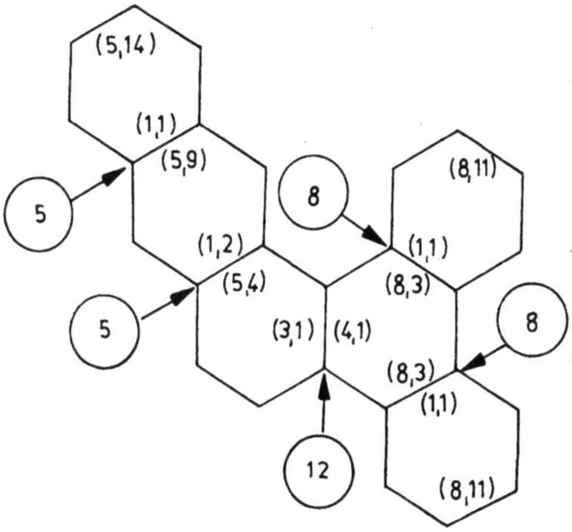

Fig. 18

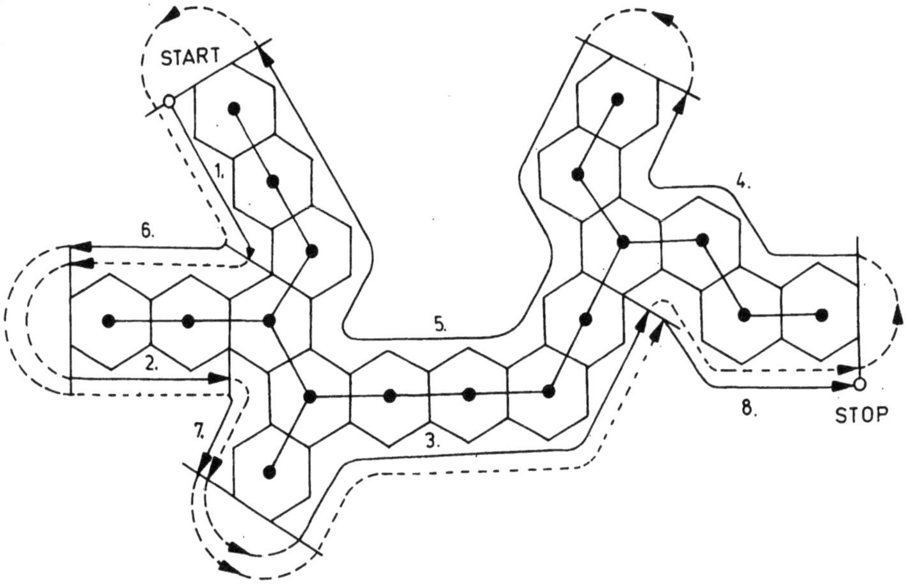

Fig. 19

6) After the second run we find an internal edge e marked on both sides, say the left side marked $(\bar{r}(e), \bar{b}(e))$ in the first run and the right side marked $(\bar{r}'(e), \bar{b}'(e))$ in the second run; then

$$\bar{r}(e) \cdot \bar{r}'(e) = r(e) = r(B, e) \tag{8}$$

(In Fig. 18, the $r(B, e)$ are the encircled numbers).

The most convenient way of applying this part of the algorithm consists in following the peripheral circuit of B; is is easy to see that after no more than two circumambulations all internal edges are marked on both sides (see Fig. 19) so that, by (8), for these edges the r-values can be calculated. To complete the algorithm, we once more follow the peripheral circuit starting from an arbitrary end edge; applying Eqs. (7) and (1) we obtain in order the r-values also of all external edges (Fig. 20).

Remark: The graphical procedure can be simplified by drawing just the dualist graph instead of B (Fig. 21).

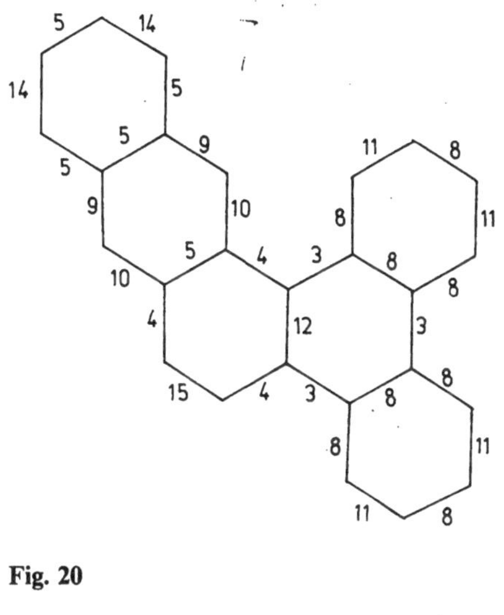

Fig. 20

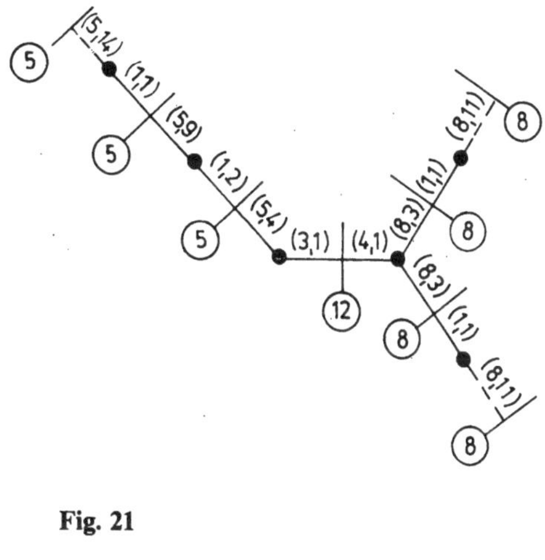

Fig. 21

4 Calculating the Characteristic Polynomial and the Eigenvectors of a Benzenoid System

4.1 Preliminaries

Let G be a graph on n vertices $v_1, v_2, \ldots, v_n$, $\mathbf{A}$ its 0,1-adjacency matrix, and $\mathbf{I}$ the $n \times n$ unit matrix. The characteristic polynomial

$$F_A(\lambda) = \det(\lambda \cdot \mathbf{I} - \mathbf{A})$$

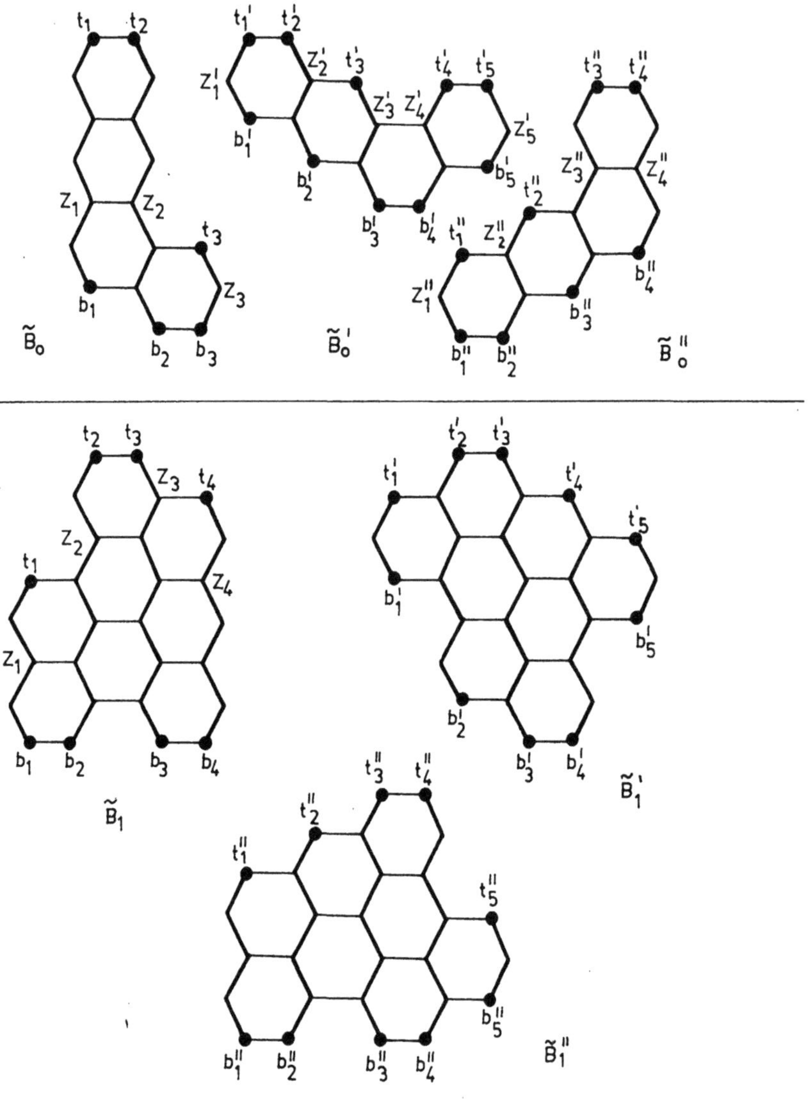

$\tilde{B}_0$, $\tilde{B}_0'$, $\tilde{B}_0''$ and $\tilde{B}_1$, $\tilde{B}_1'$ $\tilde{B}_1''$ are three different drawings of the same hexagonal system B_0 or B_1, respectively.

Fig. 22

and the eigenvalues of **A** (i.e., the roots of the equation $F_A(\lambda) = 0$) are called the characteristic polynomial of G, denoted $F_G(\lambda)$, and the eigenvalues of G, respectively.

Let λ^0 be an eigenvalue of G and let **0** denote the zero vector on n components. The set $\mathbf{E}(\lambda^0)$ of all solutions **x** of the equation

$$(\lambda^0 \cdot \mathbf{I} - \mathbf{A})\,\mathbf{x} = \mathbf{0}$$

forms the eigenspace of G belonging to λ^0; every non-zero $\mathbf{x}^0 \in \mathbf{E}(\lambda^0)$ (with $|\mathbf{x}^0| = 1$) is a (normalized) eigenvector of G belonging to λ^0.

4.2 The General Case

Let B be a BS on n vertices. Assume B to be drawn in the plane in such a way that some of its edges are horizontal, this can be done in six ways (where some of the drawings may be equal); cf. Fig. 22. In such a drawing $\tilde{B}$ of B we find a set of "zigzag lines", $Z_1, Z_2, \dots, Z_p$, say, where Z_i is a maximal monotonic path (non-interrupted zigzag line) in $\tilde{B}$ connecting a top point t_i with a bottom point b_i, as indicated in Fig. 22. (An example of an "interrupted" zigzag line, consisting of two different Z_is, is given in Fig. 23.) The points $t_1, t_2, \dots, t_p$ are called the *top vertices* of $\tilde{B}$. From the six drawings of B we select one in which p, the number of maximal non-interrupted zigzag lines, is minimum; so, considering the examples of Fig. 22, we select $\tilde{B}_0$ (or the drawing obtained from $\tilde{B}_0$ by turning it 180°), and $\tilde{B}_1$, respectively.

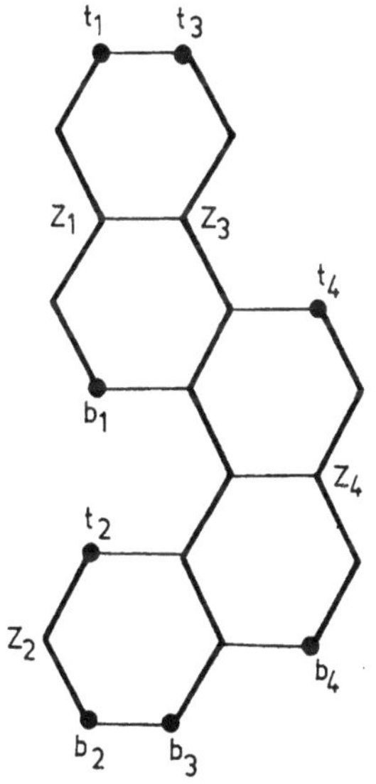

Fig. 23

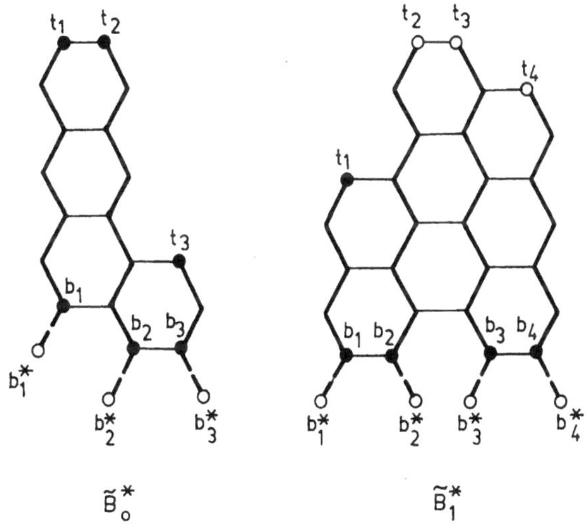

Fig. 24

Every zigzag line Z_i of $\tilde{B}$ is prolonged beyond its bottom point b_i by one unit segment connecting b_i with an additional "virtual" vertex b_i^*; thus $\tilde{B}$ is turned into a figure which we call $\tilde{B}^*$; see Fig. 24.

For any vertex v of $\tilde{B}^*$, let v^1 denote its unique upper neighbour (if it exists), let

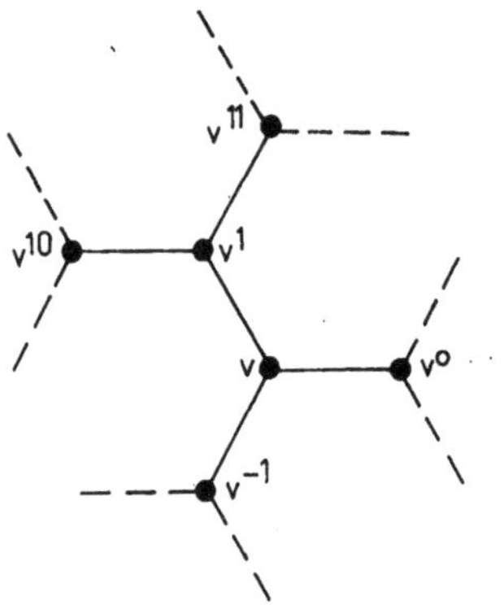

Fig. 25

v^0 denote its unique neighbour which lies on the same level as v (if it exists), and let v^{-1} denote its unique lower neighbour (if it exists); Fig. 25.

The algorithm which we shall now describe, is very similar to algorithm **A**.

Algorithm C

To every vertex v of $\tilde{B}^*$ assign a vector $\mathbf{w}(v; \lambda) = (w_1(v, \lambda), w_2(v, \lambda), \ldots, w_p(v, \lambda))$ by means of the following recursive procedure.

(i) For top vertex t_k, put

$$\mathbf{w}(t_k, \lambda) = (\delta_{1k}, \delta_{2k}, \ldots, \delta_{pk}), \quad \text{where} \quad \delta_{ik} = \begin{cases} 0 & \text{if} \quad i \neq k \\ 1 & \text{if} \quad i = k \end{cases}$$

$(i, k = 1, 2, \ldots, p)$;

(ii) for any vertex v of $\tilde{B}^*$ which is not a top vertex, put

$$\mathbf{w}(v, \lambda) = \lambda \cdot \mathbf{w}(v^1, \lambda) - \sum_{r=0,1} \mathbf{w}(v^{1r}, \lambda)$$

where the convention is adopted that, if v^{10} or v^{11} does not exist, then the corresponding (meaningless) term of the sum is to be neglected.

It is easy to see that, running through $\tilde{B}^*$ from top to bottom, we have no difficulty in successively calculating the vectors $\mathbf{w}(v, \lambda)$, which by (i) and (ii) are uniquely determined. Form the $n \times p$ matrix

$$\begin{aligned} \mathbf{W}(\tilde{B}, \lambda) &= (\mathbf{w}^{\mathsf{T}}(v_1, \lambda), \mathbf{w}^{\mathsf{T}}(v_2, \lambda), \ldots, \mathbf{w}^{\mathsf{T}}(v_n, \lambda))^{\mathsf{T}} \\ &= (w_k(v_i, \lambda)) \end{aligned} \tag{9}$$

$(i = 1, 2, \ldots, n, k = 1, 2, \ldots, p)$

and the $p \times p$ matrix

$$\begin{aligned} \mathbf{W}^*(\tilde{B}, \lambda) &= (\mathbf{w}^{\mathsf{T}}(b_1^*, \lambda), \mathbf{w}^{\mathsf{T}}(b_2^*, \lambda), \ldots, \mathbf{w}^{\mathsf{T}}(b_p^*, \lambda))^{\mathsf{T}} \\ &= (w_k(b_j^*, \lambda)) \end{aligned} \tag{10}$$

$(j, k = 1, 2, \ldots, p)$.

Theorem 3 [19, 45].

$$F_B(\lambda) = \varepsilon \cdot \det \mathbf{W}^*(\tilde{B}, \lambda) \quad \text{where} \quad \varepsilon \in \{1; -1\}. \tag{11}$$

Let λ^0 be an eigenvalue of B (i.e., a root of $F_B(\lambda)$, see (11)). Consider the system of linear equations

$$\mathbf{W}^*(\tilde{B}; \lambda^0) \cdot \mathbf{y} = \mathbf{0} \tag{12}$$

where $\mathbf{0}$ is the zero vector on p components. The space $\mathbf{S}(\lambda^0)$ of the solutions of (12) has the same dimension as the eigenspace $\mathbf{E}(\lambda^0)$. Furthermore, there is a simple 1,1-correspondence between $\mathbf{S}(\lambda^0)$ and $\mathbf{E}(\lambda^0)$ which we will now describe.

Theorem 4 [19, 45].

Let λ^0 be an eigenvalue of B, let $\mathbf{y}' \neq \mathbf{0}$ be a solution of

$$\mathbf{W}^*(\tilde{B}, \lambda^0) \cdot \mathbf{y} = \mathbf{0} \tag{13}$$

and put

$$\mathbf{x}' = \mathbf{W}(\tilde{B}, \lambda^0) \cdot \mathbf{y}' \tag{14a}$$

or, in other words, put

$$\mathbf{x}' = (x'_1, x'_2, \ldots, x'_n)^{\mathrm{T}} \quad \textit{where} \quad x'_i = \mathbf{w}(v_i, \lambda^0) \cdot \mathbf{y}'. \tag{14b}$$

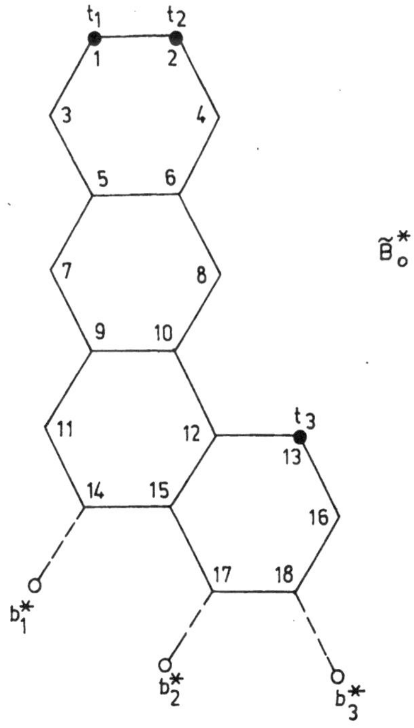

Fig. 26

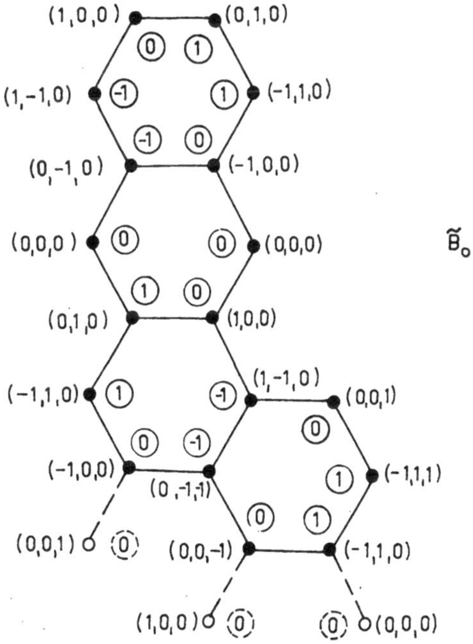

Fig. 27

Then the vector $\mathbf{x}'$ *is an eigenvector of* B *belonging to* λ^0 *and every eigenvector of* B *belonging to* λ^0 *can be obtained this way. Furthermore, if* $\mathbf{y}^1, \mathbf{y}^2, \ldots, \mathbf{y}^q$ *are non-zero solutions to (12) and* $\mathbf{x}^1, \mathbf{x}^2, \ldots, \mathbf{x}^q$ *are the corresponding eigenvectors of* B *(obtained from* $\mathbf{y}^1, \mathbf{y}^2, \ldots, \mathbf{y}^q$ *by virtue of (14a) or (14b)), then* $\mathbf{x}^1, \mathbf{x}^2, \ldots, \mathbf{x}^q$ *are linearly independent if and only if* $\mathbf{y}^1, \mathbf{y}^2, \ldots, \mathbf{y}^q$ *are linearly independent. In particular,* $\mathbf{x}^1, \mathbf{x}^2, \ldots, \mathbf{x}^q$ *form a basis of* $\mathbf{E}(\lambda^0)$ *if and only if* $\mathbf{y}^1, \mathbf{y}^2, \ldots, \mathbf{y}^q$ *form a basis of the space* $\mathbf{S}(\lambda^0)$ *of solutions to (13).*

As a consequence, no eigenvalue of B can have a multiplicity larger than p.

Example 1: For the BS depicted in Fig. 26 we obtain the following scheme.

$w_k(v_i, \lambda)$

i	k = 1	2	3
1	1	0	0
2	0	1	0
3	λ	-1	0
4	-1	λ	0
5	$\lambda^2 - 1$	$-\lambda$	0
6	$-\lambda$	$\lambda^2 - 1$	0
7	$\lambda^3 - \lambda$	$-2\lambda^2 + 2$	0
8	$-2\lambda^2 + 2$	$\lambda^3 - \lambda$	0
9	$\lambda^4 - 2\lambda^2 + 1$	$-2\lambda^3 + 3\lambda$	0
10	$-2\lambda^3 + 3\lambda$	$\lambda^4 - 2\lambda^2 + 1$	0
11	$\lambda^5 - \lambda^3 - \lambda$	$-3\lambda^4 + 7\lambda^2 - 3$	0
12	$-3\lambda^4 + 7\lambda^2 - 3$	$\lambda^5 - \lambda^3 - \lambda$	0
13	0	0	1
14	$\lambda^6 - 2\lambda^4 + \lambda^2 - 1$	$-3\lambda^5 + 9\lambda^3 - 6\lambda$	0
15	$-3\lambda^5 + 9\lambda^3 - 6\lambda$	$\lambda^6 - 2\lambda^4 + \lambda^2 - 1$	-1
16	$3\lambda^4 - 7\lambda^2 + 3$	$-\lambda^5 + \lambda^3 + \lambda$	λ
17	$-4\lambda^6 + 14\lambda^4 - 14\lambda^2 + 4$	$\lambda^7 - 7\lambda^3 + 6\lambda$	$-\lambda$
18	$3\lambda^5 - 7\lambda^3 + 3\lambda$	$-\lambda^6 + \lambda^4 + \lambda^2$	$\lambda^2 - 1$

The matrix $\mathbf{W}(\tilde{B}_0, \lambda)$

$w_k(b_j^*, \lambda)$

j	k = 1	2	3
1	$\lambda^7 - 7\lambda^3 + 6\lambda$	$-4\lambda^6 + 14\lambda^4 - 14\lambda^2 + 4$	1
2	$-4\lambda^7 + 14\lambda^5 - 16\lambda^3 + 7\lambda$	$\lambda^8 - 6\lambda^4 + 4\lambda^2 + 1$	$-2\lambda^2 + 2$
3	$7\lambda^6 - 24\lambda^4 + 24\lambda^2 - 7$	$-2\lambda^7 + 2\lambda^5 + 7\lambda^3 - 7\lambda$	$\lambda^3 - \lambda$

The matrix $\mathbf{W}^*(\tilde{B}_0, \lambda)$

Furthermore,

$$F_{B_0}(\lambda) = \varepsilon \cdot \det \mathbf{W}^*(\tilde{B}_0, \lambda)$$
$$= \lambda^{18} - 21\lambda^{16} + 180\lambda^{14} - 823\lambda^{12} + 2203\lambda^{10} - 3558\lambda^{8} +$$
$$+ 3430\lambda^{6} - 1868\lambda^{4} + 505\lambda^{2} - 49\,. \qquad (15)$$

$\lambda^1 = 1$ is simple root of $F_{B_0}(\lambda)$, i.e. simple eigenvalue of B_0. In Fig. 27 the vector $\mathbf{w}(v_i, \lambda^1)$ and the components of the (non-normalized) eigenvector corresponding to λ^1 (see Theorem 4) are written in brackets and in circles, respectively, close to vertex v_i $(i = 1, 2, \ldots, 18)$. (The zeros in the dotted circles only serve for checking the correctness of our calculations.)

4.3 Catacondensed Benzenoid Systems

For catacondensed benzenoid systems we can do even better.

Let B be a CBS with β bifurcation cells, $\beta \in \{0, 1, 2, \ldots\}$; note that B has precisely $(\beta + 2)$ end cells. For every bifurcation cell c^*, arbitrarily distinguish exactly one of the three external edges which lie on the boundary of c^*. Delete all internal edges, end edges, and distinguished edges of B; what remains is a set of $2(\beta + 1)$ disjoint paths P lying on the periphery, and covering all vertices, of B. Put $2(\beta + 1) = q$ (Fig. 28.1). Arbitrarily specify one of the paths P as P_1 and, following the periphery in the positive sense, number the paths P consecutively from 1 to q. Direct paths $P_1, P_3, \ldots, P_{q-1}$ in the positive and paths $P_2, P_4, \ldots, P_q$ in the negative sense; denote the source vertex and the mouth vertex of path P_j – i.e., the vertices from which P_j issues and in which it terminates – by s_j and m_j, respectively; denote the resulting figure by $\tilde{\tilde{B}}$ (see Fig. 28.2). To each path P_j from $\tilde{\tilde{B}}$, add a new "virtual" vertex m_j^* and a directed edge from m_j to m_j^*: thus P_j is turned into a directed path P_j^* from s_j to m_j^*, and $\tilde{\tilde{B}}$ is turned into a figure which we will call $\tilde{\tilde{B}}^*$ (Fig. 28.3). Note

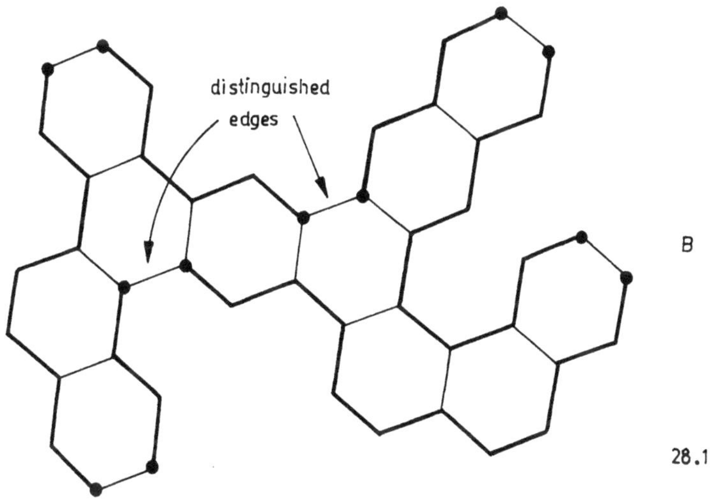

28.1

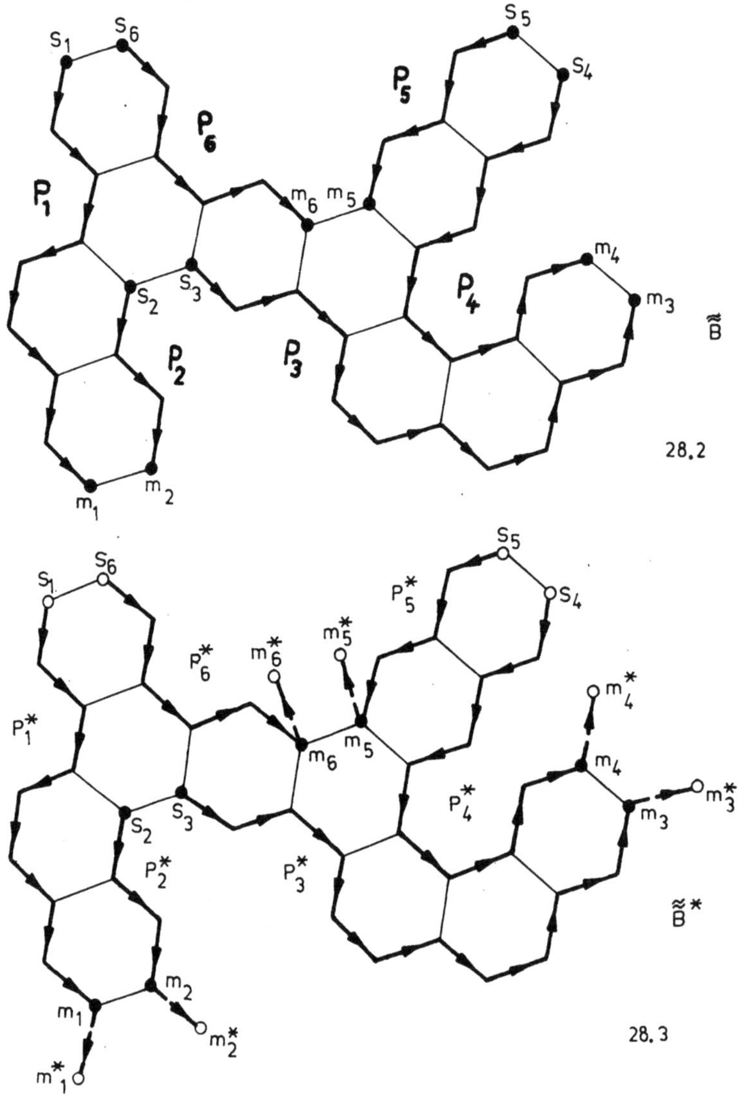

Fig. 28

that each end edge and each distinguished edge of B connects two source vertices s_j, s_{j+1} or two mouth vertices m_j, m_{j+1} (the subscript is to be reduced modulo q). Further, with respect to the system of directed paths $P_1^*, P_2^*, \ldots, P_q^*$, every vertex v of $\tilde{\tilde{B}}^*$ which is not a source vertex, has a unique immediate predecessor which we denote by v^+; let $N^+(v) := N(v^+) - \{v\}$ be the set of neighbours of v^+ in $\tilde{\tilde{B}}^*$ which are different from v. (Note that $v \notin N^+(v)$, $v^+ \notin N^+(v)$.)

Algorithm D

To every vertex v of B^* assign a vector

$$\mathbf{w}(v, \lambda) = (w_1(v, \lambda), w_2(v, \lambda), \dots, w_q(v, \lambda))$$

by means of the following recursive procedure.

(i) For $k = 1, 2, \dots, q$, mark source vertex s_k and put

$$\mathbf{w}(s_k, \lambda) = (\delta_{1k}, \delta_{2k}, \dots, \delta_{qk}) \quad \text{where} \quad \delta_{jk} = \begin{cases} 0 & \text{if} \quad j \neq k \\ 1 & \text{if} \quad j = k \end{cases}$$

$$(j = 1, 2, \dots, q).$$

(ii) Let v be a vertex of $\tilde{\tilde{B}}^*$ not yet marked and suppose that v^+ and all vertices of $N^+(v)$ have already been marked: mark v and put

$$\mathbf{w}(v, \lambda) = \lambda \cdot \mathbf{w}(v^+, \lambda) - \sum_{v' \in N^+(v)} \mathbf{w}(v', \lambda).$$

Running through $\tilde{\tilde{B}}^*$ following the directed paths P_j^* from the source vertices s_j to be virtual vertices m_j^*, we have no difficulty in successively (in some order) calculating the vectors $\mathbf{w}(v, \lambda)$ (which, by (i) and (ii), are uniquely determined) for all vertices v of $\tilde{\tilde{B}}^*$.

Form the $n \times q$ matrix

$$\mathbf{W}(\tilde{\tilde{B}}, \lambda) = (\mathbf{w}^\mathsf{T}(v_1, \lambda), \mathbf{w}^\mathsf{T}(v_2, \lambda), \dots, \mathbf{w}^\mathsf{T}(v_n, \lambda))^\mathsf{T}$$
$$= (w_k(v_i, \lambda))$$
$$(i = 1, 2, \dots, n;\ k = 1, 2, \dots, q)$$

and the $q \times q$ matrix

$$\mathbf{W}^*(\tilde{\tilde{B}}, \lambda) = (\mathbf{w}^\mathsf{T}(m_1^* \cdot \lambda), \mathbf{w}^\mathsf{T}(m_2^*, \lambda), \dots, \mathbf{w}^\mathsf{T}(m_q^*, \lambda))^\mathsf{T}$$
$$= (w_k(m_j^*, \lambda))$$
$$(j, k = 1, 2, \dots, q).$$

Using these matrices, calculate the characteristic polynomial $F_B(\lambda)$ and the eigenspaces of B in precisely the same way as described in Theorems 3 and 4 (Section 4.2) with $\tilde{B}$ replaced by $\tilde{\tilde{B}}$.

4.4 Remarks

Remark 1: Algorithm **D** is particularly efficient if applied to a CBS whose dualist is a path (i.e., satisfying $\beta = 0$); for such a system Π, the resulting determinant for $F_\Pi(\lambda)$ and the linear systems to be solved for obtained the eigenspaces are of order 2 (see Example 2 below).

As a consequence, no eigenvalue of Π can have a multiplicity larger than 2.

Remark 2: For the elements of many subclasses of CBSs it can easily be proved by induction using Algorithm **D** that ± 1 are eigenvalues (in particular this is true for all linear CBSs, i.e., CBSs without bifurcation or knee cells; in addition it can be shown that for all linear CBSs which have an odd number of cells the numbers ± 2 are simple eigenvalues).

Remark 3: Handling polynomials on a computer – especially determinants of polynomials – requires much time and memory. Therefore, in implementing Algorithm **C** and **D**, it seems to be favourable to proceed according to the following instructions.

(i) Write $F_B(\lambda) = \lambda^\varrho \cdot f_B(\lambda^2)$, $\varrho \in \{0, 1\}$, $\varrho \equiv n$, mod 2.

(ii) For $k + 1$ distinct values of x^* (the nodes), arbitrarily specified but all non-negative or all positive, respectively (e.g., for $x^* = 0, 1, \ldots, k$ or $x^* = 1, 2, \ldots, k + 1$), calculate the values $f_B(x^*)$ applying Algorithm **C** or **D** to the $k + 1$ values $\lambda^* = \sqrt{x^*}$.

(iii) From the $k + 1$ nodes x^* and the $k + 1$ nodal values $f_B(x^*)$, calculate $f_B(x)$ by means of the Lagrange or Newton Interpolation Formula.

(iv) Calculate the zeros $x_1, x_2, \ldots, x_k$ of $f_B(x)$. From these the spectrum $\{\lambda_1, \lambda_2, \ldots, \lambda_n\}$ of B is obtained as $\{\pm\sqrt{x_1}, \pm\sqrt{x_2}, \ldots, \pm\sqrt{x_k}\}$ or $\{0, \pm\sqrt{x_1}, \pm\sqrt{x_2}, \ldots, \pm\sqrt{x_k}\}$, respectively.

(v) In order to obtain the eigenspace $\mathbf{E}(\lambda_i)$, calculate the vectors $\mathbf{w}(v_j, \lambda_i)$ $(j = 1, 2, \ldots, n)$ using algorithm **C** or **D** (which now operates on numbers, not on polynomials) and proceed as described above.

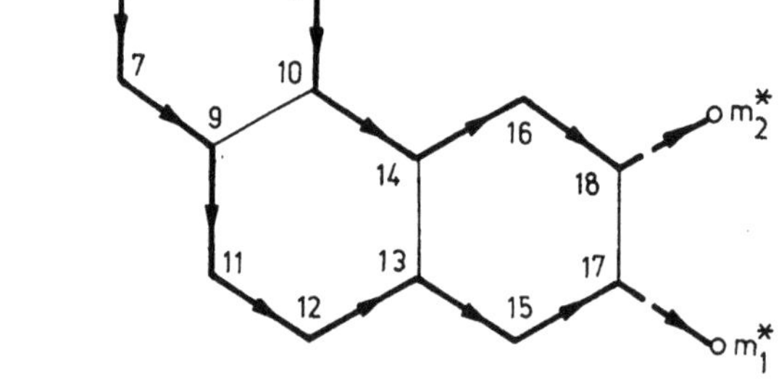

Fig. 29

Example 2: For the BS depicted in Fig. 29 we obtain

$w_k(v_i, \lambda)$

i	k = 1	2
1	1	0
2	0	1
3	λ	-1
4	-1	λ
5	$\lambda^2 - 1$	$-\lambda$
6	$-\lambda$	$\lambda^2 - 1$
7	$\lambda^3 - \lambda$	$-2\lambda^2 + 2$
8	$-2\lambda^2 + 2$	$\lambda^3 - \lambda$
9	$\lambda^4 - 2\lambda^2 + 1$	$-2\lambda^3 + 3\lambda$
10	$-2\lambda^3 + 3\lambda$	$\lambda^4 - 2\lambda^2 + 1$
11	$\lambda^5 - \lambda^3 - \lambda$	$-3\lambda^4 + 7\lambda^2 - 3$
12	$\lambda^6 - 2\lambda^4 + \lambda^2 - 1$	$-3\lambda^5 + 9\lambda^3 - 6\lambda$
13	$\lambda^7 - 3\lambda^5 + 2\lambda^3$	$-3\lambda^6 + 12\lambda^4 - 13\lambda^2 + 3$
14	$-3\lambda^4 + 7\lambda^2 - 3$	$\lambda^5 - \lambda^3 - \lambda$
15	$\lambda^8 - 4\lambda^6 + 7\lambda^4 - 8\lambda^2 + 4$	$-3\lambda^7 + 14\lambda^5 - 21\lambda^3 + 10\lambda$
16	$-\lambda^7 + 7\lambda^3 - 6\lambda$	$4\lambda^6 - 14\lambda^4 + 14\lambda^2 - 4$
17	$\lambda^9 - 5\lambda^7 + 10\lambda^5 - 10\lambda^3 + 4\lambda$	$-3\lambda^8 + 17\lambda^6 - 33\lambda^4 + 23\lambda^2 - 3$
18	$-\lambda^8 + 10\lambda^4 - 13\lambda^2 + 3$	$4\lambda^7 - 15\lambda^5 + 15\lambda^3 - 3\lambda$

The matrix $\mathbf{W}(\tilde{\tilde{B}}_0, \lambda)$

and

$w_k(m_j^*, \lambda)$

j	k = 1	2
1	$\lambda^{10} - 5\lambda^8 + 14\lambda^6 - 27\lambda^4 + 25\lambda^2 - 7$	$-3\lambda^9 + 16\lambda^7 - 32\lambda^5 + 29\lambda^3 - 10\lambda$
2	$-2\lambda^9 + 6\lambda^7 - 10\lambda^3 + 5\lambda$	$7\lambda^8 - 36\lambda^6 + 62\lambda^4 - 40\lambda^2 + 7$

The matrix $\mathbf{W}^*(\tilde{\tilde{B}}_0, \lambda)$

$F_{B_0}(\lambda) = \varepsilon \cdot \det \mathbf{W}^*(\tilde{\tilde{B}}_0, \lambda)$ is identical with the polynomial given in (15). $\lambda^1 = 1$ is a simple root of $F_{B_0}(\lambda)$. In Fig. 30, the vectors $\mathbf{w}(v_i, \lambda^1)$ and the components of the eigenvector corresponding to λ^1 are written in brackets and in circles, respectively, close to vertex v_i ($i = 1, 2, \ldots, 18$); see also Fig. 27.

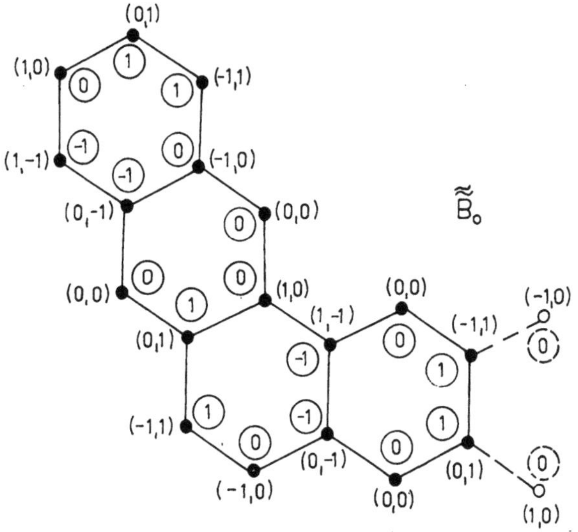

Fig. 30

5 Calculating the Number of Spanning Trees in Benzenoid Systems

Let G be a graph. A spanning tree of G is a tree of G which covers all vertices. Denote by $t(G)$ the number of all spanning trees of G.

Let B be a BS with h cells, $D = D(B)$ the dualist graph of B, $\mathbf{A} = \mathbf{A}(D)$ the 0,1-adjacency matrix of D, and $\mathbf{I}$ the $h \times h$ unit matrix.

In 1983, I. Gutman, R. B. Mallion and J. W. Essam proved the following theorem.

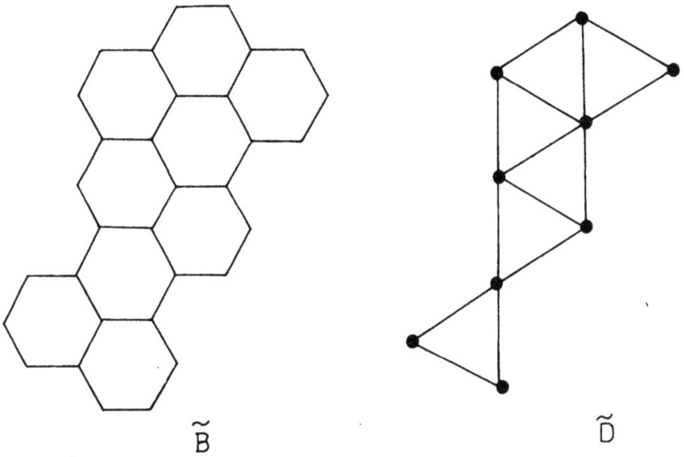

Fig. 31

Theorem 5 [42].

$$t(B) = \det(6 \cdot \mathbf{I} - \mathbf{A}). \tag{16}$$

Let $\tilde{B}$ be a drawing of B (see 4.1) and $\tilde{D}$ the inner dual of $\tilde{B}$ (see Fig. 31). In $\tilde{D}$ we find a set of disjoint "vertical segments" covering all vertices, $\tilde{P}_1, \tilde{P}_2, \ldots, \tilde{P}_{p-1}$, say, where $\tilde{P}_k$ is a maximal vertical path (non-interrupted vertical line which may reduce to a single point) in $\tilde{D}$ connecting a top point $\tilde{t}_k$ with a bottom point $\tilde{b}_k$ ($k = 1, 2, \ldots, p-1$), as indicated in Fig. 32. The points $\tilde{t}_1, \tilde{t}_2, \ldots, \tilde{t}_{p-1}$ are called the top vertices of $\tilde{D}$. Every $\tilde{P}_k$ of $\tilde{D}$ is prolonged beyond its bottom point $\tilde{b}_k$ by one short vertical segment connecting $\tilde{b}_k$ with an additional "virtual" vertex $\tilde{b}_k^*$; let $\tilde{P}_k^*$ denote the prolonged path. Direct all paths $\tilde{P}_1^*, \tilde{P}_2^*, \ldots, \tilde{P}_{p-1}^*$ from top to bottom; thus $\tilde{D}$ is turned into a figure which we call $\tilde{D}^*$; see Fig. 33.

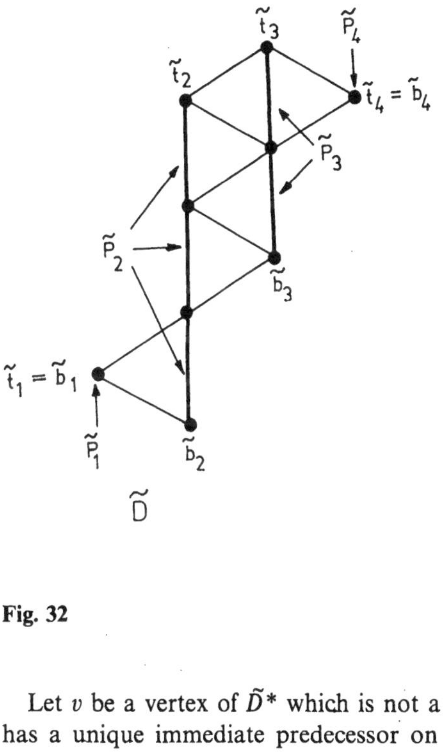

Fig. 32

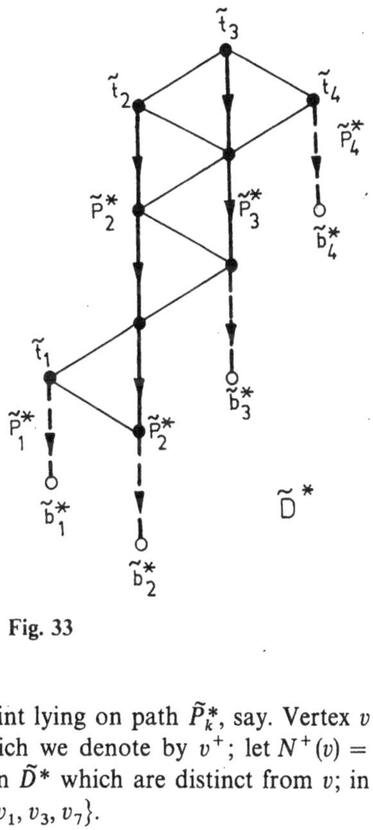

Fig. 33

Let v be a vertex of $\tilde{D}^*$ which is not a top point lying on path $\tilde{P}_k^*$, say. Vertex v has a unique immediate predecessor on $\tilde{P}_k^*$ which we denote by v^+; let $N^+(v) = N(v^+) - \{v\}$ be the set of neighbours of v^+ in $\tilde{D}^*$ which are distinct from v; in Fig. 34, $N^+(v_1) = \{v_3, v_4, v_5, v_6\}$ and $N^+(v_2) = \{v_1, v_3, v_7\}$.

Algorithm E₅

To every vertex v of $\tilde{D}^*$assign a vector $\mathbf{w}(v) = (w_1(v), w_2(v), \ldots, w_{p-1}(v))$ by means of the following recursive procedure.

(i) For $k = 1, 2, \ldots, p-1$ mark top vertex $\tilde{t}_k$ and put $\mathbf{w}(\tilde{t}_k) = (\delta_{1k}, \delta_{2k}, \ldots, \delta_{p-1,k})$ where

$$\delta_{lk} = \begin{cases} 0 & \text{if} \quad l \neq k \\ 1 & \text{if} \quad l = k, \end{cases}$$

$(l = 1, 2, \ldots, p-1)$.

(ii) Let v be a vertex of $\tilde{D}^*$ not yet marked and suppose that v^+ and all vertices of $N^+(v)$ have already been marked; mark v and put

$$\mathbf{w}(v) = 6 \cdot \mathbf{w}(v^+) - \sum_{v' \in N^+(v)} \mathbf{w}(v').$$

Running through $\tilde{D}^*$ from top to bottom, we have no difficulty in successively calculating the vectors $\mathbf{w}(v)$ (which, by (i) and (ii), are uniquely determined).

Form the $(p-1) \times (p-1)$ matrix $\mathbf{W}^*(\tilde{D}) = (\mathbf{w}^\mathsf{T}(\tilde{b}_1^*), \mathbf{w}^\mathsf{T}(\tilde{b}_2^*), \ldots, \mathbf{w}^\mathsf{T}(\tilde{b}_{p-1}^*))^\mathsf{T} = (w_l(\tilde{b}_k^*))$; $l, k = 1, 2, \ldots, p-1$.

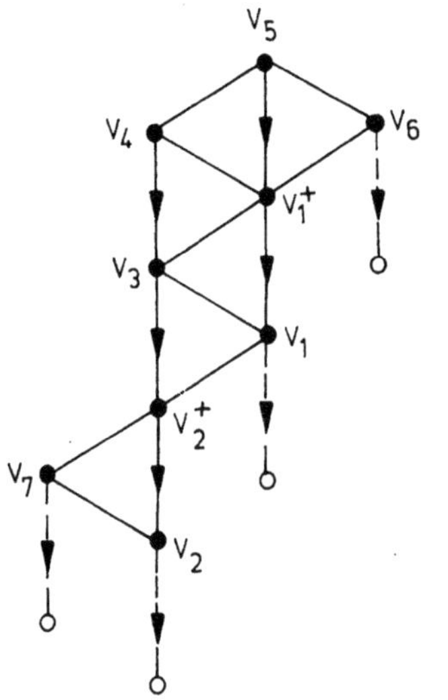

Fig. 34

Theorem 6 [19, 45].

$$t(B) = |\det \mathbf{W}^*(\tilde{D})| \quad \text{where} \quad D = D(B). \tag{17}$$

Example: For the BS depicted in Fig. 35 we obtain

$$t(B_2) = \begin{vmatrix} 42 & -14 \\ -48 & 42 \end{vmatrix} = 1092.$$

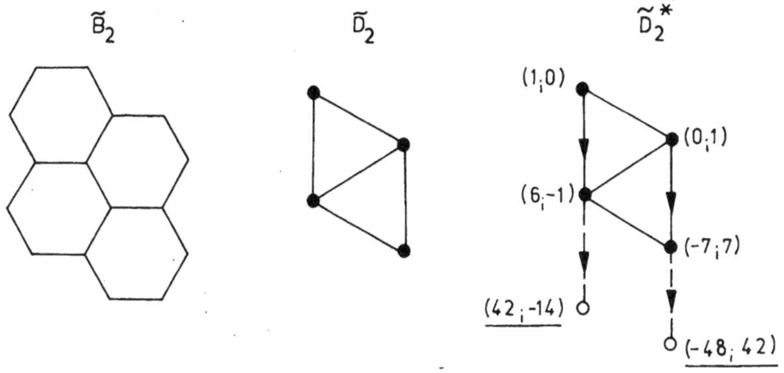

Fig.35

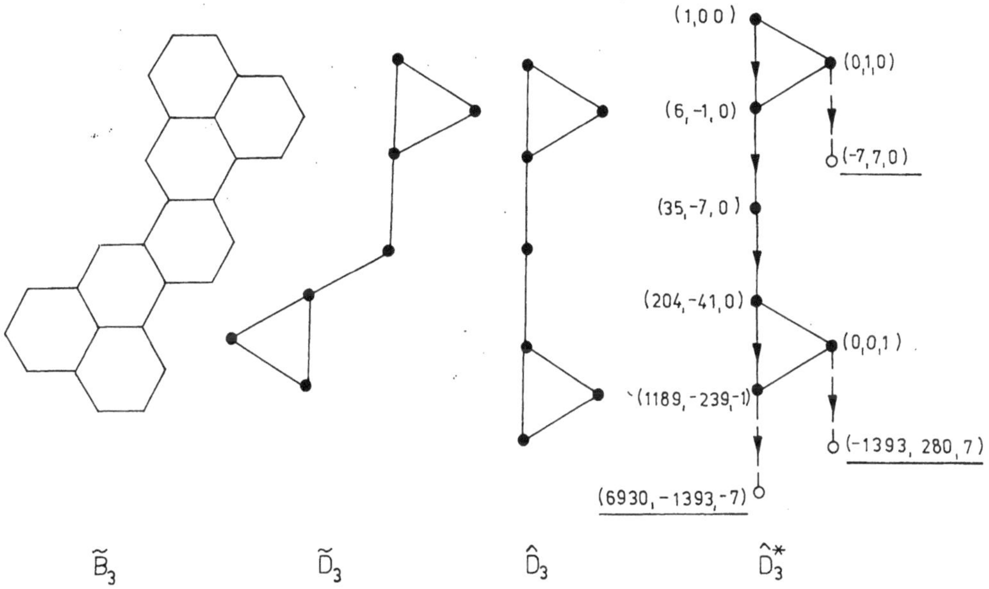

$$t(B_3) = \varepsilon \cdot \begin{vmatrix} 6930 & -1393 & -7 \\ -7 & 7 & 0 \\ -1393 & 280 & 7 \end{vmatrix} = \underline{\underline{216776}}$$

Fig. 36

If D has cut points, then the order of det $\mathbf{W}^*(\tilde{D})$ can, possibly, be further reduced by successively rotating parts of $\tilde{D}$ around these points such that, in the resulting figure $\widehat{D}$,

- every line segment is parallel to one of the three original directions;
- the number of maximal vertical paths is minimum (Figs. 36, 37, 38).

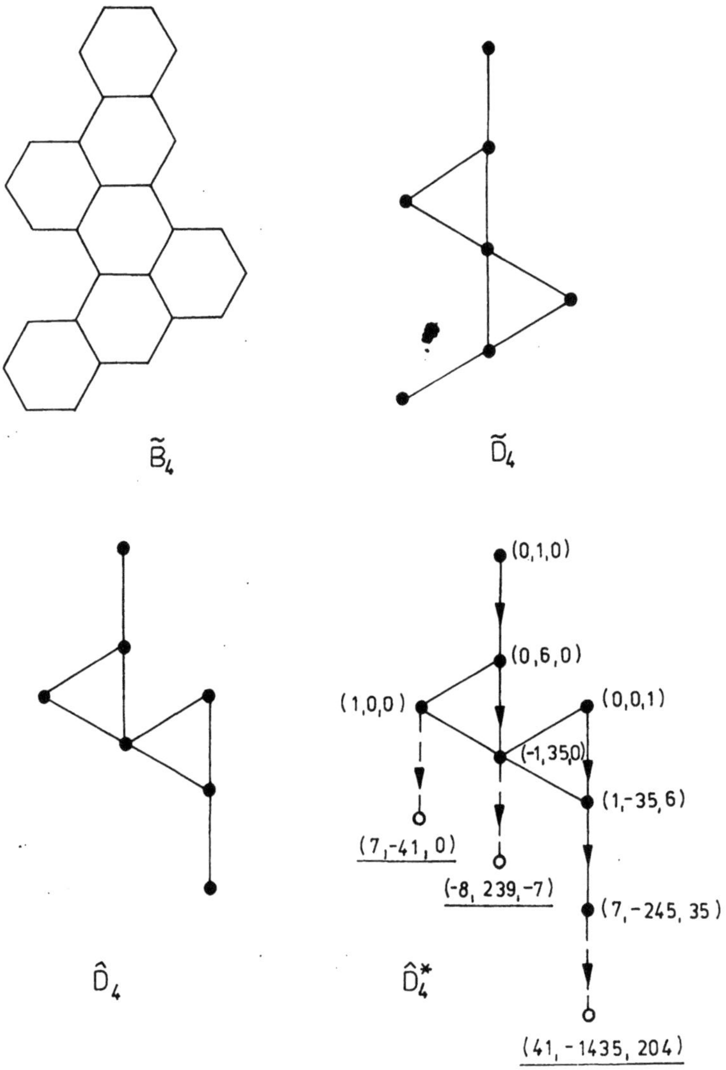

$$t(B_4) = \varepsilon \cdot \begin{vmatrix} 7 & -41 & 0 \\ -8 & 239 & -7 \\ 41 & -1435 & 204 \end{vmatrix} = \underline{\underline{215832}}$$

Fig. 37

In particular, if B is a CBS which has exactly β bifurcation cells, then the order of the resulting det $\mathbf{W}^*(\hat{D})$ lies between $(\beta + 3)/2$ and $(\beta + 1)$. Also, in this case all non-vertical segments may be replaced by horizontal ones (see $\check{D}$ in Fig. 38).

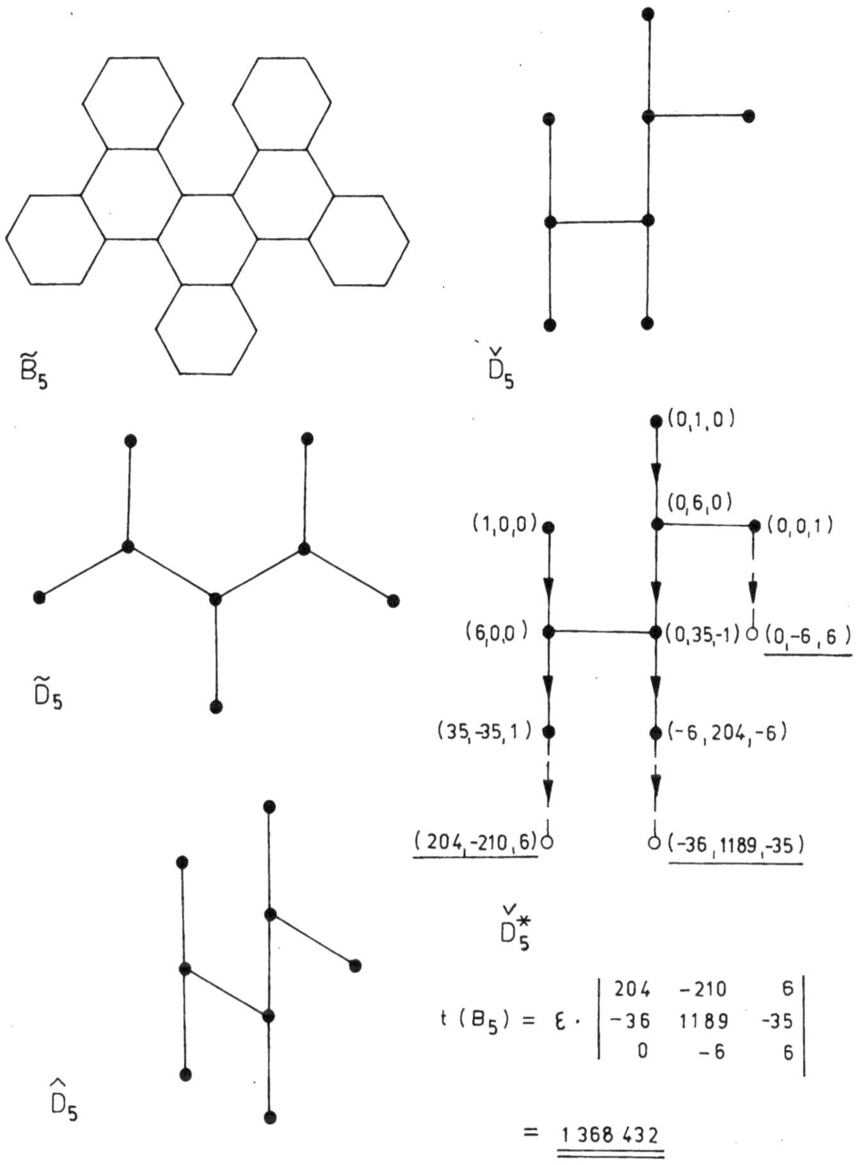

Fig. 38

6 Concluding Remark

It is worth mentioning that the algorithms described here can – in a suitably extended form – be applied to classes of graphs much larger than just the class of benzenoid systems (see, e.g. Ref. [44]).

7 References

1. Pauling L, Brockway LO, Beach JY (1935) J. Am. Chem. Soc. 57: 2705
2. Gordon M, Davison WHT (1952) J. Chem. Phys. 20: 428
3. Dewar MJS, Longuet-Higgins HC (1952) Proc. Roy. Soc. A 214: 482
4. Kasteleyn PW (1961) Physica 27: 1209
5. Cvetković D, Gutman I, Trinajstić N (1972) Chem. Phys. Letters 16: 614
6. Sachs H (1986) Algebra und Graphentheorie – Beiträge zur Jahrestagung „Algebra und Grenzgebiete" in Siebenlehn, 28. Oktober–1. November 1985. Bergakademie Freiberg, Sektion Mathematik, Freiberg, p 65
7. Yen TF (1971) Theor. Chim. Acta 20: 399
8. Herndon WC (1973) Tetrahedron 29: 3
9. Cvetković D, Gutman I (1974) Croat. Chem. Acta 46: 15
10. Randić M: J. Chem. Soc. Faraday Trans. II 1976: 232
11. Polansky OE, Gutman I (1980) Match (Mülheim) 8: 269
12. Džonova-Jerman-Blažić B, Trinajstić N (1982) Croat. Chem. Acta 55: 347
13. Trinajstić N (1983) Chemical Graph Theory, vols. I, II, CRC Press, Boca Raton, FL
14. Gutman I, Polansky OE (1986) Mathematical concepts in organic chemistry, Springer, Berlin Heidelberg New York
15. Cyvin SJ, Gutman I (1988) Kekulé structures in benzenoid hydrocarbons, Springer, Berlin Heidelberg New York (Lect. Notes Chem. 46)
16. Ham NS (1958) J. Chem. Phys. 29: 1229
17. Ruedenberg K (1954) J. Chem. Phys. 22: 1878
18. John P, Sachs H: J. Chem. Soc. Faraday Trans. II (in press)
19. Al-Khnaifes KA (1988) Thesis TH Ilmenau
20. John P, Rempel J (1985) Graphs, hypergraphs and applications, Proc. Conf. Graph Theory, Eyba, Oct. 1984, (Teubner-Texte zur Mathematik, Band 73) Teubner, Leipzig, p 72
21. John P, Sachs H (1985) Graphs, hypergraphs and applications, Proceedings of the Conf. on Graph Theory, Eyba, Oct 1984, (Teubner-Texte zur Mathematik, Band 73), Teubner, Leipzig, p 80
22. John P, Sachs H (1985) In: Bodendiek R, Schumacher H, Walter G (eds) Graphen in Forschung und Unterricht, Verlag Barbara Franzbecker, Bad Salzdetfurth, p 85
23. Heilbronner E (1953) Helv. Chim. Acta 36: 170
24. Heilbronner E (1954) Helv. Chim. Acta 37: 913
25. Lihtenbaum LM (1956) Trudy 3-go Vses. Matem. S"ezda, Tom 1, p 135
26. Collatz L, Sinogowitz U (1957) Hambg. math. Abhandlg. 21: 63
27. Milić M (1964) IEEE Trans. Circuit Theory CT-11: 423
28. Sachs H (1964) Publ. Math. Debrecen 11: 119
29. Spialter L (1964) J. Chem. Docum. 4: 269
30. Coulson CA, Streitwieser A Jr (1965) Dictionary of π-Electron Calculations, Pergamon, Oxford
31. Hosoya H (1972) Theor. Chim. Acta 25: 215
32. Mallion RB, Schwenk AJ, Trinajstić N (1975) In: Fiedler M (ed) Recent advances in graph theory, Proceedings of the Second Czechoslovak Symposium on Graph Theory, Academia, Prague, p 345
33. Aihara JI (1976) J. Amer. Chem. Soc. 98: 6840
34. Balasubramanian K (1982) Int. J. Quant. Chem. 21: 581
35. Collatz L (1978) Resultate Math. 1: 42
36. Cvetković DM, Doob M, Sachs H (1980) Spectra on graphs-theory and application, Academic, New York
37. Cvetković DM, Doob M, Gutman I, Torgašev A (1988) Recent results in the theory of graph spectra, North-Holland, Amsterdam (Annals of Discrete Mathematics 36)
38. Kirchhoff G (1847) Ann. Phys. Chem. 72: 497
39. Brooks RL, Smith CBA, Stone AH, Tutte WT (1940) Duke Math. J. 7: 312
40. Trent H (1954) Proc. Nat. Acad. Sci. USA 40: 1004
41. Hutschenreuther H (1967) Wiss. Z. d. TH Ilmenau, 13: 403
42. Gutman I, Mallion RB, Essam JW (1983) Mol. Phys. 50: 859

43. Gutman I, Cyvin SJ (1987) Chem. Phys. Letters 136: 137
44. John P, Sachs H, Zernitz H (1987) Zastosowania Matematyki/Applicationes Mathematicae 19: 465
45. in preparation
46. Sachs H (1984) Combinatorica 4: 89

The Existence of Kekulé Structures in a Benzenoid System

Zhang Fuji[1], Guo Xiaofeng[1] and Chen Rongsi[2]

[1] Department of Mathematics, Xinjiang University, Wulumuqi, Xinjiang, The People's Republic of China
[2] College of Finance and Economics, Fuzhou University, Fuzhou, Fujian, The People's Republic of China

Table of Contents

The existence of Kekulé structures in a benzenoid system is the first fundamental problem in the topological theory of benzenoid systems. It was considered as one of the most difficult open problems in this theory. Many investigations have been made in order to find necessary and sufficient conditions for the existence of Kekulé structures in a benzenoid system. Some fairly simple conditions which are both necessary and sufficient have been given in the last few years. In this chapter we review the main results and give a rigorous proof for some necessary and sufficient conditions for the existence of Kekulé structures in a benzenoid system. In addition, by using the above results, a construction method of some concealed non-Kekuléan benzenoid systems is given.

1 Introduction

A Kekulé structure or a 1-factor of a benzenoid system H is an independent edge set in H such that every vertex in H is incident with an edge in the edge set. A benzenoid system is said to be Kekuléan if it possesses a Kekulé structure, otherwise it is said to be non-Kekuléan. It was first pointed out by Clar et al. [1, 2] that Kekulé structures are of paramount importance for the stability of benzenoid systems. In fact, up to now, no non-Kekuléan benzenoid system has been synthesized by chemists. Therefore the existence of Kekulé structures in a benzenoid system is a fundamental problem in the topological theory of benzenoid systems.

Many investigations have been made in order to find necessary and sufficient conditions for the existence of Kekulé structures in a benzenoid system. On the other hand, some algorithms for determining whether or not a given benzenoid system has Kekulé structures have been developed.

2 Some Early Results

Since a benzenoid system *H* is a bipartite graph, the existence of Kekulé structures of H is equivalent to the existence of 1-factors (perfect matchings) of a bipartite graph. In 1935, P. Hall found the following necessary and sufficient conditions.

Theorem 1 [3]. Let G be a bipartite graph with bipartition (X, Y). Then G contains a matching that saturates every vertex in X if, and only if $|N(S)| = |S|$ for all $S \subseteq X$.

Here N(S) is the set of all vertices adjacent to the vertices in S and called the neighbour set of S.

Clearly, if $|X| = |Y|$ a matching which saturates every vertex in X is a 1-factor (perfect matching) of G. Hence Theorem 1 can be used to decide whether or not a given benzenoid system H has Kekulé structures. But, using Theorem 1, we have to examine all subsets of X, where (X, Y) is the bipartition of H. This is evidently tedious.

Let H be a benzenoid system drawn in a plane such that one of the three edge directions is vertical. A peak (valley) of H is a vertex in H which lies above (below) its all adjacent vertices. A monotonic path (P – V path) in H is such a path starting from a peak, running monotonously downwards and terminating in a valley. A monotonic path system of H is a selection of independent monotonic paths which contain all peaks and valleys of H. We denote by $p(H)$ and $v(H)$ the number of peaks and valleys of H, respectively, and $\Delta = p(H) - v(H)$. For convenience we colour the vertices of H black and white so that any two adjacent vertices have different colours, and peaks of H are coloured black (see Fig. 1). Let n_b and n_w be the number of black and white vertices, respectively. Then $n_b - n_w = p(H) - v(H) = \Delta$.

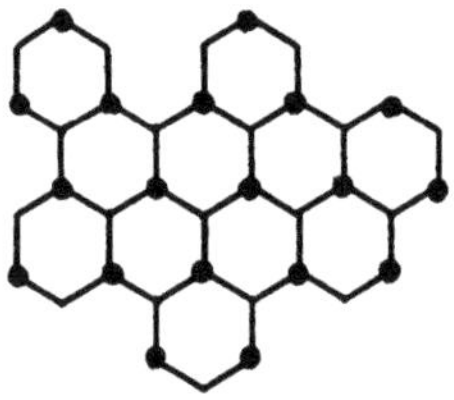

Fig. 1

In 1952 the following theorem was discovered by Gordon and Davison [4] and rigorously proved by Sachs [5].

Theorem 2. (a) The number of Kekulé structures in a benzenoid system is equal to the number of selections of independent monotonic paths. (b) There is a one-one correspondence between the systems of independent monotonic paths and the Kekulé structures.

Theorem 2 implies the following.

Corollary 1. A benzenoid system H has Kekulé structures if, and only if H has monotonic path systems.

The above theorem and corollary tell us that the existence of Kekulé structures of H is equivalent to the existence of monotonic path systems in H. However, it is also difficult to decide whether or not a given benzenoid system has monotonic path systems.

Another early result was given by Dewar and Longuet-Higgins [6] in 1952.

Theorem 3. Let $\mathbf{A}$ be the adjacency matrix of a benzenoid system H with n vertices and K the number of Kekulé structures of H. Then $\det \mathbf{A} = (-1)^{n/2} K^2$.

An evident corollary is the following.

Corollary 2. A benzenoid system has Kekulé structures if, and only if $\det \mathbf{A} \neq 0$.

Theorem 3 and Corollary 2 transform the question of existence of Kekulé structures in benzenoid systems to an algebraic problem. But it is also of little practical use, since the calculation of determinants of high orders is troublesome.

After this chemists hoped to find some fairly simple necessary and sufficient conditions. This is why until 1982–1983 I. Gutman and N. Trinajstić still pointed out several times [7–9] that the problem of recognizing Kekuléan benzenoid systems was an open problem, and it was thought to be one of the most difficult open problems in the topological theory of benzenoid systems.

We mention that though the above results are inconvenient for practical use, Theorem 1 can be used to prove stronger results, and Corollary 1 can lead us to find an algorithm for determining whether or not a given benzenoid system has Kekulé structures or monotonic path systems. Recently a simple and efficient algorithm of

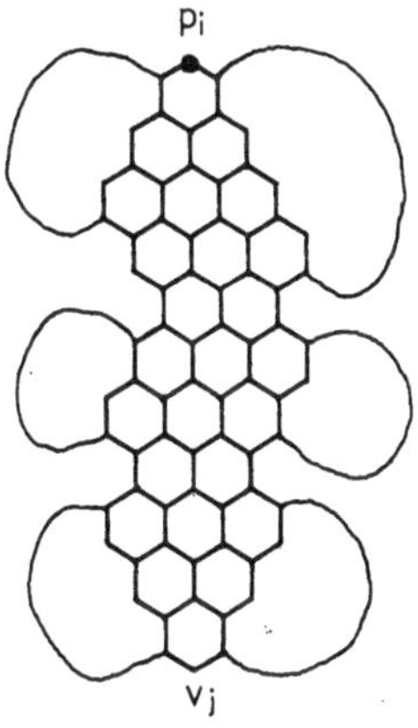

Fig. 2

relevance to this problem, called canonical P − V path elimination, has been found by Guo Xiaofeng and Zhang Fuji [10]. A result based on Corollary 2 was developed by John, Sachs and Rempel [11, 12] in 1985. They defined a new matrix $\mathbf{W} = (w_{ij})$ such that w_{ij} denotes the number of monotonic paths which start at the i-th peak and end at the j-th valley in a benzenoid system, and proved the following theorem.

Theorem 4. Let H be a benzenoid system with $\Delta = 0$, and K the number of Kekulé structures of H. Then $K = \pm \det \mathbf{W}$.

Corollary 3. A benzenoid system has Kekulé structure if, and only if $\Delta = 0$ and $\det \mathbf{W} \neq 0$.

Note that the order of **W** is smaller than the order of **A**, and w_{ij} is equal to the number of Kekulé structures of the subsystem induced by the edges contained in monotonic paths from the i-th peak p_i to the j-th valley v_j (see Fig. 2). A method for calculating the matrix element w_{ij} has been found by I. Gutman, L. X. Su, and S. J. Cyvin [13–15].

3 Some Necessary or Sufficient Structural Requirements for the Existence of Kekulé Structures in Benzenoid Systems

It is well known that a Kekuléan benzenoid system must have the same number of black and white vertices, and consequently $\Delta = 0$. The benzenoid systems with $\Delta \neq 0$ are said to be obvious non-Kekuléan. In the following, we focus our attention on the benzenoid systems with $\Delta = 0$.

The conditions $\Delta = 0$ is necessary but not sufficient for the existence of Kekulé structures. The smallest non-Kekuléan benzenoid systems with $\Delta = 0$ (called concealed non-Kekuléan) were found during the years 1972–1986; see, e.g. Brunvoll et al. [16]. Two of them are shown in Fig. 3.

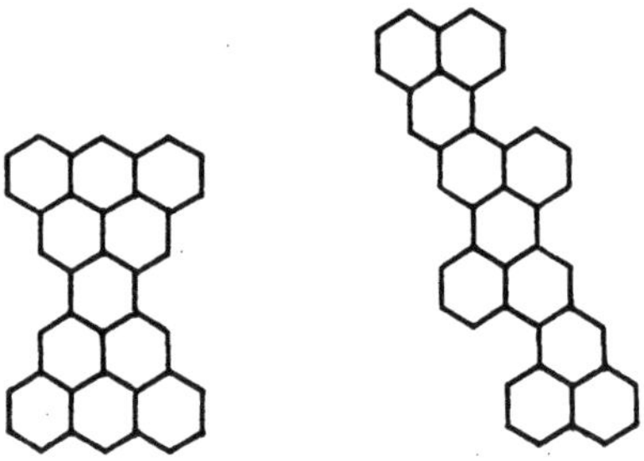

Fig. 3

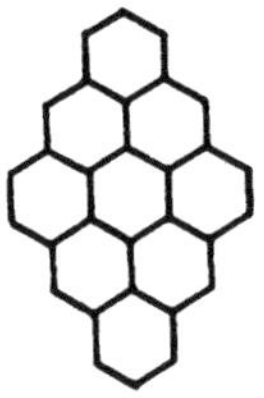

Fig. 4

If the number of vertices is even, then an evidently sufficient condition for the existence of Kekulé structures of a benzenoid system is the existence of a Hamiltonian path [17]. As a corollary, all catacondensed benzenoid systems are Kekuléan [18]. But the condition is not necessary. A Kekuléan benzenoid system with no Hamilton path is shown in Fig. 4.

Some other necessary or sufficient (but not necessary and sufficient) criteria for the differentiation between Kekuléan and non-Kekuléan benzenoid systems can be found in a number of recently published papers [5, 19–23]. In particular, Sachs gave some important results [5]. He introduced the concepts of a cut segment and a cut in a benzenoid system.

Let H be a benzenoid system drawn in the plane. A straight line segment C with end points P_1, P_2 is called a cut segment if:
(1) C is orthogonal to one of the three edge directions,
(2) each of P_1, P_2 is the centre of an edge,
(3) every point of C is either an interior or a boundary point of some hexagon of H,
(4) the graph obtained from H by deleting all edges intersected by C has exactly two components.

Let $\mathbb{C}$ denote the set of edges of H intersected by C; $\mathbb{C}$ is called a cut of H. If C is in the horizontal position, the component of H $-$ $\mathbb{C}$ lying at the upper bank of $\mathbb{C}$ is denoted by U($\mathbb{C}$), and the component at the lower bank is denoted by L($\mathbb{C}$); see Fig. 5. Let $p(H/U(\mathbb{C}))$, $p(H/L(\mathbb{C}))$, $(v(H/U(\mathbb{C}))$, $v(H/L(\mathbb{C})))$ denote the number of peaks (valleys) of H which belong to U($\mathbb{C}$) or L($\mathbb{C}$), respectively.

Sachs gave the following necessary conditions, and conjectured them also to be sufficient.

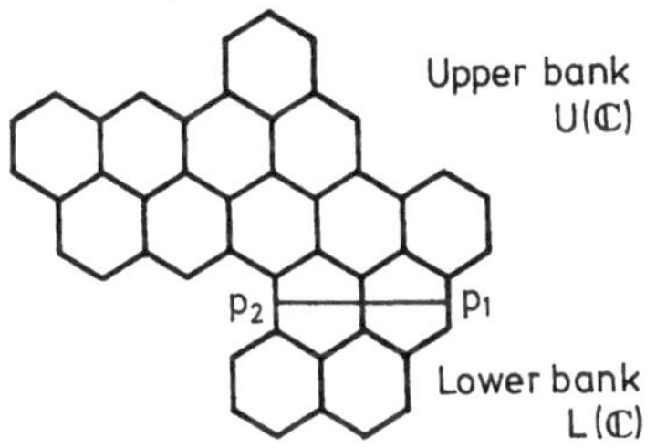

Fig. 5

Fig. 6

Lemma. Let H be a benzenoid system which has a Kekulé structure. Then for each of the six possible positions of H,
(i) $p(H) = v(H)$,
(ii) $0 \leqq p(H/U(\mathbb{C})) - v(H/U(\mathbb{C})) \leqq |\mathbb{C}|$,where $\mathbb{C}$ runs through all horizontal cuts.

Conjecture. Let H be a benzenoid system satisfying, for each of its possible positions,
(i) $p(H) = v(H)$,
(ii) $0 \leqq p(H/U(\mathbb{C})) - v(H/U(\mathbb{C})) \leqq |\mathbb{C}|$, where $\mathbb{C}$ runs through all horizontal cuts. Then H has a Kekulé structure.

Many chemists and mathematicians have attempted to prove Sachs' conjecture, but without success. Later on several counterexamples of Sachs' conjecture have been found; see e.g. Zhang Fuji, Chen Rongsi and Guo Xiaofeng [24]; cf. Fig. 6.

4 Necessary and Sufficient Structural Requirements for the Existence of Kekulé Structures in Benzenoid Systems

The failure of Sachs' conjecture made us consider some further necessary and sufficient conditions for the existence of Kekulé structures. The first result can be formulated in the following way [24].

Consider a benzenoid system H whose vertices are coloured in the usual (previously described) manner. An edge-cut (EC)* is a collection $\{e_1, e_2, \dots, e_t\}$ of edges of H such that:

(i) the graph obtained from H by deleting the edges in (EC)* has exactly two components H′ and H″;

(ii) for each edge e_i, $i = 1, 2, \dots, t$, its black end vertex belongs to H″, and so its white end vertex belongs to H′;

(iii) each pair of edges e_i, e_{i+1}, $i = 1, 2, \dots, t-1$, belongs to the same hexagon, and e_1 and e_t belong to the boundary of H.

Theorem 5. A benzenoid system H has Kekulé structures if, and only if $\Delta = 0$ and $n_b(H') \leqq n_w(H')$ for every edge-cut (EC)* of H.

The benzenoid system shown in Fig. 6 has no Kekulé structure, since the edge-cut indicated by dash lines satisfies the three conditions of (EC)*, but $n_b(H') > n_w(H')$.

Theorem 5 is somewhat stronger than Theorem 1. But, using Theorem 5, we would have to examine every possible edge-cut (EC)* of a benzenoid system. This is also tedious.

After this a stronger result was given by Kostochka [25]. He showed that in Theorem 5 it is sufficient to consider only those edge-cuts which are realized by straight lines or by two straight lines forming an angle of 60°. These lines may intersect the boundary only twice.

Some fairly simple necessary and sufficient conditions, which are similar to that of Kostochka, were independently found by Zhang Fuji and Chen Rongsi [26]. In the following we give their rigorous proof.

In order to get some necessary and sufficient conditions for benzenoid systems to have Kekulé structures, the concepts of the horizontal cut segment and the horizontal cut of H were generalized to the horizontal g-cut segment and the horizontal g-cut, respectively.

Let H be a benzenoid system drawn in the plane such that one of the three edge directions is vertical. A broken line segment $C = P_1P_2P_3$ (possibly, $P_2 = P_3$) is called a horizontal g-cut segment of H if:

(1) P_1P_2 is horizontal,

(2) each of P_1, P_3 is the centre of an edge lying on the boundary of H; and if $P_2 \neq P_3$, P_2 is the centre of a hexagon of H,

(3) every point of C is either an interior or a boundary point of some hexagon of H,

(4) if $P_2 \neq P_3$, the angle $P_1P_2P_3$ is $\pi/3$.

Please see Fig. 7. In particular, if $P_2 = P_3$, C becomes a horizontal cut segment.

For a horizontal g-cut segment $C = P_1P_2P_3$, let $\mathfrak{C}_{12}$ and $\mathfrak{C}_{23}$ denote the set of edges of H intersected by straight line segment $C_{12} = P_1P_2$ and $C_{23} = P_2P_3$, respectively. Let $\mathfrak{C} = \mathfrak{C}_{12} \cup \mathfrak{C}_{23}$. $\mathfrak{C}$ is called a horizontal g-cut. In particular, if C is a horizontal cut segment, we have $\mathfrak{C}_{23} = \emptyset$ and $\mathfrak{C} = \mathfrak{C}_{12}$, and $\mathfrak{C}$ is called a

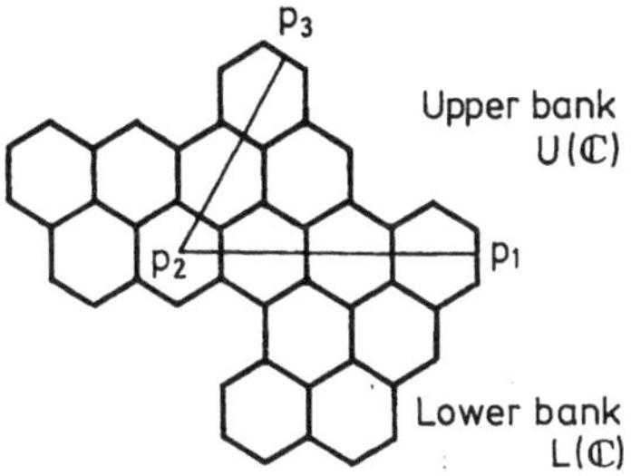

Fig. 7

horizontal cut. Clearly, $H - \mathfrak{C}$ has exactly two components, which are called the upper bank and lower bank, respectively, denoted by $U(\mathfrak{C})$ and $L(\mathfrak{C})$.

Theorem 6 [26]. Let H be a benzenoid system. Then H has a Kekulé structure if, and only if, for each of its six possible positions,
(1) $p(H) = v(H)$,
(2) $p(H/U(\mathfrak{C})) - v(H/U(\mathfrak{C})) \leqq |\mathfrak{C}_{12}|$,
where $\mathfrak{C}$ runs through all horizontal g-cuts.

Proof. First we prove the necessity. Suppose that H has a Kekulé structure. Then $\Delta = p(H) - v(H) = 0$, that is, $p(H) = v(H)$. Let $\mathfrak{C}$ be a horizontal g-cut of H. Let S be the set of black vertices in $U(\mathfrak{C})$. Then $|S| - |N(S)| = p(H/U(\mathfrak{C})) - v(H/U(\mathfrak{C})) - |\mathfrak{C}_{12}|$ (see Fig. 7). By Theorem 1, we have $|S| \leqq |N(S)|$. Hence $p(H/U(\mathfrak{C})) - v(H/U(\mathfrak{C})) = |\mathfrak{C}_{12}| + |S| - |N(S)| \leqq |\mathfrak{C}_{12}|$.

Then we prove the sufficiency by contradiction. Suppose that H satisfies the conditions in the theorem, but has no Kekulé structure.

Let B(H) and W(H) be the set of black and white vertices of H, respectively, and V(H) the set of vertices of H. For $S \subset V(H)$, the subgraph of H induced by S is denoted by $\langle S \rangle$. If, for all $S^* \subset B(H)$ such that the induced subgraph $\langle S^* \cup N(S^*) \rangle$ is connected, $|S^*| \leqq |N(S^*)|$, then $|S| \leqq |N(S)|$ for all $S \subset B(H)$, and by Theorem 1 and $\Delta = 0$, H has a Kekulé structure, a contradiction. Hence we can take a subset S of B(H) with maximum cardinality such that $|S| > |N(S)|$ holds and the induced subgraph $G = \langle S \cup N(S) \rangle$ is connected. Note that $S \neq B(H)$ since $|B(H)| = |W(H)| = |N(B(H))|$. By the maximality of S, we can deduce that $|S| = |N(S)| + 1$. (Otherwise, if $|S| > |N(S)| + 1$, then we can take a vertex v which is not in S and is adjacent to a vertex in N(S). Put $S' = S \cup \{v\}$. Since $|N(\{v\})| \leqq 3$, we have $|N(S')| \leqq |N(S)| + 2 < |S| + 1 = |S'|$, contradicting the maximality of S). From $S \neq B(H)$, we have that $G' = \langle V(H) - S \cup N(S) \rangle$ is not an empty graph. Furthermore, we can claim that G′ is connected. Let G_i', $i = 1, 2, \ldots, m$, be the components of G′. We have

$$|W(G')| - |B(G')| = \sum_{i=1}^{m} (|W(G_i')| - |B(G_i')|) = |B(G)| - |W(G)|$$
$$= |S| - |N(S)| = 1 .$$

On the other hand, $|W(G_i')| - |B(G_i')| \geqq 1$, by the maximality of S. Hence $m = 1$, and G′ is connected.

We show that G has some edges lying on the boundary of H.

Note that every black vertex of G has in G the same valency (the number of adjacent vertices) as in H and that every vertex of valency 2 in H lies on the boundary of H. Let m_i, $i = 2, 3$, denote the number of black vertices of G which have valency i and let r be the number of edges of G. Then $m_2 + m_3 = |S|$, $2m_2 + 3m_3 = r \leqq 3\,|N(S)| = 3\,|S| - 3$, thus $3\,|S| - m_2 = r \leqq 3\,|S| - 3$, implying $m_2 \geqq 3$. This proves this assertion.

Put $W(G') = W(H) - N(S) = T$. Then $B(G') = B(H) - S = N(T)$, i.e., $G' = \langle T \cup N(T) \rangle$, and $|T| = |W(G')| = |B(G')| + 1 = |N(T)| + 1$.

Note the analogy between G, S and G′, T. By the same argument, G′ has also some edges lying on the boundary of H. Thus we can be sure that any closed Jordan curve in the plane separating G and G′ (which are both connected) must traverse the external region of H and, therefore, intersect exactly two edges which lie on the boundary of H.

We denote the set of those edges with one end vertex in G and the other in G′ by (G, G′). It is easy to see that for each edge in (G, G′) the end in G is white, and the one in G′ is black. We connect (G, G′) with a broken line segment (BLS) $L = h_1h_2 \ldots h_{n+1}h_{n+2}$ satisfying

(1) h_ih_{i+1} is orthogonal to one of the three directions of H for $i = 1, 2, \ldots, n + 1$,
(2) each of h_1 and h_{n+2} is the centre of edge lying on the boundary of H,
(3) h_i is the centre of a hexagon of H for $i = 2, \ldots, n + 1$,
(4) every point of L is either an interior or a boundary point of some hexagon of H,
(5) the set of edges of H intersected by L is just the set (G, G′).

Since the vertices in G which are incident with the edges in (G, G′) are all white, the directed angle $\angle h_{i-1}h_ih_{i+1}$ at each turning point $h_i (2 \leqq i \leqq n + 1)$ is $+\pi/3$ or $-\pi/3$. Nevertheless, by the maximality of $S = B(G)$, all angles $\angle h_{i-1}h_ih_{i+1}$ are equal and we may suppose that they all equal to $\pi/3$ (see Fig. 8).

For $n = 0$ or $n = 1$, the BLS L corresponding to G is just a horizontal g-cut segment for some of the six possible positions of H. Now let $\mathbb{C} = L$. Since $|S| = |B(G)| > |N(S)| = |W(G)|$, we deduce that $p(H/U(\mathbb{C})) - v(H/U(\mathbb{C})) > |\mathbb{C}_{12}|$, contradicting (2) in the theorem.

For $n \geqq 2$, we can show that there is a connected subgraph G^* of G such that the corresponding BLS of G^* is just a horizontal g-cut segment for some of the six possible positions.

We denote G^1 the subgraph of G obtained from G by deleting the vertices between line segments h_2h_3 and $h'_2h'_3$ together with their incident edges, where $h'_2(h'_3)$ is the centre of a neighbour of the hexagon with centre $h_2(h_3)$ and lies on $h_1h_2(h_3h_4)$. G^1 is said to be obtained from G by a shift of h_2h_3. One can see that $|B(G^1)| - |W(G^1)| > |B(G)| - |W(G)| = 1$ (see Fig. 8). Repeat shifting h_2h_3 until we first get one or several disjoint connected subgraphs $G_1, \ldots, G_t$ $(t \geqq 1)$ such that the BLS $L^i = h_1^i \ldots h_{n(i)}^i$ corresponding to G_i is a BLS with $n(i) < n$. It is easy to see that $\sum_{i=1}^{t} |B(G_i)| - \sum_{i=1}^{t} |W(G_i)| > |B(G)| - |W(G)| = 1$, and there is at least one $j (1 \leqq j \leqq t)$ such that G_j satisfies $|B(G_j)| > |W(G_j)|$. Let $S_j = B(G_j)$, then $N(S_j) = W(G_j)$, and $|S_j| > |N(S_j)|$. If $n(j) \leqq 1$, let $\mathbb{C}^j$ be the horizontal g-cut corresponding to L^j, and deduce as before that $p(H/U(\mathbb{C}^j)) - v(H/U(\mathbb{C}^j)) > |\mathbb{C}_{12}^j|$, a contradiction. If $n(j) \geqq 2$, then we can repeat the above argument for G_j by induction and eventually get a G^* such that the BLS

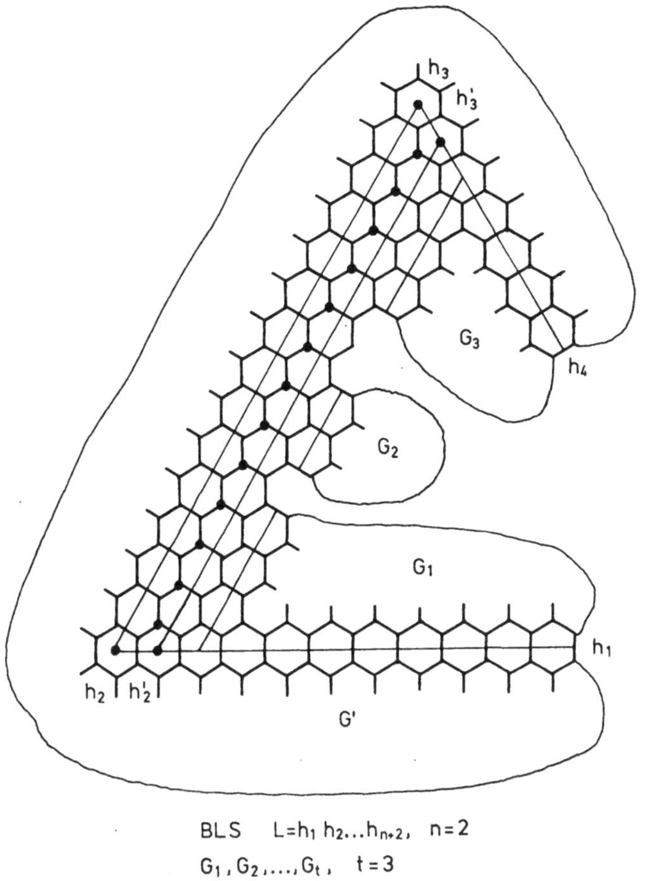

Fig. 8

corresponding to G* is just a horizontal g-cut segment for some of the six positions, and deduce a contradiction.

Now the theorem is proved. ■

From Theorem 6 and Lemma we see that there exist some concealed non-Kekuléan benzenoid systems which satisfy the conditions of the lemma but not (2) of the theorem. We call them the concealed non-Kekuléan benzenoid systems of type I. Such an example is shown in Fig. 6.

It is natural to investigate the smallest concealed non-Kekuléan benzenoid systems of type I. It may enable us to find some simpler necessary and sufficient conditions for benzenoid systems with small number of hexagons to have Kekulé structures.

The smallest concealed non-Kekuléan benzenoid system of type I was first found by Zhang Fuji and Guo Xiaofeng [27] (see Fig. 9).

Theorem 7 [27]. Let H be a smallest concealed non-Kekuléan benzenoid system of type I. Then (i) $h = 14$, (ii) H is unique as shown in Fig. 9.

Proof. Since H is of type I, there is a horizontal g-cut $\mathbb{C} = P_1P_2P_3(P_2 \neq P_3)$ such that $p(H/U(\mathbb{C})) - v(H/U(\mathbb{C})) > |\mathbb{C}_{12}|$.

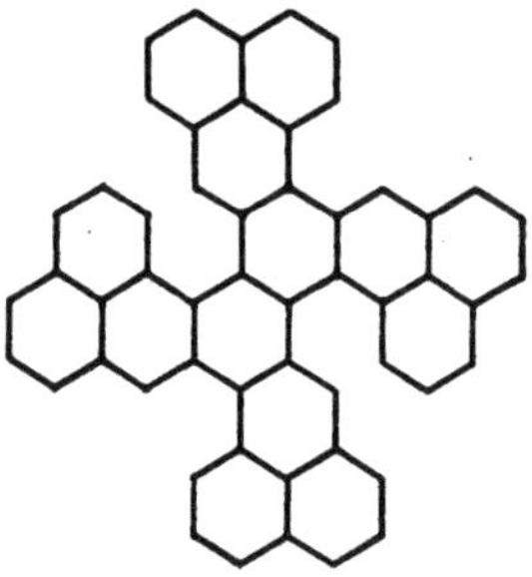

Fig. 9

If $|\mathbb{C}_{12}| = |\mathbb{C}_{23}| = 1$, then $U(\mathbb{C})$ has only one vertex, and $p(H/U(\mathbb{C})) - v(H/U(\mathbb{C})) = 0 < |\mathbb{C}_{12}|$, a contradiction. Hence either $|\mathbb{C}_{12}|$ or $|\mathbb{C}_{23}|$ is greater than one. Suppose that $|\mathbb{C}_{12}| = 2$, $|\mathbb{C}_{23}| = 1$; see Fig. 10(1). Then the hexagon s_1 belongs to H. Otherwise, $p(H/U(\mathbb{C})) - v(H/U(\mathbb{C})) = 1 < |\mathbb{C}_{12}|$, again a contradiction. We assert also that the hexagon s_2 belongs to H. Otherwise, the horizontal cut segment C′ passing through the centre of s_1 would satisfy $p(H/U(\mathbb{C}')) - v(H/U(\mathbb{C}')) = |B(U(\mathbb{C}'))| - |W(U(\mathbb{C}'))| + |\mathbb{C}'| = [|B(U(\mathbb{C}))| - 2] - [|W(U(\mathbb{C}))| - 2] + |\mathbb{C}_{12}| = p(H/U(\mathbb{C})) - v(H/U(\mathbb{C})) > |\mathbb{C}_{12}| = |\mathbb{C}'|$, contradicting that H is of type I; see Fig. 10(2). Moreover, the hexagon s_3 must belong to H. Otherwise there is a cut segment C″ as shown in Fig. 10(3). Put C″ in the horizontal position; then $v(H/U(\mathbb{C}'')) - p(H/U(\mathbb{C}'')) = |B(U(\mathbb{C}''))| - |W(U(\mathbb{C}''))| - |\mathbb{C}''| = [|B(U(\mathbb{C}))| - 1] - [|W(U(\mathbb{C}))| - 3] - |\mathbb{C}_{12}| = |B(U(\mathbb{C}))| - |W(U(\mathbb{C}))| = p(H/U(\mathbb{C})) - v(H/U(\mathbb{C})) - |\mathbb{C}_{12}| > 0$. This contradicts that H is of type I.

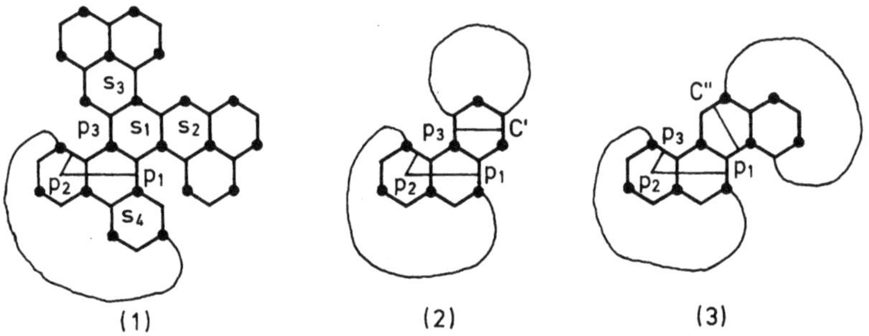

Fig. 10

Now, since $p(H/U(\mathbb{C})) - v(H/U(\mathbb{C})) = v(H/L(\mathbb{C})) - p(H/L(\mathbb{C})) \geqq |\mathbb{C}_{12}| + 1 = 3$, it is easy to see that $U(\mathbb{C})$ contains at least seven hexagons, and if $U(\mathbb{C})$ contains exactly seven hexagons, $U(\mathbb{C})$ can only be as shown in Fig. 10(1). By the same reason, the hexagon s_4 – see Fig. 10(1) – must belong to H, and $L(\mathbb{C})$ and H can only be as shown in Fig. 9. It is easy to verify that H is of type I.

In other cases, $|\mathbb{C}_{12}| + |\mathbb{C}_{23}| \geqq 4$. We prove that $h > 14$.

Let X(Y) be the set of the hexagons in $U(\mathbb{C})$ $(L(\mathbb{C}))$, and let H[X] be the benzenoid system induced by X.

If $|\mathbb{C}_{23}| = 1$, $|\mathbb{C}_{12}| \geqq 3$, then $v(H/L(\mathbb{C})) - p(H/L(\mathbb{C})) = p(H/U(\mathbb{C})) - v(H/U(\mathbb{C})) \geqq |\mathbb{C}_{12}| + 1 \geqq 4$. We can see that $|Y| \geqq 5$. If $h \leqq 14$, then $|X| = h - |Y| - |\mathbb{C}| + 1 \leqq 6$. On the other hand, $p(H[X]) - v(H[X])$ must be greater than one, so $|X| \geqq 6$. Hence we have that $|X| = 6$, $|Y| = 5$, and $|\mathbb{C}_{12}| = 3$. Then H[X] consists of either one triangulene or two phenalenes (see Fig. 11), and H can only be as shown in Fig. 12. But in H there is a horizontal cut $\mathbb{C}'$ such that $p(H/U(\mathbb{C}')) - v(H/U(\mathbb{C}')) > |\mathbb{C}'|$, contradicting that H is of type I.

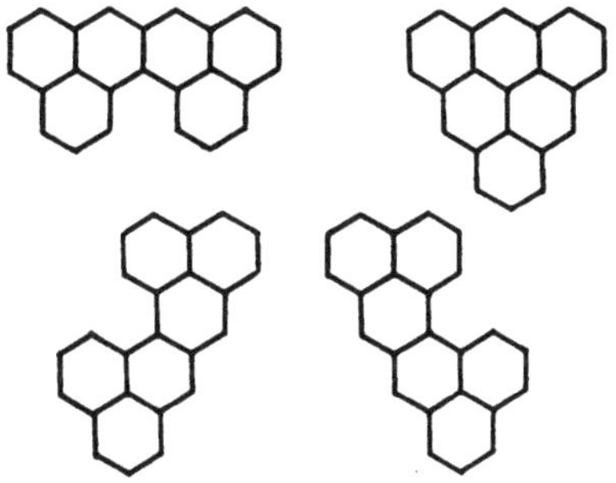

Fig. 11

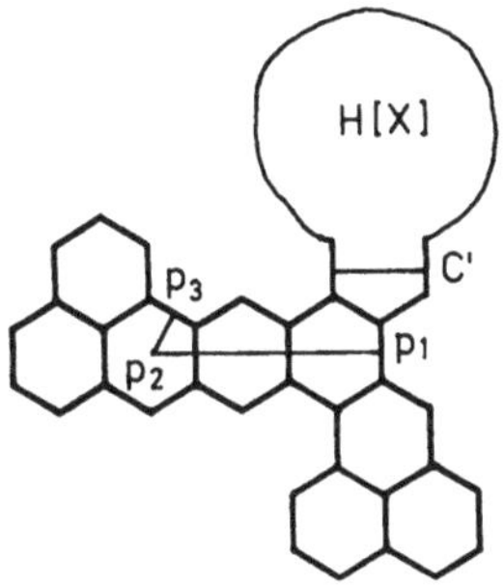

Fig. 12

If $|\mathbb{C}_{12}| \geqq 2$, $|\mathbb{C}_{23}| \geqq 2$, and $h \leqq 14$, it is easy to see that $|Y| \geqq 6$, so that $|X| \leqq 5$. Clearly, this is impossible.

Now it follows that the benzenoid system shown in Fig. 9 is the unique smallest concealed non-Kekuléan benzenoid system of type I. ■

By Theorem 7 for a benzenoid system with $h < 14$, the necessary conditions in Lemma are also sufficient. In fact, we have the stronger result.

Theorem 8 [27]. Let H be a benzenoid system with $h < 14$. Then H has a Kekulé structure if, and only if, for each of its six possible positions and every horizontal cut $\mathbb{C}$,
(i) $p(H) = v(H)$,
(ii) $p(H/U(\mathbb{C})) - v(H/U(\mathbb{C})) \leqq |\mathbb{C}|$.

Proof. The necessity is obvious. We need only to prove the sufficiency. Suppose that conditions (i) and (ii) are satisfied, but H has no Kekulé structure.

Since $h < 14$, by Theorem 7, H is not of type I. So there is a horizontal cut $\mathbb{C}$ in H which does not satisfy the conditions in Lemma. But H satisfies the conditions (i) and (ii) by our assuption. Hence $p(H/U(\mathbb{C})) - v(H/U(\mathbb{C})) < 0$.

Let X(Y) be the set of the hexagons in $U(\mathbb{C})$ $(L(\mathbb{C}))$, and let H[X] be the benzenoid system induced by X.

If $|\mathbb{C}| = 2$, then $v(U(\mathbb{C})) - p(U(\mathbb{C})) \geqq 1$, and $v(H[X]) - p(H[X]) \geqq 2$. So $|X| \geqq 6$. Similarly, $|Y| \geqq 6$. From $h \leqq 13$, we have that $|X| = |Y| = 6$. Thus $v(H[X]) - p(H[X]) = 2, v(H/U(\mathbb{C})) - p(H/U(\mathbb{C})) = 1$, and $U(\mathbb{C})$ can only be one of the benzenoid systems as shown in Fig. 13. Put the cut segment C′ in horizontal position, then $p(H/U(\mathbb{C}')) - v(H/U(\mathbb{C}')) = 3 > |\mathbb{C}'|$. This contradicts our assumption.

If $|\mathbb{C}| > 2$, then either $|X|$ or $|Y|$, say $|X|$, is less than six. So $v(H[X]) - p(H[X]) \leqq 1$, and $p(H/U(\mathbb{C})) - v(H/U(\mathbb{C})) \geqq 0$. This is also a contradiction.

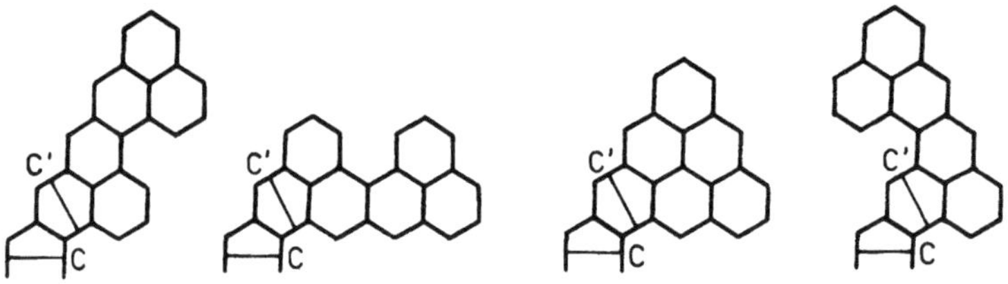

Fig. 13

The theorem is now proved. ■

For the benzenoid systems with $h < 11$, the conditions of Theorem 8 can be further simplified. As stated above, a smallest concealed non-Kekuléan benzenoid system has eleven hexagons (see Fig. 3). It implies that, for benzenoid systems with $h < 11$, (i) of Theorem 8 is also sufficient. By Theorem 8 we can give it a simple proof.

Theorem 9. Let H be a benzenoid system with $h < 11$. Then H has a Kekulé structure if, and only if $p(H) = v(H)$.

Proof. The necessary is obvious. Suppose therefore that $p(H) = v(H)$, but H has no Kekulé structure.

By Theorem 8, there is a horizontal cut $\mathbb{C}$ in H such that $p(H/U(\mathbb{C})) - v(H/U(\mathbb{C})) \geqq |\mathbb{C}| + 1 \geqq 3$. Since $h < 11$, one of $U(\mathbb{C})$ and $L\mathbb{C})$, say $U(\mathbb{C})$, contains at most 4 hexagons. Then we have that $p(H/U(\mathbb{C})) - v(H/U(\mathbb{C})) \leqq 2$, a contradiction. ■

By computer-aided generation the authors of Ref. [16] found that there are exactly eight smallest concealed non-Kekuléan benzenoid systems. In addition, He Wenchen et al. [28] asserted that there are exactly ninety-eight concealed non-Kekuléan benzenoid systems with $h = 12$.

Using Theorem 8, we can give a construction method for concealed non-Kekuléan benzenoid systems with $h < 14$, which does not depend on computer-aided generation. By this construction method, we proved the above results from References [16 and 28]. Furthermore, we have proved that there are exactly 1097 concealed non-Kekuléan benzenoid systems with $h = 13$ [27, 29].

5 Acknowledgements

We would like to thank Prof. I. Gutman and S. J. Cyvin for their manuscript "Kekuléan and Non-Kekuléan Benzenoid Hydrocarbons" prior to publication [30].

We would also like to acknowledge support from the NNSFC.

6 References

1. Clar E, Stewart DG (1953) J. Amer. Chem. Soc. 75: 2667
2. Clar E, Kemp W, Stewart DG (1958) Tetrahedron 3: 325
3. Hall P (1935) J. London Math. Soc. 10: 26
4. Gordon M, Davison WHT (1952) J. Chem. Phys. 20: 428
5. Sachs H (1984) Combinatorica 4: 89

6. Dewar MJS, Longuet-Higgins HC (1952) Proc. Roy. Soc. A214: 482
7. Gutman I (1982) Bull. Soc. Chim. Beograd 47: 453
8. Gutman I (1983) Croat. Chem. Acta 56: 365
9. Trinajstić N (1983) Chemical graph theory, CRC Press, Boca Raton, FL, vol 1, p 27
10. Guo XF, Zhang FJ: J. Math. Chem. (submitted)
11. John P, Sachs H (1985) In: Bodendiek R, Schumacher H, Walter G (eds) Graphen in Forschung und Unterricht, Verlag Barbara Franzbecker, Bad Salzdetfurth, p 85
12. John P, Rempel J (1985) Proc. Internat. Conf. Graph Theory, Eyba, Oct. 1984, Teubner, Leipzig, p 72 (Teubner – Texte zu Mathematik, Band 73)
13. Gutman I, Su LX, Cyvin SJ (1987) J. Serb. Chem. Soc. 52: 263
14. Cyvin SJ, Gutman I (1988) J. Mol. Struct. (Theochem) 164: 183
15. Gutman I, Cyvin SJ (1987) Chem. Phys. Letters 136: 137
16. Brunvoll J, Cyvin SJ, Cyvin BN, Gutman I, He WJ, He WC (1987) Match 22: 105
17. Gutman I (1974) Croat. Chem. Acta 46: 209
18. Cyvin SJ, Gutman I (1988) Kekulé structures in benzenoid hydrocarbons, Springer, Berlin Heidelberg New York
19. Dias JR (1985) Nouv. J. Chim. 9: 125
20. Dias JR (1986) J. Mol. Struct. (Theochem) 149: 213
21. Dias JR (1987) Handbook of polycyclic hydrocarbons, Part A: Benzenoid hydrocarbons, Elsevier, Amsterdam, Chapt 2
22. Hosoya H (1986) Croat. Chem. Acta 59: 583
23. Cyvin DJ, Gutman I (1987) J. Mol. Struct. (Theochem) 150: 157
24. Zhang FJ, Chen RS, Guo XF (1985) Graphs and combinatorics 1: 383
25. Kostochka AV (1985) Proc. 30 Internat. Wiss. Koll. TH Ilmenau 1985, Vortragsreihe F, p 49
26. Zhang FJ, Chen RS (1987) Nature J. (in Chinese) 10: 163; Acta Math. Appl. Sinica (1989) 1: 1 (Engl. Ser.)
27. Zhang FJ, Guo XF (1988) Match 23: 229
28. He WC, He WJ, Cyvin BN, Cyvin SJ, Brunvoll J (1988) Match 23: 201
29. Guo XF, Zhang FJ: Match 24: 85
30. Gutman I, Cyvin SJ (1988) J. Serb. Chem. Soc. 53: 391

Peak-Valley Path Method on Benzenoid and Coronoid Systems

He Wenchen[1] and He Wenjie[2]

[1] Hebei Chemical Engineering Institute, Shijiazhuang, The People's Republic of China
[2] Hebei Academy of Sciences, Sciences, Shijiazhiang, The People's Republic of China

Table of Contents

The present paper is a review of the P–V path method. The concept of the P–V path was proposed by Gordon and Davison in 1952. In the last few years this method has been greatly developed and has become one of the important approaches for investigating Kekulé structures of benzenoid hydrocarbons. The superiority of this method is its simplicity and visualization. According to the properties of the P–V path, some algorithmic approaches have been developed for deciding whether a benzenoid or a coronoid hydrocarbon is Kekuléan. In 1985, John and Sachs introduced the concept of the P–V matrix and deduced the John–Sachs theorem which states that the absolute value of the determinant of the P–V matrix of a benzenoid or a coronoid hydrocarbon G is equal to the number of Kekulé structures of G.

1 Introduction

This article is a review of the P–V path method, which is used for investigating Kekulé structures of benzenoid and coronoid (single and multiple) hydrocarbons and has been developed in the last few years.

In the present review we follow the definitions, symbols, notation and terminology of a recent book [1]. Benzenoid hydrocarbons are represented by benzenoid systems.

Only Kekuléan benzenoid and coronoid hydrocarbons are known to exist. Non-Kekuléan benzenoid and coronoid systems have never been synthesized [2–6], they should be polyradicals and have very low chemical stability.

In the classical work of Gordon and Davison [7], the concept of peak-valley path was proposed for the first time. In the last few years, the peak-valley path method (P–V path method) has been developed greatly [8–12] and has become one of the most important approaches used for investigating Kekulé structures of benzenoid and coronoid hydrocarbons.

Consider a benzenoid or a coronoid system G drawn so that some of its edges are vertical. A peak is defined as a vertex lying above all its first neighbours and a valley is a vertex lying below all its first neighbours. A monotonic path (peak-valley path or P–V path) in G is a path connecting a peak and a valley, such that starting at the peak one always goes downwards. Two paths are independent when they have no vertices in common (see Fig. 1). The vertices of G are coloured by two colours (say black and white) so that adjacent vertices never have the same colour, and the peaks and valeys are coloured white and black respectively. We denote the numbers of peaks and valleys by $n_\wedge$ and $n_\vee$; the numbers of white vertices and black vertices by n_w and n_b respectively.

Theorem 1. $\Delta = |n_\wedge - n_\vee| = |n_w - n_b|$ (Ref. [11]) (1)

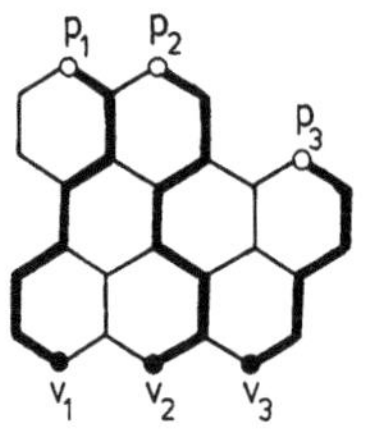

Fig. 1. Peaks and valleys

Gordon and Davison [7] observed the following result.

In every Kekulé structure of a benzenoid system there is a unique monotonic alternating path (alternating P–V path or conjugated P–V path [11, 12] connecting a peak with a valley and starting with a double bond. This is true for all peaks and all valleys, and the pertinent alternating P–V paths are mutually independent. The set of the mutually independent alternating P–V paths is called an alternating P–V path system.

The following results are true for coronoid systems as well as benzenoids.

Theorem 2. There is a one-to-one correspondence between an alternating P–V path system and a Kekulé structure [7, 8, 13, 14].

Theorem 3. For a Kekuléan benzenoid or coronoid system the number of peaks is equal to the number of valleys [11].

$\Delta = 0$ is a necessary but not sufficient condition for Kekuléan benzenoids and coronoids. The systems with $\Delta \neq 0$ are obvious non-Kekuléan systems and those non-Kekuléan systems with $\Delta = 0$ are called concealed non-Kekuléan systems. (see Fig. 2)

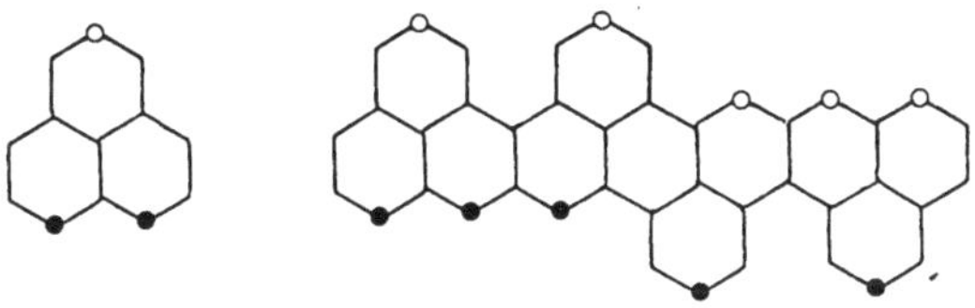

Fig. 2. Obvious and concealed non-Kekuléan systems

2 Algorithmic Approaches for Deciding Whether a Benzenoid or Coronoid Hydrocarbon is Kekuléan or Non-Kekuléan

(1) A simple method for recognizing an obvious non-Kekuléan system is by counting its peaks and valleys.

(2) *The Peeling Algorithm*

Gutman and Cyvin proposed and algorithm [13] based on Theorem 2. We illustrate the peeling algorithm on the example shown in Fig. 3. Every step of the algorithm is to delete the extreme left P–V path along the perimeter of the system. They thought originally that the fragment obtained in this way would be Kekuléan if and only if the initial benzenoid system was Kekuléan.

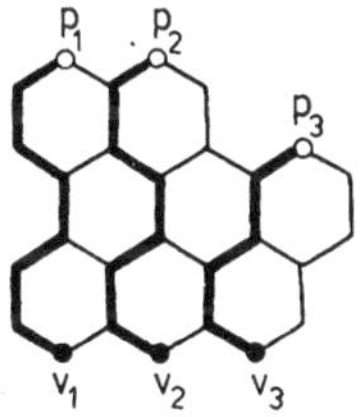

Fig. 3. Illustration of the peeling algorithm

But it needs not always be so. A counterexample of the peeling algorithm is shown in Fig. 4 [15, 16]. For the Kekuléan system in the figure, the P–V path along the extreme left perimeter doesn't belong to any alternating P–V path system of the Kekuléan hydrocarbon.

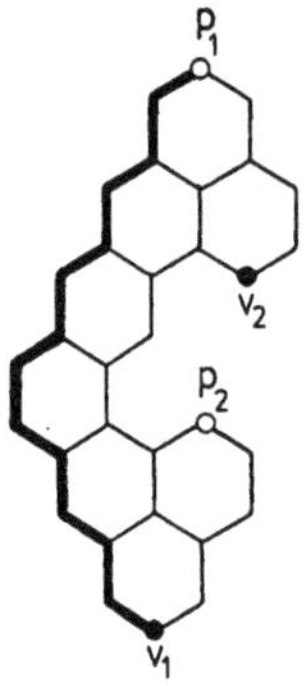

Fig. 4. Failure of the peeling algorithm

(3) *Sachs' Algorithm*

Using the P–V path method, Sachs [8] obtained a number of topological properties for a benzenoid system. He proposed an algorithm, based on the topological properties, for recognizing the existence of Kekulé structures.

A generalized benzenoid system B is defined as a system which has the following properties:

(i) Every vertex and every edge of B are on the regular hexagonal lattice;

(ii) every finite region of B is a congruent regular hexagon with edge length 1. The system contains no "holes" with a size of at least two hexagons.

Consider a general benzenoid system B whose vertices are colored with two colors (say black and white) in the usual manner so that adjacent vertices never have the same colour, and the peaks and the valleys are coloured white and black respectively.

An edge-cut of B is an edge set $\{e_1, e_2, \ldots, e_t\}$, such that:

(i) by deleting $e_1, e_2, \ldots, e_t$ from B, it decomposes exactly into two parts B′ and B″;

(ii) for each edge e_i ($i = 1, 2, \ldots, t$), its black end vertex belongs to B′ (while its white end vertex belongs to B″);

(iii) e_1 and e_t belong to the perimeter.

We denote the number of peaks and valleys in B′ by $n'_\wedge$ and $n'_\vee$ respectively. Similary, $n''_\wedge$ and $n''_\wedge$ are the numbers of peaks and valleys in B″ respectively. Clearly, $n'_\wedge + n''_\wedge = n_\wedge$ and $n'_\vee + n''_\vee = n_\vee$.

Let

$$s' = n'_\wedge - n'_\vee; \qquad s'' = n''_\vee - n''_\wedge \tag{2}$$

In the case of $\Delta = 0$, for any edge-cut, $s' = s''$.

An elementary edge-cut is defined so that in addition to the preceding conditions (i)–(iii) we have the fourth requirement:

(iv) all the edges e_i ($i = 1, 2, \ldots, t$) are mutually parallel.

For convenience, we perform the elementary edge-cuts so that the edges e_i ($i = 1, 2, \ldots, t$) are vertical. Then B′ lies above the edge-cut and is called the upper bank of the edge-cut, and B″ is called the lower bank.

A broken line segment through all the centres of the cut edges of an edge-cut is called a cut segment. For an elementary edge-cut, the cut segment is horizontal. A cut segment $C = a_1a_2a_3$ (possibly, $a_2 = a_3$) is called a horizontal g-cut segment if

a_1a_2 is horizontal and the angle $a_1a_2a_3$ is $\pi/3$; see Refs. [17, 18]. Then the edge-cut is called a horizontal g-cut. For a horizontal g-cut segment $C = a_1a_2a_3$, let t_{12} and t_{23} denote the numbers of edges of B intersected by straight line segments a_1a_2 and a_2a_3, respectively. For the horizontal g-cut, B′ lies above the horizontal segment a_1a_2 and is called the upper bank of the cut, while B″ is the lower bank. (see Fig. 5)

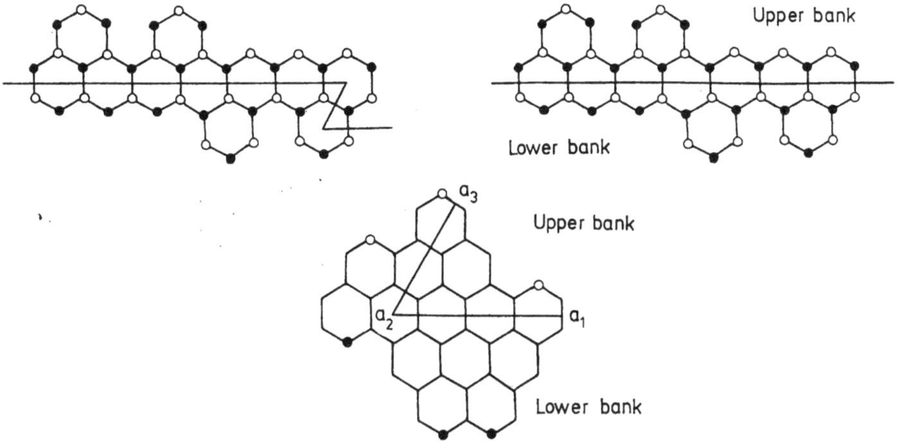

Fig. 5. Edge-cut, elementary edge-cut and horizontal g-cut

Sachs proved the following theorem [8].

Theorem 4. In a connected Kelulėan benzenoid system B for every elementary edge-cut

$$0 \leqq s' = s'' \leqq t \tag{3}$$

He conjectured that conversly, if (3) holds for all edge-cuts (and for all positions), then B is Kekulėan [8].

But this conjecture is not correct [17, 18]. A counterexample is shown in Fig. 6.

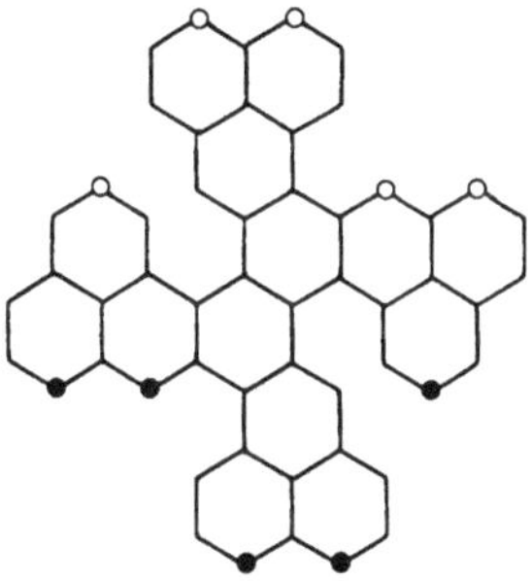

Fig. 6. A counterexample of Sachs' conjecture

Zhang and Guo [17] proved the following theorem.

Theorem 5. A benzenoid system B with $\Delta = 0$ is Kekuléan, if, and only if for each of six positions and for every horizontal g-cut of B, $s' \leqq t_{12}$.

Now let us give the definition of free edge.

An edge of a benzenoid system, connecting vertices of degree two is called a free edge. They, of course, belong to the perimeter.

Sachs proved that a benzenoid system contains at least six free edges [8, 19]. In fact, in each of the three edge directions there are at least two free edges [19].

Sachs [8] proposed the other two theorems:

Theorem 6. Let $\{e_1, e_2, \ldots, e_t\}$ be an edge-cut of a uléan benzenoid system B. For any Kekulé structure of B, the number of the double bond edges in the edge-cut is equal to $t - s$, where $s = s' = s''$.

Hence if $s = 0$, then all the edges e_i ($i = 1, 2, \ldots, t$) are fixed double bond edges; if $s = t$, then all the edges e_i are fixed single bond edges.

Theorem 7. Let B be a Kekuléan benzenoid system, an edge-cut is made through one of its free edges, e. If $t - s > 0$, then there exists a Kekulé structure of B in which e is a double bond edge.

Based on the above two theorems Sachs proposed the following algorithm for recognizing Kekulé structures:

Make an elementary edge-cut through a vertical free edge e of B.

(a) If $\Delta \neq 0$, then B is obvious non-Kekuléan.

For the case of $\Delta = 0$,

(b) If $s < 0$ or $s > t$, then B is concealed non-Kekuléan;

(c) if $0 < s < t$, then delete the free edge e together with its two vertices;

(d) if $s = t$, then delete all the edges of the edge-cut;

(e) if $s = 0$, then delete all the edges of the edge-cut together with their ends.

In all the cases (b)–(e), B is Kekuléan, if, and only if the obtained new system $\bar{B}$ (probably disconnected) is Kekuléan.

In a benzenoid systems, an elementary edge-cut $\{e_1, e_2, \ldots, e_t\}$ contains two external edges e_1 and e_t. If e_1 is a free edge, and at least one of the two vertices of e_t is of degree two, then the two ends u, v of e_1 form a convex pair, denoted by (u, v) [19].

Based on Sachs' work, Sheng [19] obtained the following two theorems.

Theorems 8. If B is a Kekuléan benzenoid system, (u, v) is a convex pair of B, and e is the free edge whose ends are u and v, then B possesses a Kekulé structure in which e is a double bond.

Theorem 9. A benzenoid system possesses at least two convex edges respectively in each of three edge directions.

From the properties of P–V paths, Theorem 8 is easily proved.

According to Theorem 8, B is Kekuléan if, and only if after deleting a convex pair, the remaining system $\bar{B}$ is Kekuléan.

Sachs' algorithm can be used on coronoid systems, but the deletion method of convex pair can not. An example is shown in Fig. 7. (u, v) is a convex pair in the coronoid system, but in any Kekulé structure of the system, the edge uv is not a double bond edge.

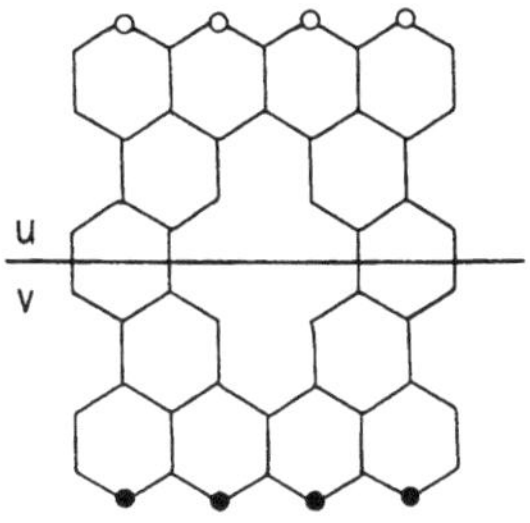

Fig. 7. A coronoid system

3 P–V Path Network Flow Method

In 1987, He and He used the network method to investigating P–V path structures of benzenoid and coronoid system [11].

Consider a benzenoid or a coronoid system G. A network N is constructed from G as follows [11, 20, 21].

(i) Direct all the edges of G downwards (vertically or diagonally);
(ii) add a source A and a directed edge from A to each peak of G;
(iii) add a sink B and a directed edge from each valley of G to B.

Denote the flow value along the directed edge qr (from q to r) by $f(q, r)$.

$$f(q, r) = 0 \text{ or } 1 . \tag{4}$$

The capacity $c(q, r)$ of edge qr is the maximum possible flow value along the edge. If the direction from q to r is downwards, then $c(q, r) = 1$, otherwise $c(q, r) = 0$. We always have $f(q, r) \leqq c(q, r)$.

For all vertices other than A and B, we have the conservation of flow

$$\sum_r f(q, r) = \sum_r f(r, q) . \tag{5}$$

The value of the net flow of a network N is defined as the total flow leaving the source A (and arriving at the sink B) and is denoted by $F(N)$.

$$F(N) = \sum_p f(A, p) = \sum_v f(v, B) . \tag{6}$$

We immediately have

$$F(N) \leqq \min(n_\wedge(G), n_\vee(G)) . \tag{7}$$

Consider a Kekuléan system G shown in Fig. 8. All the edges leaving A have flow value 1 so that

$$F(N) = \max F(N) = n_\wedge(G) = n_\vee(G) .$$

As seen in the figure, the flow is along all the disjoint alternating P–V paths (an alternating P–V path system).

Theorem 10. [11, 12]. A benzenoid or a coronoid system G is Keluléan if, and only if

$$\max F(\mathrm{N}) = n_{\wedge}(\mathrm{G}) = n_{\vee}(\mathrm{G}) . \tag{8}$$

According to the algorithm for maximizing the flow in a network called the labelling method or augmenting method, which is proposed by Ford and Fulkerson [22–24] in 1957, revised by Edmonds and Karp [25] in 1972, and improved by Karzanov [26] and Malhotra et al. [27] in 1978, we can calculate max F(N) by computer.

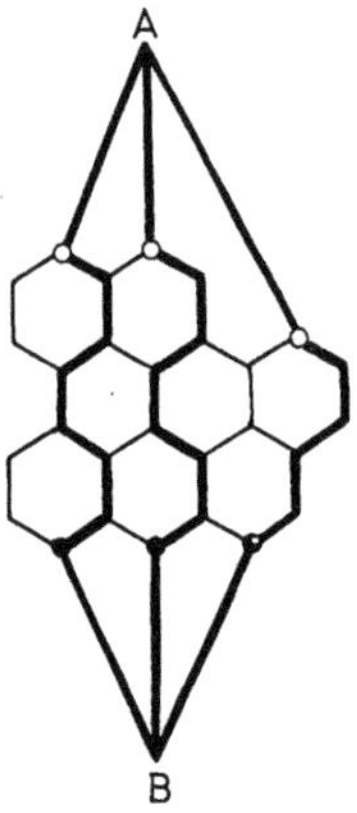

Fig. 8. Maximum net flow

With the augmentation of the net flow, the disjoint P–V paths are determined step by step. And, finally, by (8), we can decide whether G is Kekuléan or not.

The method is very simple. Starting with the zero flow (i.e. all edges in N have a flow value 0), in each step, choose a flow augmenting path (a path from A to B in which the flow of each forward-edge (downward-edge)/(upward-edge) is 0 and the flow of each reverse-edge is 1), and change the flow of each forward-edge of the path from 0 to 1, and the flow of each reverse-edge of the path from 1 to 0. Thus, we can augment the net flow step by step until there is no augmenting paths in G. Finally, the net flow reaches its maximum value max F(N).

An example is shown in Fig. 9. As seen in the figure, using this method we can overcome the difficulty of the peeling method.

Now we give the definition of edge-cut of a network N as follows [11].

An edge-cut of N is an edge set $\{e_1, e_2, \ldots, e_t\}$, such that:

(a) by deleting $e_1, e_2, \ldots, e_t$ from N, it decomposes exactly into two parts N′ and N″;

(b) for each cut edge e_i ($i = 1, 2, \ldots, t$), one of its end vertices belongs to N′, and the other belongs to N″.

(c) A belongs to N′, and B belongs to N″.

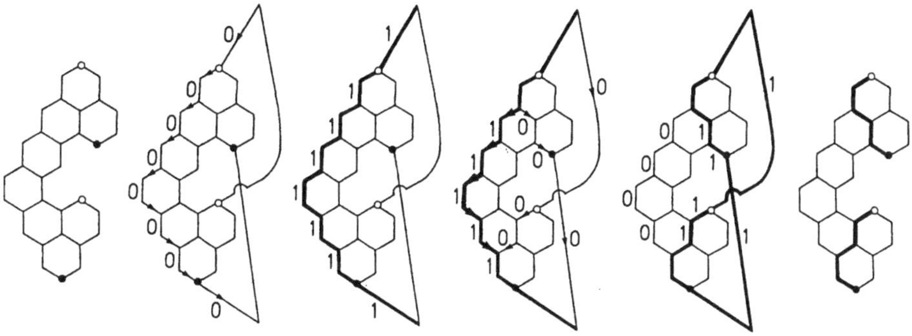

Fig. 9. Flow augmenting method

The capacity of an edge-cut of N denoted by Cap(N′, N″), is defined to be the sum of the capacities of the cut edges incident from vertices in N′ and incident to vertices in N″.

$$\mathrm{Cap}(\mathrm{N}', \mathrm{N}'') = \sum_{\mathrm{q} \in \mathrm{N}', \mathrm{r} \in \mathrm{N}''} c(\mathrm{q}, \mathrm{r}) . \tag{9}$$

Denote the number of the edges in the edge-cut by t. We have

$$\min \mathrm{Cap}(\mathrm{N}', \mathrm{N}'') \leqq \mathrm{Cap}(\mathrm{N}', \mathrm{N}'') \leqq t \tag{10}$$

Among all the edge-cuts of N, the one having minimum capacity min Cap(N′, N″) is called the minimum cut.

According to the max-flow min-cut theorem [20–22],

$$\max F(\mathrm{N}) = \min \mathrm{Cap}(\mathrm{N}', \mathrm{N}'') \tag{11}$$

Obviously,

$$\max F(\mathrm{N}) \leqq \min(n_\wedge(\mathrm{G}), n_\vee(\mathrm{G})) . \tag{12}$$

So

$$0 \leqq \min \mathrm{Cap}(\mathrm{N}', \mathrm{N}'') < \min(n_\wedge(\mathrm{G}), n_\vee(\mathrm{G})) . \tag{13}$$

We immediately have the following theorem:

Theorem 11 [11]. A benzenoid or coronoid system G is Kekuléan if and only if

$$\min \mathrm{Cap}(\mathrm{N}', \mathrm{N}'') = n_\wedge(\mathrm{G}) = n_\vee(\mathrm{G}) \tag{14}$$

For the example shown in Fig. 10, min Cap(N′, N″) = 2 < $n_\wedge$(G) = 3. So the system is non-Kekuléan.

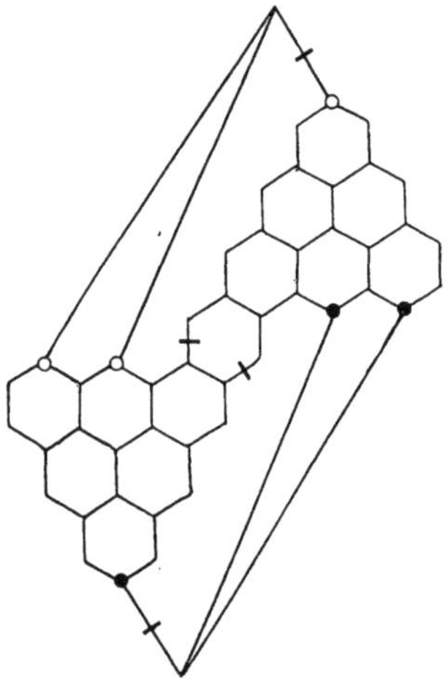

Fig. 10. A non-Kekuléan system

Consider a Kekuléan benzenoid G whose edges are directed downwards (vertically or diagonally). A generalized edge-cut $\{e_1, e_2, \ldots, e_t\}$ of G is defined such that:

(a) by deleting all the cut edges $e_1, e_2, \ldots, e_t$ from G, it decomposes exactly into two segments G′ and G″.

(b) for each cut edge $e_i = q_i r_i$ ($i = 1, 2, \ldots, t$), one of its end vertex belongs to G′, and the other belongs to G″.

$$\mathrm{Cap}(G', G'') = \sum_{q \in G', r \in G''} c(q, r) \leqq t . \tag{15}$$

Theorem 12. For an arbitrary generalized edge-cut of a Kekuléan benzenoid (or a coronoid) G, $s' = s'' \leqq t$

Proof: For any vertex q ∈ G′ other then $n'_\wedge$ peaks and $n'_\vee$ valleys in G′:

$$0 = \sum_r f(q, r) - \sum_r f(r, q) .$$

G is Kekuléan. So for all the peaks p ∈ G′ and valleys v ∈ G′,

$$n'_\wedge - n'_\vee = \sum_{p,r} f(p, r) - \sum_{r,v} f(r, v) .$$

Summing these equations over all q ∈ G′, including the peaks and valleys, we obtain:

$$n'_\wedge - n'_\vee = \sum_{q \in G'} \left(\sum_r f(q, r) - \sum_r f(r, q) \right)$$

Now:

$$\sum_{q\in G'}\sum_{r} f(q, r) = \sum_{q\in G', r\in G'} f(q, r) + \sum_{q\in G', r\in G''} f(q, r)$$

and

$$\sum_{q\in G'}\sum_{r} f(r, q) = \sum_{q\in G', r\in G'} f(r, q) + \sum_{q\in G', r\in G''} f(r, q)\,.$$

Clearly,

$$\sum_{q\in G', r\in G'} f(r, q) \text{ is the same as } \sum_{q\in G', r\in G'} f(q, r) \text{ and so}$$

$$\begin{aligned} n'_{\wedge} - n'_{\vee} &= \sum_{q\in G', r\in G''} f(q, r) - \sum_{q\in G', r\in G''} f(r, q) \\ &= \sum_{q\in G', r\in G''} f(q, r) - \sum_{q\in G'', r\in G'} f(q, r) \\ &\leqq \sum_{q\in G', r\in G''} c(q, r) - \sum_{q\in G'', r\in G'} f(q, r) \\ &= \mathrm{Cap}(G', G'') - \sum_{q\in G'', r\in G'} f(q, r) \\ &\leqq \mathrm{Cap}(G', G'') \leqq t \end{aligned}$$

Q.E.D.

Theorem 4 is the special case of this theorem.

Now, we give the following theorem which can be derived from Theorems 5 and 11.

Theorem 13. A benzenoid system G with $\Delta = 0$ is Kekuléan if, and only if $s' = n'_{\wedge} - n'_{\vee} \leqq \mathrm{Cap}(N', N'')$ holds for every generalized edge-cut and for each position of G.

Proof: Now, we only need to prove the sufficiency of the conditions in this theorem.

If for every position and for every generalized edge-cut of G, $s' \leqq \mathrm{Cap}(N', N'')$ holds, but G is a concealed non-Kekuléan system, then by using Theorem 4, we may find a horizontal g-cut, for which $s' > t_{12} = \mathrm{Cap}_g(N', N'')$. It is a contradiction. Q.E.D.

4 P–V Matrix Method

The P–V matrix method is an algebraic (or a combinatorial) method used to determine the number of Kekulé structures, K.

Theorem 14. For a benzenoid or a coronoid G,

$$K(G) = |\det \mathbf{A}(G)|^{1/2}\,, \tag{16}$$

where $\mathbf{A}(G)$ is the adjacent matrix of the vertices of G.

By theorem 14, the following known result [28] is obtained immediately.

Theorem 15. For a benzenoid or a coronoid G with $\Delta = 0$,

$$K(G) = |\det \mathbf{M}(G)|\,, \tag{17}$$

where $\mathbf{M}(G)$ is an $h \times h$ matrix ($h = n_w = n_b$) with the elements:

$$m_{ij} = \begin{cases} 1 & \text{(if white vertex } w_i \text{ is adjacent to black vertex } b_j) \\ 0 & \text{(if } w_i \text{ is not adjacent to } b_j). \end{cases}$$

John, Sachs and Rempel observed that by using the P–V matrix of G, the above results can be simplified substantially [9, 10].

Consider a benzenoid or a coronoid system G. Denote by w_{ij} the number of P–V paths which start at the i-th peak p_i and end at the j-th valley v_j. These numbers define a matrix $\mathbf{W}$ called the P–V matrix of G. If $\Delta = 0$, then $\mathbf{W}$ is a $g \times g$ ($g = n_\wedge = n_\vee$) square matrix.

John, Sachs and Rempel gave the following theorem [9, 10]:

Theorem 16. For a benzenoid or a coronoid G with $\Delta = 0$,

$$K(G) = |\det \mathbf{W}(G)|\,. \tag{18}$$

The matrix elements w_{ij} can very easily be determined either by computers or by hands, with the following method [12].

Let the valleys have the values:

$$v_s = \begin{cases} 1 & (s = j) \\ 0 & (s \neq j) \end{cases} \tag{19}$$

and every other vertex have a value equal to the sum of the values of the vertices which are below and adjacent to it. The obtained peak values are merely the value w_{ij} of the elements in the j-th column of $\mathbf{W}(G)$.

For the example shown in Fig. 11,

$$\mathbf{W}(G) = \begin{bmatrix} 16 & 1 & 6 \\ 2 & 2 & 1 \\ 1 & 0 & 2 \end{bmatrix}$$

Below we outline the proof of Theorem 16. This proof was given by He and He [12].

Lemma 1. For a benzenoid or a coronoid G with $\Delta = 0$,

$$|\det \mathbf{W}(G)| = |\det \mathbf{W}'(G)|\,, \tag{20}$$

where $\mathbf{W}'(G)$ is a $g \times g$ matrix with elements

$$w'_{ij} = (-1)^{I_{ij}} \times w_{ij}\,, \qquad (i = 1, 2, \ldots, g;\ \ j = 1, 2, \ldots, g)\,, \tag{21}$$

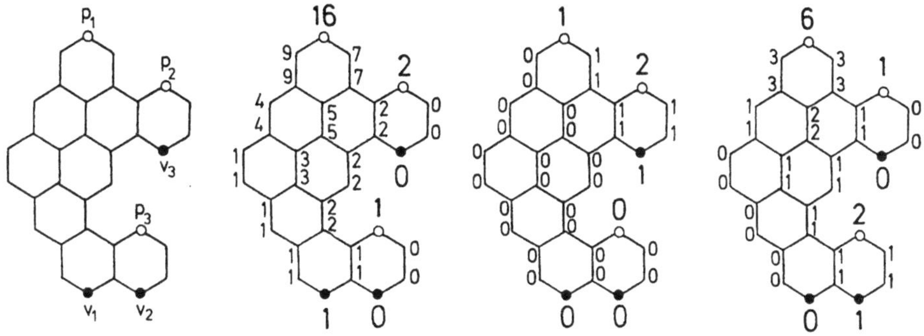

Fig. 11. Determination of w_{ij}

where w_{ij}'s are the P–V matrix elements and I_{ij} is the number of the diagonal edges in the P–V path issuing from the i-th peak p_i and terminating in the j-th valley v_j (in the case of $w_{ij} = 0$, I_{ij} is arbitrary).

Proof: The length difference of two P–V paths p_iv_j and $p_{i'}v_j$ with a common end v_j does not depend on j ($j = 1, 2, \ldots, g$). Neither does the value $I_{ij} - I_{i'j}$: the corresponding elements in any two rows (say the i-th and the i'-th rows) of $\mathbf{W}'(G)$ have the same sign (if $(-1)^{I_{ij}-I_{i'j}} = 1$) or the opposite sign (if $(-1)^{I_{ij}-I_{i'j}} = -1$).

Thus, we immediately obtain that (20) holds.

The proof of Theorem 16 is as follows.

For convenience, we will adopt the following conventions.

(i) The peaks as well as valleys are labelled $1, 2, \ldots, g$ and the other white (black) vertices are labelled $g + 1, g + 2, \ldots, h$.

(ii) The labels $g + 1, g + 2, \ldots, h$ of the white vertices are given by sweeping from the left to the right and from top to bottom. Every black vertex which is not a valley is given the same label as the white vertex immediately beneath it.

Thus, the $h \times h$ matrix $\mathbf{M}(G)$ has the following form:

$$\mathbf{M}(G) = \begin{bmatrix} m_{11} & m_{12} & \cdots & m_{1g} & m_{1,g+1} & m_{1,g+2} & m_{1,g+3} & \cdots & m_{1h} \\ m_{21} & m_{22} & \cdots & m_{2g} & m_{2,g+1} & m_{2,g+2} & m_{2,g+3} & \cdots & m_{2h} \\ \cdots & \cdots & \cdots & \cdots & \cdots & \cdots & \cdots & \cdots & \cdots \\ m_{g1} & m_{g2} & \cdots & m_{gg} & m_{g,g+1} & m_{g,g+2} & m_{g,g+3} & \cdots & m_{gh} \\ m_{g+1,1} & m_{g+1,2} & \cdots & m_{g+1,g} & 1 & m_{g+1,g+2} & m_{g+1,g+3} & \cdots & m_{g+1,h} \\ m_{g+2,1} & m_{g+2,2} & \cdots & m_{g+2,g} & & 1 & m_{g+2,g+3} & \cdots & m_{g+2,h} \\ \cdots & \cdots & \cdots & \cdots & & & 1 & \cdots & m_{g+3,h} \\ \cdots & \cdots & \cdots & \cdots & & 0 & & \cdots & \\ m_{h1} & m_{h2} & \cdots & m_{hg} & & & & & 1 \end{bmatrix}$$

Now let us transform the determinant $\det \mathbf{M}(G)$. To begin with, consider the h-th column (supposing $h \geqq g + 1$). In this column, there are other nonzero elements than m_{hh}, say, m_{ph} and m_{qh} (i.e. $m_{ph} = m_{qh} = 1$). By subtracting the corresponding element values of the h-th row from those of the p-th row and from those of the q-th row, we can make all the elements of the h-th column, except the diagonal element m_{hh}, become zero. Using the same method, we can then make all the elements of the $(h-1)$-th column except $m_{h-1,h-1}$ transform into zero. Continuing with such transformations will finally transform all the elements m_{ij} $(i = 1, 2, \ldots, h;\ j = g+1, g+2, \ldots, h)$ except for m_{jj} $(j \geqq g+1)$ into zero. Thus,

$$\det \mathbf{M}(G) = \begin{vmatrix} m'_{11} & m'_{12} & \cdots & m'_{1g} & 0 & 0 & \cdots & \cdots & 0 \\ m'_{21} & m'_{22} & \cdots & m'_{2g} & 0 & 0 & \cdots & \cdots & 0 \\ \cdots & \cdots & \cdots & \cdots & \cdots & \cdots & \cdots & \cdots & \cdots \\ m'_{g1} & m'_{g2} & \cdots & m'_{gg} & 0 & 0 & \cdots & \cdots & 0 \\ m'_{g+1,1} & m'_{g+1,2} & \cdots & m'_{g+1,g} & 1 & & & 0 & \\ m'_{g+2,1} & m'_{g+2,2} & \cdots & m'_{g+2,g} & & 1 & & & \\ \cdots & \cdots & \cdots & \cdots & & & & \cdots & \\ \cdots & \cdots & \cdots & \cdots & 0 & & & & \\ m'_{h1} & m'_{h2} & \cdots & m'_{hg} & & & & & 1 \end{vmatrix}$$

where the magnitude of m'_{ij} $(i = 1, 2, \ldots, h;\ j = 1, 2, \ldots, g)$ is equal to the number of possible paths which run down monotonously from the white vertex w_i $(i = 1, 2, \ldots, h)$ to the valley v_j $(j = 1, 2, \ldots, g)$ and the sign of m'_{ij} is equal to $(-1)^{I_{ij}}$, where I_{ij} is the number of the diagonal edges in each path $w_i v_j$. Hence,

$$\det \mathbf{M}(G) = \begin{vmatrix} m'_{11} & m'_{12} & \cdots & m'_{1g} \\ m'_{21} & m'_{22} & \cdots & m'_{2g} \\ \cdots & \cdots & \cdots & \cdots \\ m'_{g1} & m'_{g2} & \cdots & m'_{gg} \end{vmatrix} = \begin{vmatrix} w'_{11} & w'_{12} & \cdots & w'_{1g} \\ w'_{21} & w'_{22} & \cdots & w'_{2g} \\ \cdots & \cdots & \cdots & \cdots \\ w'_{g1} & w'_{g2} & \cdots & w'_{gg} \end{vmatrix}$$

Theorem 16 holds.

Theorems 15 and 16 look quite similar. One should, however, observe that the order of **W** is equal to a half number of vertices (which is usually a large number) whereas the order of **W** is equal to the number of peaks (which is often quite small). In addition to this, the matrix elements W_{ij} are easily calculated [12, 29].

5 References

1. Cyvin SJ, Gutman I (1988) Kekulé structures in benzenoid hydrocarbons, Springer, Berlin Heidelberg New York
2. Clar E, Stewart DG (1953) J. Amer. Chem. Soc. 75: 2667
3. Clar E, Kemp W, Stewart DG (1958) Tetrahedron 3: 325
4. Staab HA, Diederich F (1983) Chem. Ber. 116: 3487
5. Diederich F, Staab HA (1978) Angew. Chem. 90: 383
6. Funhoff DJH, Staab HA (1986) Angew. Chem. 98: 757
7. Gordon M, Davison WHT (1952) J. Chem. Phys. 20: 428
8. Sachs H (1984) Combinatorica 4: 89
9. John P, Sachs H (1985) In: Bodendiek R, Schumacher H, Walter G (eds) Graphen in Forschung und Unterricht, Verlag Barbara Franzbecker, Bad Salzdetfurth, p 85
10. John P, Rempel J (1985) In: Sachs H (ed) Proc. Int. Conf. Graph Theory, Eyba, October 1984, Teubner, Leipzig, 72
11. He WJ, He WC (1987) In: King RB, Rouvray DH (eds) Graph theory and topology in chemistry, Elsevier, Amsterdam, p 476

12. He WJ, He WC (1989) Theor. Chim. Acta. 75: 389
13. Gutman I, Cyvin SJ (1986) J. Mol. Struct. (Theochem) 138: 325
14. Cyvin SJ, Gutman I (1987) J. Mol. Struct. (Theochem) 150: 157
15. Cyvin SJ, Gutman I (1988) J. Mol. Struct. (Theochem) 164: 183
16. Gutman I, Cyvin SJ (1988) J. Serb. Chem. Soc. 53: 391
17. Zhang FJ, Chen RS, Guo XF (1985) Graphs and combinatorics 1: 383
18. Zhang FJ, Guo XF (1988) Match 23: 229
19. Sheng RQ: This volume
20. Bondy JA, Murty USR (1976) Graph theory with applications, Macmillan, New York
21. Gibbons A (1985) Algorithmic graph theory, Cambridge University Press, Cambridge
22. Ford LR Jr, Fulkerson DR (1956) Canad. J. Math. 8: 399
23. Ford LR Jr, Fulkerson DR (1957) Canad. J. Math. 9: 210
24. Ford LR Jr, Fulkerson DR (1962) Flows in networks, Princeton University Press, Princeton
25. Edmonds J, Karp RM (1972) J. Assoc. Comput. Math. 19: 248
26. Karzanov AV (1974) Soviet. Math. Dokl. 15: 434
27. Malhotra VM, Pramodh Kumar M, Maheswari SN (1978) An $0(V^3)$ Algorithm for finding maximum flows in networks, Computer Science Program, India Institute of Technology, Kanpur
28. Dewar MJS, Longuet-Higgins HC (1952) Proc. Roy. Soc. A214: 482
29. Gutman I, Cyvin SJ (1987) Chem. Phys. Letters 136: 137

Rapid Ways to Recognize Kekuléan Benzenoid Systems

Rong-qin Sheng

Shanghai Institute of Computer Technology, 546 Yu Yuan Road, Shanghai, The People's Republic of China

Table of Contents

Rong-qin Sheng

This chapter presents rapid ways to determine whether or not a given benzenoid system possesses a Kekulé structure. Two simple were discovered by Sachs and the present author, respectively. Both algorithms operate by deleting vertices and edges from the benzenoid system examined. Emphasis is given to the algorithm proposed by the present author. The algorithm seems to be the simplest of the presently known ones. Its simplicity consists of the fact that is performed diagrammatically without the need for arithmetic operations. Also, simple procedures are described for examining two special classes of benzenoid systems; (1) the benzenoid systems with consecutively located peaks on the perimeters, (2) R : S with a normal benzenoid system S. Meanwhile, some concepts, such as monotonic paths, generalized benzenoid systems, convex pairs and quasi-benzenoid systems, are reviewed. Some structural properties of benzenoid systems are described with detailed exposition.

1 Introduction

This chapter is a review of the presently known rapid techniques for recognizing Kekuléan benzenoid systems. Only the Sachs algorithm and the Sheng algorithm are simple and generally applicable. The Sheng algorithm seems to be the simplest of all the known general methods. Simple applications for special cases are included here. The algorithms in the present review involve deletion of vertices and edges.

Benzenoid molecules are represented by benzenoid systems. Only Kekuléan benzenoid molecules are known to exist chemically. Non-Kekuléan benzenoid molecules have never been synthesized [1, 2]; they should be polyradicals and therefore of very low chemical stability [3, 4].

The important problem of how to decide whether a given benzenoid system possesses Kekulé structures or not has been investigated for a long time. The first fast algorithm was given by Sachs [5]. Recently necessary and sufficient structural requirements for the existence of Kekulé structures in benzenoid systems have been discovered [6, 7], but these results do not provide a rapid method for recognizing Kekuléan benzenoid systems. Also the "P–V maximum flow method" of He and He [8, 9] should be mentioned. More recently the present author [10] put forward an algorithm which is much simpler and faster than that of Sachs.

In Sect. 2, some relevant concepts and properties will be reviewed for preparation. The Sachs algorithm and the Sheng algorithm will be described with detailed proofs in Sect. 3 and 4, respectively. In Sect. 5, simple applications for special benzenoid systems will be reported.

2 Some Relevant Concepts and Properties

2.1 Monotonic Paths and Segmentations

A benzenoid system can be defined as follows: let C be a cycle on the hexagonal lattice; then the vertices and edges lying on C and in the interior of C form a benzenoid system B.

A Kekulé structure of B is a selection of $n/2$ (n = number of vertices in B) edges in B, such that no two of them are incident. Benzenoid systems possessing Kekulé structures are called Kekuléan, otherwise non-Kekuléan. In a Kekulé structure, the selected edges are called double bonds; the edges which are not selected are single bonds. In a Kekuléan benzenoid system, an edge is called a fixed double bond if it is selected in all Kekulé structures; an edge is a fixed single bond if it is not selected in any Kekulé structure.

In the present chapter, we always draw benzenoid systems so that some of their edges are vertical. A peak is a vertex lying above its both first neighbours, a valley is a vertex lying below both its first neighbours. A monotonic path is a path connecting a peak with a valley in which, when starting at the peak, one always goes downwards. Two paths are independent when they have no vertices in common.

Theorem 2.1. There is a one-to-one correspondence between a constellation of independent monotonic paths and a Kekulé structure [5, 11–13].

An example given in Fig. 1. illustrates Theorem 2.1; the monotonic paths are marked by heavy lines, the double bonds are indicated by double lines.

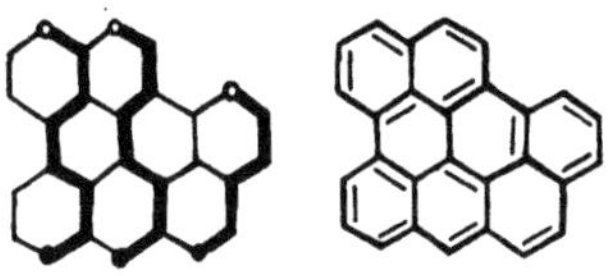

Fig. 1. A constellation of independent monotonic paths and the corresponding Kekulé structure. The peaks and valleys are indicated by white and black circles, respectively

A segmentation of a benzenoid system produces two segments (upper and lower) by cutting horizontally through a number of edges, so that two terminal edges belong to the perimeter. The cut edges are called tracks. We denote

tr = number of tracks,

s = difference between the numbers of peaks and valleys of the upper segment.

Theorem 2.2. For a segmentation of a Kekuléan benzenoid system, the number of double bonds in the tracks, denoted by d, is the same for all Kekulé structures. Furthermore, $d = tr - s$ [5, 14, 15].

Figure 2. depicts an example for the concept of segmentation.

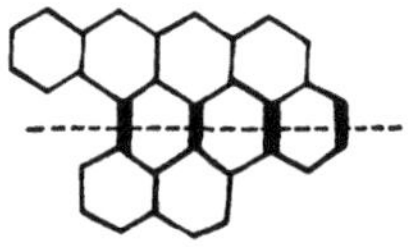

Fig. 2. A segmentation of a benzenoid system is indicated by a stippled line. The heavy lines represent tracks. Here $s = 3$, $tr = 4$

Theorem 2.3. In a Kekuléan benzenoid system, one has $0 \leqq s \leqq tr$ for every segmentation [5, 12].

Theorem 2.4. For a segmentation of a Kekuléan benzenoid system, (a) $s = 0$ holds if and only if the corresponding tracks are all fixed double bonds; (b) $s = tr$ holds if and only if the corresponding tracks are all fixed single bonds [12, 14, 15].

Two examples depicted in Fig. 3. illustrate Theorem 2.4.

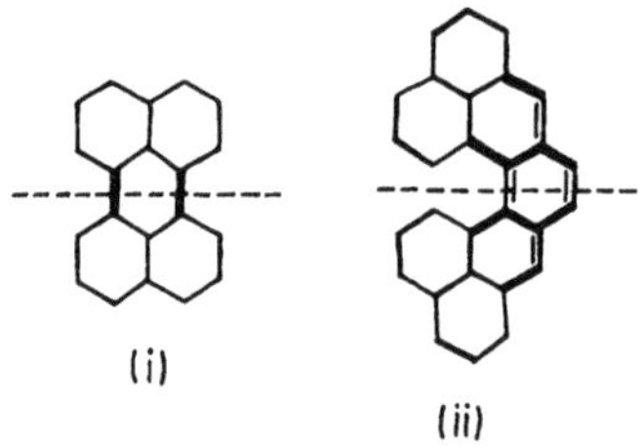

Fig. 3. Segmentations of benzenoid systems: **(i)** $s = tr = 2$ implies that the corresponding tracks are all fixed single bonds and represented by heavy lines; for **(ii)** $s = 0$, hence the tracks are all fixed double bonds and represented by double lines

The fundamental theorems 2.1–2.4 are well known results; therefore their proofs are omitted in this chapter.

In a benzenoid system the vertices can be colored by two colors (say black and white) so that adjacent vertices never have the same color.

Let the vertices in a benzenoid system with a segmentation be colored. It is easy to verify that the end vertices of the tracks belonging to the upper segment have the same color. The following result can easily be derived.

Theorem 2.5. In a Kekuléan benzenoid system with a segmentation, if the vertices are colored so that the end vertices of the tracks belonging to the upper segment are black, then the difference between the numbers of black and white vertices in the upper segment equals d. Here $d = tr - s$ as in Theorem 2.2.

2.2 Generalized Benzenoid Systems

A generalized benzenoid system is obtained by deleting some vertices and edges from a benzenoid system [14]. A generalized benzenoid system may be disconnected; then each independent conjugated subsystem is called a component. A benzenoid system is a special case of a generalized benzenoid system.

In the present chapter we always draw generalized benzenoid systems so that some of their edges are vertical.

We have the same definitions of a Kekulé structure, Kekuléan, non-Kekuléan, double and single bonds, fixed double and fixed single bonds for generalized benzenoid systems as for benzenoid systems.

It is obvious that a disconnected generalized benzenoid system is Kekuléan if and only if its components all are Kekuléan.

A peak (respectively valley) is defined as a vertex which has at least one non-vertical incident edge and lies above (respectively below) all its neighbours. The definition of the (independent) monotonic paths is the same as in Paragraph 2.1.

A segmentation of a connected generalized benzenoid system produces two segments (upper and lower) by cutting horizontally through a number of edges. The cut edges are again called tracks. The definitions of the symbols tr and s are the same as in Paragraph 2.1. Fig. 4. shows an example.

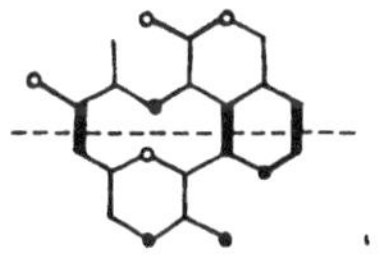

Fig. 4. A segmentation of a connected generalized benzenoid system with $s = 2$, $tr = 3$. The heavy lines represent tracks. The peaks and valleys are indicated by white and black circles, respectively

The generalized benzenoid systems have many properties similar to benzenoid systems. We outline the results similar to theorems 2.1–2.5 in the following.

Theorem 2.6. In a Kekuléan connected generalized benzenoid system with a segmentation,

(I) there is a one-to-one correspondence between a constellation of independent monotonic paths and a Kekulé structure;

(II) the number of double bonds in the tracks, denoted by d, is the same for all Kekulé structures. Furthermore, $d = tr - s$;

(III) $0 \leqq s \leqq tr$;

(IV) $s = 0$ holds if and only if the corresponding tracks are all fixed double bonds;

(V) $s = tr$ holds if and only if the corresponding tracks are all fixed single bonds;

(VI) if the vertices are colored so that the end vertices of the tracks belonging to the upper segment are black, then the difference between the numbers of black and white vertices in the upper segment is equal to d as defined in (II).

3 Sachs Algorithm

3.1 Free Edges

In a benzenoid system, the vertices and edges lying on the perimeter are called external. An external edge whose both end vertices are of degree two is called a free edge. We define

n_e = number of external vertices,

n_2 = number of the external vertices of degree two,

n'_3 = number of the external vertices of degree three.

Lemma 3.1. $n_2 = n'_3 + 6$.

Proof: Consider the sum of all internal angles of the polygon formed by the perimeter, that is $(n_2 + n'_3 - 2)\,180°$. The internal angle corresponding to an external vertex of degree two (respectively three) is 120° (respectively 240°). Hence $n_2 120° + n'_3 240° = (n_2 + n'_3 - 2)\,180°$, from which the result follows.

By means of Lemma 3.1 we can immediately get

Lemma 3.2. A benzenoid system contains at least six free edges.

A stronger statement than Lemma 3.2 is that a benzenoid system contains at least two free edges in each of the three edge directions. In fact the vertical, extreme left (and right) edge is a free edge (see Fig. 5.).

Fig. 5. The extreme left and right vertical edges (marked by heavy lines) are free edges

Theorem 3.1. Let B be a Kekuléan benzenoid system and e a vertical free edge. Assume that a segmentation is made through e. If $s < tr$, then B has a Kekulé structure in which e is a double bond.

Proof: According to the assumption $s < tr$ and Theorem 2.2, $d = tr - s > 0$, so there is at least one double bond in the tracks. Assume K is a Kekulé structure in which e is a single bond, e′ is a double bond in the tracks, and the tracks lying to the left of e′ are all single bonds. Here we assume e lies to the left of e′. Consider the benzenoid chain indicated in Fig. 6. and the notation defined therein; we get $T = \{e', p_{2i-1}, q_{2i-1}, 1 \leqq i \leqq t\} \subset K$. Let $S = \{e, p_{2i}, q_{2i}, 1 \leqq i \leqq t\}$, $K' = (K - T) \cup S$. Then K′ is another Kekulé structure in which e is a double bond. Therefore the result follows.

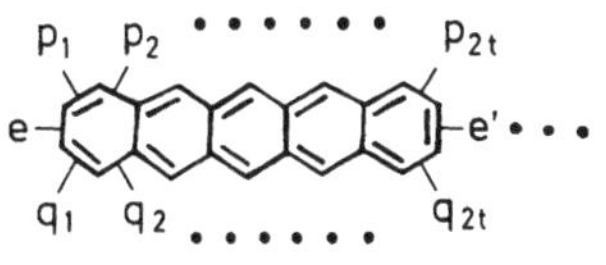

Fig. 6. A benzenoid chain (see the text)

As an immediate consequence of Theorem 2.2 and Theorem 3.1 we have

Corollary 3.1. Let B be a Kekuléan benzenoid system with a segmentation through a free edge e. If $s = tr - 1$, then B has a Kekulé structure in which e is a double bond, and the tracks different from e are all single bonds.

3.2 The Procedure of the Sachs Algorithm

Let H be a generalized benzenoid system, which satisfies that all finite regions are mutually congruent regular hexagons in the plane divided by H; or, equivalently, each of the components of H contains no "holes". We shall examine whether H is Kekuléan. Note that H is Kekuléan if and only if every component of H is Kekuléan. Therefore we may assume that H is connected.

First, we check whether H contains an equal number of peaks and valleys; if not, then H is non-Kekuléan; in particular this happens if H consists of an isolated vertex.

Suppose H has an equal number of peaks and valleys. We shall select some vertices and edges, such that deleting them does not influence the existence of Kekulé structures. There are two cases.

Case I. H contains a cut-edge (an edge not belonging to cycles) e. If e is a hanging edge (an edge joining a pendent vertex and its neighbour), we may delete two end vertices of e. When we say deleting a vertex, this means that the vertex and its incident edges are all deleted. If e is not a hanging edge, we may assume e is vertical. Consider the segmentation through e. We may assume the vertices of the upper segment are colored so that the end vertex of e lying in the upper segment is black. Then calculate d = difference between the number of black and wihte vertices in the upper segment. If $d < 0$ or $d > 1$, then, according to VI of Theorem 2.6, H is non-Kekuléan. If $d = 0$ we may delete e; if $d = 1$ we may delete two end vertices of e. Examples are depicted in Fig. 7.

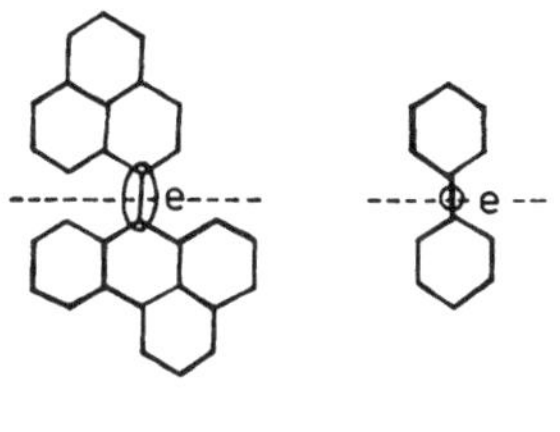

Fig. 7. Exemplification of Case I of the Sachs algorithm: (i) $d = 1$; (ii) $d = 0$

Case II. H does not have a cut-edge. Then H is a benzenoid system. According to Lemma 3.2 we can select a free edge e. We may assume e is vertical and consider the segmentation through e. We calculate s and tr.

If $s < 0$ or $s > tr$, then by Theorem 2.3, H is non-Kekuléan;

if $s = tr$, then by (b) of Theorem 2.4, we may delete all the corresponding tracks (see Fig. 8a);

if $s = tr - 1$, then according to Corollary 3.1, we may delete two end vertices of e and all the tracks different from e (see Fig. 8b);

if $0 < s < tr - 1$, then according to Theorem 3.1, we may delete two end vertices of e (see Fig. 8c);

if $s = 0$, then from (a) of Theorem 2.3, we may delete two end vertices of each of the corresponding tracks (see Fig. 8d).

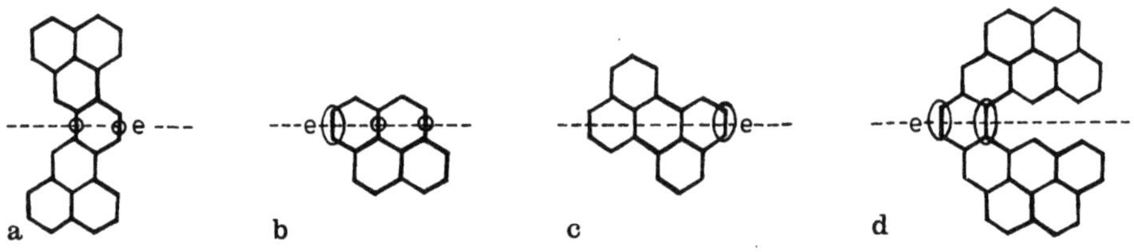

Fig. 8. Exemplification of Case II of the Sachs algorithm: **(a)** $s = tr = 2$; **(b)** $s = tr - 1 = 2$; **(c)** $s = 2$, $tr = 4$; **(d)** $s = 0$

After one step of the above algorithm we obtain the remaining part of H, which may be denoted by H′. It is easy to show that H′ also is a generalized benzenoid system whose components have no "holes" and that H is Kekuléan if and only if H′ is Kekuléan. Therefore we treat the components of H′ as described above and continue further on.

The algorithm comes to a stop when either (a) a non-Kekuléan condition is reached; then H is non-Kekuléan, or (b) all vertices of H are deleted; in this case H is Kekuléan. If H is recognized as Kekuléan, we recall the deleting process: if we delete two end vertices of an edge, then the edge is considered as a double bond, and in this way we can constitute a Kekulé structure of H.

Remark 3.1. In Case I, we can calculate $d = tr - s$ by means of (II) of Theorem 2.6 instead of coloring of vertices.

Remark 3.2. Let H be a connected generalized benzenoid system having no "holes". Furthermore, let e be an edge which belongs to a hexagon and its both end vertices are of degree two. We also call e a free edge. If H has a free edge e, we can also apply the deleting procedure described in Case II. After reading Paragraph 4.2, the reader will know that if H contains no hanging edges, then H possesses at least six free edges.

4 Sheng Algorithm

4.1 Convex Pairs

Let B be a benzenoid system. A is the polygon formed by the perimeter. A set of four external vertices {u, v, x, y} is called a rectangular set if u and x are adjacent to v and y respectively, and u, v, x, y form a rectangle which is covered by A. A convex pair is formed by two adjacent vertices of degree two in a rectangular set which contains at least three vertices of degree two.

An equivalent definition of a rectangular set and a convex pair is grasped intuitively: Let u, v be two adjacent external vertices and e = {u, v}. The cut orthogonal to the edge e going through the center of e and the interior of B intersects another external edge e′ = {x, y}. Then u, v, x, y form a rectangular set. If u, v are of degree two and at least one of x, y is of degree two, then u, v form a convex pair. Let it be denoted by (u, v). Note that (u, v) = (v, u). Fig. 9. is supposed to explain the above definitions.

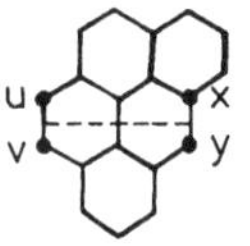

Fig. 9. A rectangular set u, v, x, y, where (u, v) is a convex pair because u, v and y are of degree two

Lemma 4.1. A benzenoid system possesses a total of $n_e/2$ rectangular sets.

Proof: The result holds because every external vertex belongs to two rectangular sets.

Theorem 4.1. A benzenoid system possesses at least six convex pairs.

Proof: Let the rectangular sets be $R_1, R_2, \ldots, R_m$, where $m = n_e/2$. $N(R_i)$ denotes the number of the vertices of degree two in R_i, $1 \leqq i \leqq m$. s_1 and s_2 denote the number of the rectangular sets which contain three and four vertices of degree two, respectively. Then the number of all convex pairs is $s_1 + 2s_2$. We have $2n_2 = \sum_{i=1}^{m} N(R_i) \leqq 3s_1 + 4s_2 + 2(m - s_1 - s_2)$. By Lemma 3.1 we get $s_1 + 2s_2 \geqq 2(n_2 - n_e/2) = n_2 - n_3' = 6$. This completes the proof.

Theorem 4.2. If B is a Kekuléan benzenoid system, (u, v) is a convex pair and e = {u, v}, then B possesses a Kekulé structure in which e is a double bond.

Proof: With reference to Fig. 10., choose the benzenoid chain with vertices $u, v, u_1, v_1, u_2, v_2, \ldots, u_{2k}, v_{2k}$, where we may assume that u_{2k} is of degree two without loss of generality. Consider the segmentation through e. If $d = 0$, i.e., the tracks are all fixed single bonds, then we deduce that $\{u_{2m-1}, u_{2m}\}$ is a single bond (see Fig. 10). Hence $\{u_{2m}, v_{2m}\}$ would have to be a double bond, which contradicts $d = 0$. Consequently $d > 0$, i.e., $s < tr$. Now the result follows from Theorem 3.1.

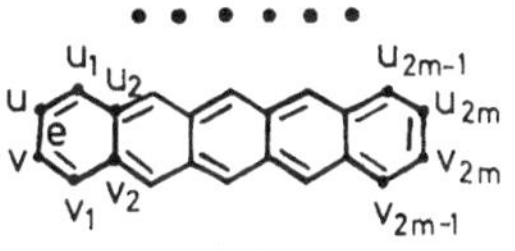

Fig. 10. A benzenoid chain (see the text)

Theorem 4.2 demonstrates the important fact that deleting a convex pair from a benzenoid system does not alter its Kekuléan/non-Kekuléan character.

4.2 Quasi-benzenoid Systems

Q is called a quasi-benzenoid system if it satisfies:

(1) Q consists of benzenoid subunits joined by acyclic lines on the hexagonal lattice;

(2) by regarding the benzenoid subunits as vertices and the acyclic lines as edges, the corresponding joining graph is a tree. It shall be called the incidence tree of Q and denoted by T(Q).

The perimeter of Q, denoted by C, comprises the perimeters of the benzenoid subunits and the acyclic lines. The vertices and edges lying on C are external, otherwise internal. A denotes the area constituted by C and the interior of C. Then we can adopt the same definitions of rectangular sets and convex pairs as in Paragrph 4.1. Fig. 11. depicts an example.

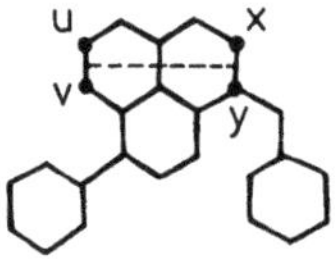

Fig. 11. A quasi-benzenoid system in which {u, v, x, y} is a rectangular set, and (u, v) is a convex pair because u, v and x are of degree two

Theorem 4.3. If Q is a quasi-benzenoid system and its incidence tree T(Q) has k leaves, then Q possesses at least $4k$ convex pairs.

Proof: Let Q* be a benzenoid subunit corresponding to a leaf of T(Q). Consider the number of the external vertices of degree two (respectively three) on the perimeter of Q*, denoted by n_2^* (respectively n_3^*), we find that $n_2^* = n_3^* + 4$. Hence by a similar proof as for Theorem 4.1, we get that Q* possesses at least four convex pairs of Q. Consequently the rest of the present proof is obvious.

T(Q) always has two leaves, therefore by Theorem 4.3 we see that a quasi-benzenoid system possesses at least eight convex pairs.

By the analogous proofs as for Theorem 4.2, together with Theorem 3.1, we get the following result.

Theorem 4.4. If Q is a Kekuléan quasi-benzenoid system, (u, v) is a convex pair and e = {u, v}, then Q possesses a Kekulé structure in which e is a double bond.

The above theorem shows that deleting a convex pair from a quasi-benzenoid system does not influence the existence of Kekulé structures.

4.3 The Procedure of the Sheng Algorithm

Let H be a connected generalized benzenoid system having no "holes". We shall recognize whether H possesses Kekulé structures. This is done by selecting some vertices, so that deleting them does not influence the existence of Kekulé structures. There are two cases.

Case 1. H contains a pendent vertex. We may delete a pendent vertex together with its first neighbour without changing the Kekuléan/non-Kekuléan nature.

Case 2. H does not contain a pendent vertex. Then H is a benzenoid or quasi-benzenoid system. According to theorems 4.1–4.4, we may choose a convex pair and delete it.

After this one step let the remaining part of H be denoted by H′. It is easy to realize that no component of H′ has any "hole" and that H is Kekuléan if and only if every component of H′ is Kekuléan. Hence we may continue the above operations for every component of H′. The process continues along the same principles.

The above procedure ends when an isolated vertex is created or when all vertices of H are pairwise deleted. In the former case H is non-Kekuléan; in the latter case it is Kekuléan, and a Kekulé structure can be constructed by regarding the edge joining the deleted pair of vertices at each step as a double bond.

Figure 12. shows two examples of the algorithm. The numbers 1, 2, 3, ... indicate the order of deletion. There are many possible orders of deletion.

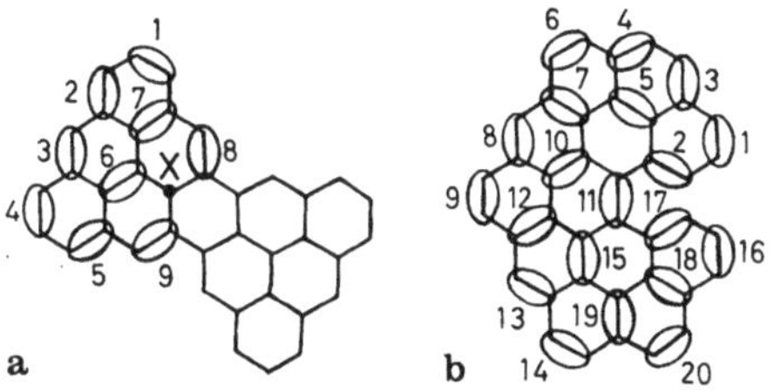

Fig. 12. Exemplification of the Sheng algorithm: **(a)** is non-Kekuléan, x is an isolated vertex; **(b)** is Kekuléan, the encircled vertex pairs correspond to the double bonds in a Kekulé structure

After reading Paragraph 4.4, the reader should be convinced that searching a convex pair is quite easy, and therefore the algorithm is fairly simple and fast. Using this algorithm, one can easily judge any large benzenoid system only from its diagram, without the need for arithmetic operations.

4.4 Distribution of Convex Pairs [16]

Consider a benzenoid system B oriented with some of its edges vertical. Assume that v is an external vertex of degree three and that its three adjacent vertices are v_1, v_2, v_3. Here $\{v, v_1\}$ is vertical and internal, while $\{v, v_2\}$ and $\{v, v_3\}$ are external. If v lies above v_1, then v is called an upper concave vertex (UCV); if v lies below v_1, then v is called an lower concave vertex (LCV). Fig. 13. indicates the two situations.

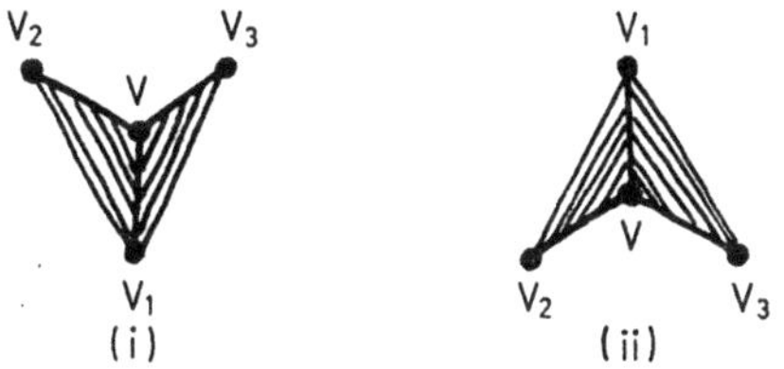

Fig. 13. (i) v is an upper concave vertex; **(ii)** v is a lower concave vertex. The hatched parts lie in the interior of B

The numbers of peaks, valleys, UVCs and LVCs in B are denoted by n_p, n_v, n_u and n_l, respectively. The following is a basic property of benzenoid systems.

Theorem 4.5. (i) $n_p = n_u + 1$. (ii) $n_v = n_l + 1$.

Proof: It is obvious that (ii) can be derived from (i) by the rotation of 180°. Hence we shall prove (i) only.

Denote by h the number of hexagons in a benzenoid system. When $h = 1$, the corresponding benzenoid system is a regular hexagon, and (i) holds in an obvious manner.

Assume that (i) holds when $h \leq k$. Then we shall verify that (i) also holds when $h = k + 1$.

Let **B** be a benzenoid system with $h = k + 1$. Denote respectively by n_p and n_u the numbers of peaks and UCVs in B. In the lowest line of hexagons, choose the first hexagon from the left and denote it by X. Then X has at most three adjacent hexagons. There are seven possible situations, as indicated in Fig. 14.

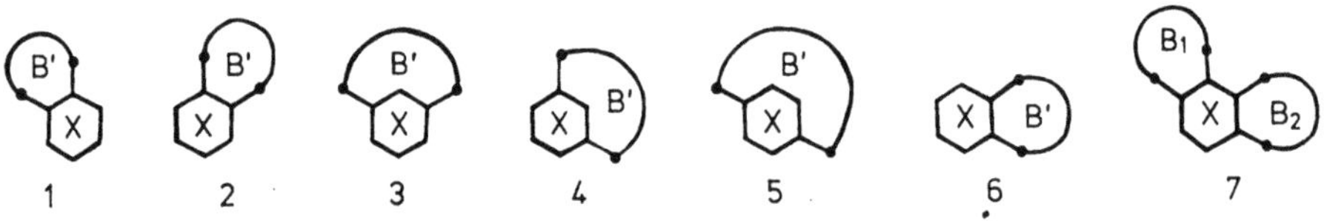

Fig. 14. Seven possible situations of the hexagon X

In each of the situations 1–5, **B** is obtained by adding X to a benzenoid system **B′** with $h = k$, and the numbers of peaks and UCVs are unchanged under the addition. Thus the assumption (i) for **B′** yields that (i) is also valid for **B**.

In the situation 6, **B** is also obtained by adding X to a benzenoid system **B′** with $h = k$. Denote respectively by n'_p and n'_u the numbers of peaks and UCVs in B′. We have

$$n_p = n'_p + 1, n_u = n'_u + 1 .$$

By assumption, one has $n'_p = n'_u + 1$. Hence also $n_p = n_u + 1$.

In the situation 7, X joints two benzenoid systems, B_1 and B_2, with $h < k$. Denote respectively by $n_p^{(i)}$ and $n_u^{(i)}$ the numbers of peaks and UVCs in B_i, $i = 1, 2$. We have

$$n_p = n_p^{(1)} + n_p^{(2)}, \qquad n_u = n_u^{(1)} + n_u^{(2)} + 1 .$$

By assumption we get $n_p^{(i)} = n_u^{(i)} + 1$, $i = 1, 2$. Hence also $n_p = n_u + 1$.

Consequently, (i) is valid for all benzenoid systems, and so is (ii).

An external edge is called a convex edge if its both end vertices form a convex pair, or equivalently, a convex edge is an edge joining two vertices of a convex pair.

Theorem 4.6. A benzenoid system possesses at least two convex edges each in the three different edge directions.

Proof: Let **B** be a benzenoid system oriented with some of its edges vertical. Because non-vertical directions can be transformed into the vertical direction by appropriate rotation, it only needs to be verified that **B** possesses at least two vertical convex edges.

Let V_e = the set of all external vertices,
V_1 = the set of the external vertices which have incident vertical edges,
V_2 = the set of all peaks, valleys, UCVs and LCVs.
Then (a) $V_1 \cap V_2 = \emptyset$, $V_1 \cup V_2 = V_e$.

Denote respectively by $n_2(S)$ and $n_3(S)$ the numbers of the vertices of degree two and three in S, where S is a set constituted by some external vertices.

By Lemma 3.1, Theorem 4.5 and (a), we obtain
(b) $n_2(V_1) - n_3(V_1) = 4$.

Assume that V_1 contains a total of k rectangular sets, denoted by $R_1, R_2, \dots, R_k$. Then we have

(c) $R_i \cap R_j = \emptyset$, $1 \leqq i < j \leqq k$; $\bigcup_{i=1}^{k} R_i = V_1$.

Hence (d) $\sum_{i=1}^{k} (n_2(R_i) - n_3(R_i)) = n_2(V_1) - n_3(V_1)$.

Assume that V_1 contains t_1 and t_2 rectangular sets which have three and four vertices of degree two, respectively. Then the number of all vertical convex edges is equal to $t_1 + 2t_2$.

By (c) we get

(e) $\sum_{i=1}^{k} (n_2(R_i) - n_3(R_i)) \leqq 2t_1 + 4t_2$.

By (b), (d) and (e), it follows that $t_1 + 2t_2 \geqq 2$.

Therefore B contains at least two vertical convex edges. The proof of Theorem 4.6 is completed.

Using Theorem 4.6, the following result can easily be derived:

Theorem 4.7. If Q is a quasi-benzenoid system and its incidence tree T(Q) has k leaves, then Q possesses at least k convex edges each in the three different edge directions.

In Theorem 4.7, $k \geqq 2$, hence a quasi-benzenoid system possesses at least two convex edges each in the three different edge directions.

5 Simple Applications for Special Cases

5.1 The Benzenoid System with Consecutively Located Peaks on the Perimeter

Sachs [5] put forward a simple algorithm for the existence of Kekulé structures in the case of benzenoid systems where all peaks follow consecutively on the perimeter. Let B be such a benzenoid system. Along the perimeter, the peaks and valleys follow the order: $p_1 \to p_2 \to \dots \to p_k \to v_k \to v_{k-1} \to \dots \to v_1$. Here we assume that B has an equal number of peaks and valleys. If B is Kekuléan, then by Theorem 2.1 B has a constellation of independent monotonic paths, each from the peak p_i to the valley v_i, $i = 1, 2, \dots, k$. We find that deleting the extreme left monotonic path does not change the existence of Kekulé structures. We assume that the extreme left monotonic path goes from p_1 to v_1. The procedure consists of the deletion of the extreme left monotonic path joining p_i and v_i at step i. The procedure ends when a monotonic path joining p_i and v_i can not be reached at step i for some i or when the step k is finished. In the former case B is non-Kekuléan; in the latter case it is Kekuléan. An example is given in Fig. 15.

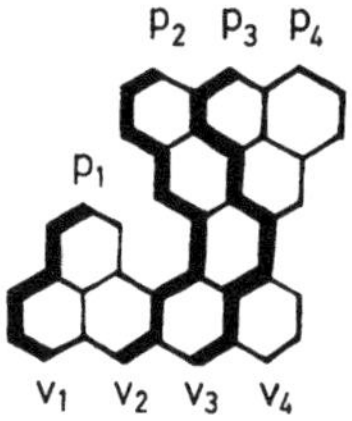

Fig. 15. This is a non-Kekuléan benzenoid system because after step 3 no paths joining p_4 and v_4 exist

In this situation we can also apply the peeling algorithm, as reported in Refs. [11, 12, 17]. The peeling algorithm consists of the deletion of the extreme left monotonic path, followed by the deletion of pendent vertices (if any) together with their first neighbours. The procedure ends when either (a) an isolated vertex is created; in this case B is non-Kekuléan, or (b) all vertices are deleted; then it is Kekuléan. An example is given in Fig. 16.

Fig. 16. By means of the peeling algorithm this benzenoid system is found to be non-Kekuléan; x is an isolated vertex

5.2 R:S with a Normal Benzenoid System S

R:S is a benzenoid system obtained by a fusion of two benzenoid units R and S which share exactly one edge (see Fig. 17).

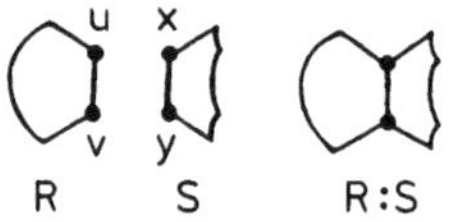

Fig. 17. R:S is obtained by identifying the vertex u of R with the vertex x of S, and the vertex v of R with the vertex y of S

Define the subgraph R′, R″, S′ and S″ as follows:

$$R' = R - u, v, \qquad R'' = R - u - v,$$

$$S' = S - x, y, \qquad S'' = S - x - y.$$

Obviously, $K\{R\} = K\{R'\} + K\{R''\}, \qquad K\{S\} = K\{S'\} + K\{S''\}$. (i)

As usual $K\{B\}$ is used to denote the Kekulé structure count of a (generalized) benzenoid system B.

Assume S is a normal benzenoid system. Then S has an even number of vertices. According to Ref. [18] we have the relation:

$$K\{R{:}S\} = K\{R\}\,K\{S\} - K\{R'\}\,K\{S'\}\,. \qquad \text{(ii)}$$

Theorem 5.1. If R and S are benzenoid systems and S is normal, then R:S is Kekuléan if and only if R is Kekuléan [15].

Proof: By (i) and (ii), we get

$$K\{R{:}S\} = K\{R\}\,K\{S''\} + K\{R''\}\,K\{S'\}\,. \qquad \text{(iii)}$$

The normal benzenoid system S obeys $K\{S'\} > 0$ and $K\{S''\} > 0$. Hence by (iii), $K\{R{:}S\} > 0$ if and only if $K\{R\} > 0$. This completes the proof.

The concept of fusion can be extended to the case of generalized benzenoid systems. Then R:S is a generalized benzenoid system obtained by a fusion of two generalized benzenoid units which share exactly one edge. By a similar proof as for Theorem 5.1, we get

Theorem 5.1'. If R is a generalized benzenoid system and S is a normal benzenoid system, then R:S is Kekuléan if and only if R is Kekuléan.

Hence, if S is a normal benzenoid system, then we may reduce R:S to R without changing the existence of Kekulé structures.

An example is shown in Fig. 18.

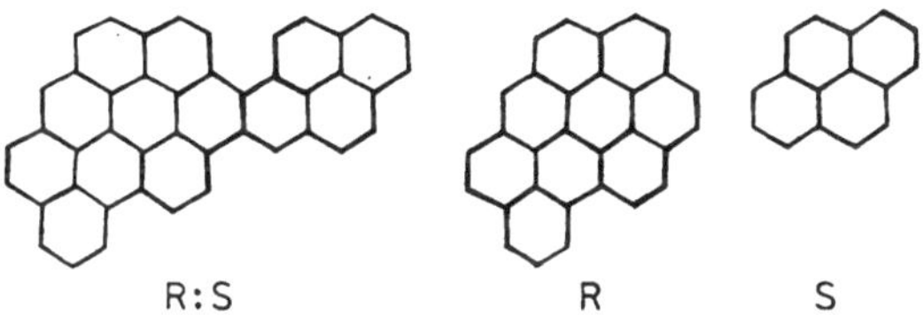

Fig. 18. S is a normal benzenoid system. Hence R:S is Kekuléan if and only if R is Kekuléan

Since all catacondensed benzenoid system are normal, we may reduce a generalized benzenoid system with fused catacondensed units without altering the Kekuléan/non-Kekuléan character. By inspecting Fig. 19. the reader may get a fair idea about how the reduction is made.

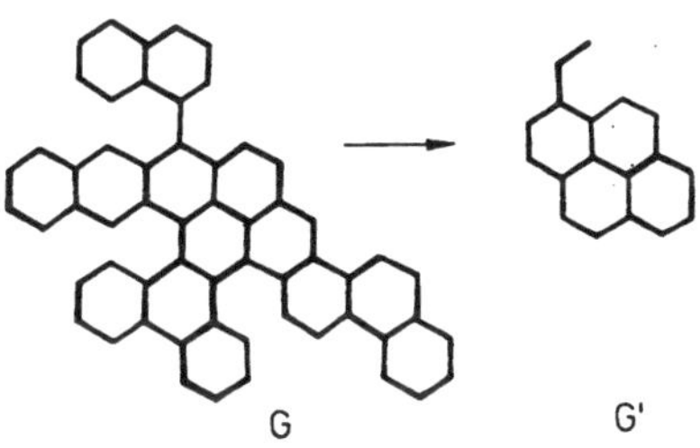

Fig. 19. Reduce G to G'. G is Kekuléan if and only if G' is Kekuléan

The fused catacondensed units can easily be recognized, and they often occur in the process of a general method of reduction. We may combine this reduction procedure with a generally applicable simple algorithm and thus raise the speed of recognition.

6 Concluding Remarks

Here we have reviewed some advances in rapid ways to recognize Kekuléan benzenoid systems. Many important properties concerning Kekuléan and non-Kekuléan (generalized) benzenoid systems have been reported.

The Sheng algorithm is to be recommended for its simplicity and rapidity. By means of this algorithm, a complicated benzenoid system can be easily reduced, and thus rapidly recognized as being Kekuléan or non-Kekuléan.

7 References

1. Clar E, Stewart DG (1953) J. Amer. Chem. Soc. 75: 2667
2. Clar E, Kemp W, Stewart DG (1958) Tetrahedron 3: 325
3. Müller E, Müller-Rodloff I (1935) Justus Liebigs Ann. Chem. 517: 134
4. Longuet-Higgins HC (1950) J. Chem. Phys. 18: 265
5. Sachs H (1984) Combinatorica 4: 89
6. Zhang F, Chen R, Guo X (1985) Graphs and Combinatorics 1: 383
7. Kostochka AV: Proc. 30th Internat. Wiss. Koll. TH Ilmenau 1985; Vortragsreihe F, pp 49–52
8. He W, He W (1985) Theor. Chim. Acta 68: 301
9. He W, He W: this volume
10. Sheng R (1987) Chem. Phys. Letters 142: 196
11. Gutman I, Cyvin SJ (1986) J. Mol. Struct. (Theochem) 138: 325
12. Cyvin SJ, Gutman I (1987) J. Mol. Struct. (Theochem) 150: 157
13. Gordon M, Davison WHT (1952) J. Chem. Phys. 20: 428
14. Sheng R (in press) Match
15. Sheng R, Cyvin SJ, Gutman I (in press) J. Mol. Struct. (Theochem)
16. Sheng R (in press) Match
17. Cyvin SJ, Gutman I (1988) J. Mol. Struct. (Theochem) 164: 183
18. Cyvin SJ, Gutman I (1988) Kekulé structures in benzenoid hydrocarbons, Springer, Berlin Heidelberg New York (Lecture Notes in Chemistry 46)

Methods of Enumerating Kekulé Structures, Exemplified by Applications to Rectangle-Shaped Benzenoids

Chen Rongsi[1], S. J. Cyvin[2], B. N. Cyvin[2], J. Brunvoll[2], and D. J. Klein[3]

Table of Contents

[1] Department of Planning and Statistics, Fuzhou University, Fuzhou, Fujian, The People's Republic of China
[2] Division of Physical Chemistry, The University of Trondheim, N-7034 Trondheim-NTH, Norway
[3] Department of Marine Sciences, Texas A & M University at Galveston, Galveston, Texas 77553-1675, U.S.A.

The enumeration of Kekulé structures for rectangle-shaped benzenoids is treated. Combinatorial formulas for K (the Kekulé structure count) are derived by several methods. The oblate rectangles, $R^j(m, n)$, with fixed values of m are treated most extensively and used to exemplify different procedures based on the method of fragmentation (chopping, summation), a fully computerized method (fitting of polynominal coefficients), application of the John-Sachs theorem, and the transfer-matrix method. For $R^j(m, n)$ with fixed values of n the relevant recurrence relations are accounted for, and general explicit combinatorial K formulas are reported. Finally a class of multiple coronoids, the "perforated" oblate rectangles, is considered in order to exemplify a "perfectly explicit" combinatorial K formula, an expression for arbitraty values of the parameters m and n.

1 Introduction

1.1 Definitions

The rectangle-shaped benzenoids (or simply rectangles) are divided into the classes of prolate and oblate rectangles. Figure 1 shows as examples one member of each of these classes. In general $R^i(m, n)$ and $R^j(m, n)$ designate regular $2m - 1$ strips [1, 2], which differ in the manner of indentation; R^i inwards and R^j outwards. The second parameter, n, indicates the number of hexagons of the top or bottom row (which have equal length).

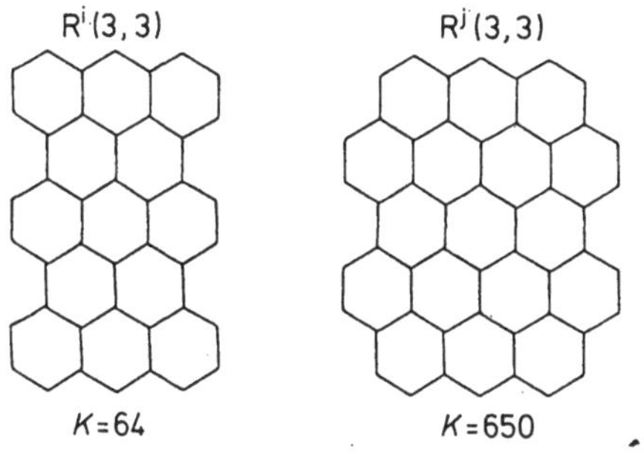

Fig. 1. A prolate (left) and oblate rectangle (right)

An oblate rectangle, $R^j(m, n)$; $m > 1$, is associated with a class of auxiliary benzenoids denoted $B(n, 2m - 2, -l)$, where $l = 0, 1, 2, \ldots, n$. Figure 2 shows the four auxiliary benzenoids associated with the right-hand system of Fig. 1. We notice at once the symmetry properties for the auxiliary benzenoids under consideration. In our example (Fig. 2) we find on one hand that $B(3, 4, 0)$ and $B(3, 4, -3)$ are isomorphic, while on the other hand $B(3, 4, -1)$ and $B(3, 4, -2)$ are isomorphic. In general: $B(n, 2m - 2, -l) = B(n, 2m - 2, l - n)$.

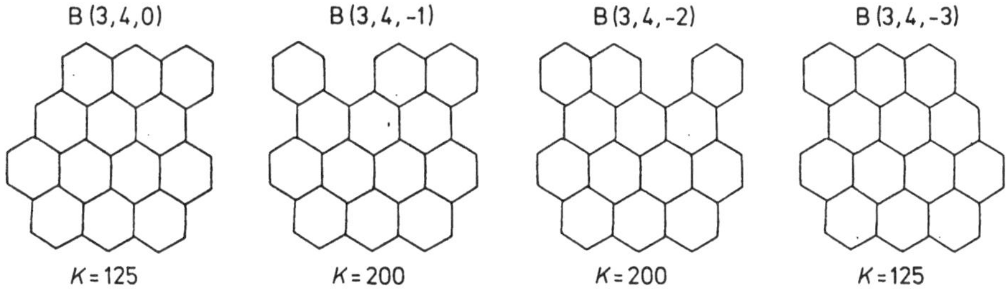

Fig. 2. Members of an auxiliary benzenoid class

The Kekulé structure count (or number of Kekulé structures) is denoted by K. More specifically, when B is a benzenoid, then $K\{B\}$ is used to identify its Kekulé structure count. Thus, for instance, $K\{R^i(3, 3)\} = 64$ and $K\{R^j(3, 3)\} = 650$; cf. Fig. 1.

Also in Fig. 2 the K numbers for the depicted benzenoids are specified. By virtue of the symmetry properties of the auxiliary benzenoids one has of course

$$K\{B(n, 2m-2, l-n)\} = K\{B(n, 2m-2, -l)\} \tag{1}$$

1.2 Prolate Rectangles

A prolate rectangle, $R^i(m, n)$, is an essentially disconnected benzenoid [1–3]. Hence the Kekulé structure counts are easily obtained by

$$K\{R^i(m, n)\} = (n+1)^m \tag{2}$$

a formula which already was derived by Yen [4].

1.3 Previous Work

The special case of Eq. (2) for $m = 2$, pertaining to the 3-tier prolate rectangles $R^i(2, n)$, was first given in the classical paper of Gordon and Davison [5]. The systematic studies of regular 3-tier [6] and 5-tier [1] strips include the appropriate classes of prolate rectangles. The general formula (2) has been re-derived in different ways [7, 8].

The studies of oblate rectangles, $R^j(m, n)$, turned out to be far more difficult and therefore more challenging. A survey of these studies is at the same time an excursion through several methods of computing K numbers and deriving K formulas, which have general importance far beyond their applications to the oblate rectangles only.

The investigations are sharply divided into the derivation of K formulas for $R^j(m, n)$ with fixed values of m on one hand and fixed values of n on the other.

In the first category come the relatively easy derivations of K formulas for $R^j(2, n)$ [4–6] and $R^j(3, n)$ [1, 4–6, 8], the 3-tier and 5-tier oblate rectangles, respectively. The significantly more difficult problem for the 7-tier oblate rectangles, $R^j(4, n)$, was solved later [9] and followed by the solution for $R^j(5, n)$ [10]. The race continued with the derivation of the K formula for $R^j(6, n)$ [11], followed by $K\{R^j(7, n)\}$ [12], and has so far culminated with the K formula for the 15-tier system $R^j(8, n)$ [13, 14].

In the second category, the K formulas for $R^j(m, n)$ with fixed values of n, the pioneering work is due to Gutman [15], who produced the explicit K formulas for $R^j(m, 1)$ and $R^j(m, 2)$. Shortly thereafter the corresponding solutions for $R^j(m, 3)$ [9, 16] and $R^j(m, 4)$ [9, 17] were reported, and eventually for $R^j(m, 5)$ [18]. These studies are dominated by recurrence relations, for which we give some relevant supplementary references [19–21].

The K enumeration of rectangles was the topic for one of the contributions at an International Conference on Graph Theory and Topology in Chemistry, Athens, Georgia, USA, 16–20 March 1987 [22]. Finally we mention that the research area in question has been summarized in details and supplemented in the book of Cyvin and Gutman [2]. Some of the latest developments [21] are not included therein. The present review provides a new twist of the presentation, which is supposed to be suited as an introduction to the fairly complicated analyses of this topic. Because of space limitations we have mainly confined the present contribution to the classes of oblate rectangles, $R^j(m, n)$, with fixed values of m. The available formulas for $R^j(m, n)$ with fixed values of n are reviewed very briefly. In supplement, a new general K formula for $R^j(m, n)$ is presented. It may be used to derive explicit formulas of $K\{R^j(m, n)\}$ for any fixed value of n.

2 Oblate Rectangles with Fixed Values of m

2.1 Trivial and Degenerate Cases

The formulas of $K\{R^j(m, n)\}$ with fixed values of m are polynomials in n. Here $m \geqq 1$. The case of $m = 1$ is trivial; here the systems degenerate to the single linear chains, $L(n)$, and the K formula reads [2, 5]

$$K\{R^j(1, n)\} = K\{L(n)\} = n + 1 \tag{3}$$

For $n = 0$ and arbitrary m the systems degenerate to no hexagons with $K = 1$. The mentioned polynomials should be consistent with this value, i.e.

$$K\{R^j(m, 0)\} = 1 \tag{4}$$

2.2 The 3-Tier and 5-Tier Oblate Rectangles

The 3-tier oblate rectangle, $R^j(2, n)$, is identical with the dihedral hexagon $O(2, 2, n)$ [2, 6]. The general K formula for hexagon-shaped benzenoids (or hexagons) yields

$$K\{R^j(2, n)\} = \frac{\binom{n+2}{2}\binom{n+3}{3}}{n+1} = \frac{1}{12}(n+1)(n+2)^2(n+3) \tag{5}$$

Also the K formula for $R^j(3, n)$, the 5-tier oblate rectangle, may be obtained by means of the general K formula for hexagons [2, 6] when it is combined with the stripping [6], a procedure based on the method of fragmentation due to Randić [23]. An example is shown in Fig. 3, where a hexagon is subjected to stripping, whereby it is split into two fragments: the hexagon without corner, and a hexagon with one unit less in the last parameter. In general for the dihedral 5-tier hexagon (with arbitrary n) one arrives at

$$K\{O(3, 3, n)\} = K\{Oa(3, 3, n)\} + K\{O(3, 3, n-1)\}; \qquad n \geqq 1 \tag{6}$$

On subjecting $Oa(3, 3, n)$ to a stripping in the same way one obtains

$$K\{Oa(3, 3, n)\} = K\{Ob(3, 3, n)\} + K\{Oa(3, 3, n-1)\}; \qquad n \geqq 1 \tag{7}$$

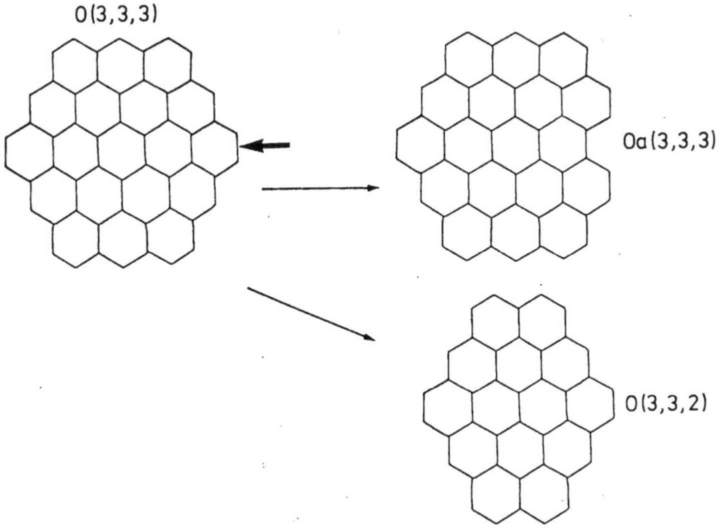

Fig. 3. Example of stripping: the method of fragmentation applied to a hexagon. The *arrow* indicates the bond which successively is assumed to be double and single

where Ob(3, 3, n) signifies the appropriate oblate rectangle without two opposite corners. A combination of (6) and (7) yields

$$K\{\mathrm{Ob}(3,3,n)\} = K\{\mathrm{O}(3,3,n)\} - 2K\{\mathrm{O}(3,3,n-1)\} + K\{\mathrm{O}(3,3,n-2)\}\,; \qquad n \geqq 2 \tag{8}$$

The system Ob(3, 3, n) is identical with the 5-tier oblate rectangle. On the right-hand side of (8) all quantities are the K numbers for different hexagons, for which the formulas are known. One obtains

$$K\{\mathrm{R}^{\mathrm{j}}(3,n)\} = \frac{\binom{n+3}{3}\binom{n+4}{4}\binom{n+5}{5}}{(n+1)\binom{n+2}{2}} - \frac{2\binom{n+2}{3}\binom{n+3}{4}\binom{n+4}{5}}{n\binom{n+1}{2}} + \frac{\binom{n+1}{3}\binom{n+2}{4}\binom{n+3}{5}}{(n-1)\binom{n}{2}}$$

$$= \frac{1}{120}(n+1)(n+2)^3(n+3)(n^2+4n+5) \tag{9}$$

The polynomial expression in (9) is valid for all n.

2.3 The Advanced Method of Chopping

The so-called advanced method of chopping [2] was employed in the derivation of the K formula for the 7-tier oblate rectangles, viz. [9]

$$K\{R^{j}(4, n)\} = \frac{1}{20160}(n+1)(n+2)^4(n+3) \times (17n^4 + 136n^3 + 439n^2 + 668n + 420) \qquad (10)$$

Here we shall illustrate the method on the simpler case of the 5-tier oblate rectangles; cf. the preceding paragraph.

In preparation of the treatment of the present paragraph it is expedient to demonstrate first the ordinary method of chopping [2, 24]. Figure 4 shows the chopping of a 5-tier prolate pentagon, $D^{i}(2, 4, n)$, a member of another class of regular strips. In the depicted example (Fig. 4) one has $n = 3$. A repeated application of the method of fragmentation [23] as indicated yields $n + 1$ fragments, each consisting of two disconnected benzenoids. One set of these benzenoids consists of the 2-tier parallelograms, $L(2, l)$, for which [2, 5]

$$K\{L(2, l)\} = \binom{l+2}{2} \qquad (11)$$

The degenerate case of no hexagons for $l = 0$ must here be incorporated. The second partners of the disconnected benzoids are members of an auxiliary benzenoid class, viz. $B(n, 2, -l)$; see the introductory definitions (Sect 1.1). For the pertinent K numbers it has been found [2, 9, 11]

$$K\{B(n, 2, -l)\} = \binom{n+2}{2}(l+1) - (n+2)\binom{l+1}{2} = \frac{1}{2}(n+2)(l+1)(n-l+1) \qquad (12)$$

As a result of the chopping of the 5-tier prolate pentagon $D^{i}(2, 4, n)$ one obtains

$$K\{D^{i}(2, 4, n)\} = \sum_{i=0}^{n} K\{L(2, i)\}\, K\{B(n, 2, -i)\} \qquad (13)$$

The summation in (13) may be evaluated by means of (11) and (12), but this issue is not pursued here, as it would lead us away from the main topic. Instead we give a formula similar to (13) for the simpler class of 4-tier pentagons [2, 6], viz. $D(2, 3, n)$. We can visualize a chopping of this system by deleting the top row in all the drawings of Fig. 4. The result is

$$K\{D(2, 3, n)\} = \sum_{i=0}^{n} K\{L(i)\}\, K\{B(n, 2, -i)\} \qquad (14)$$

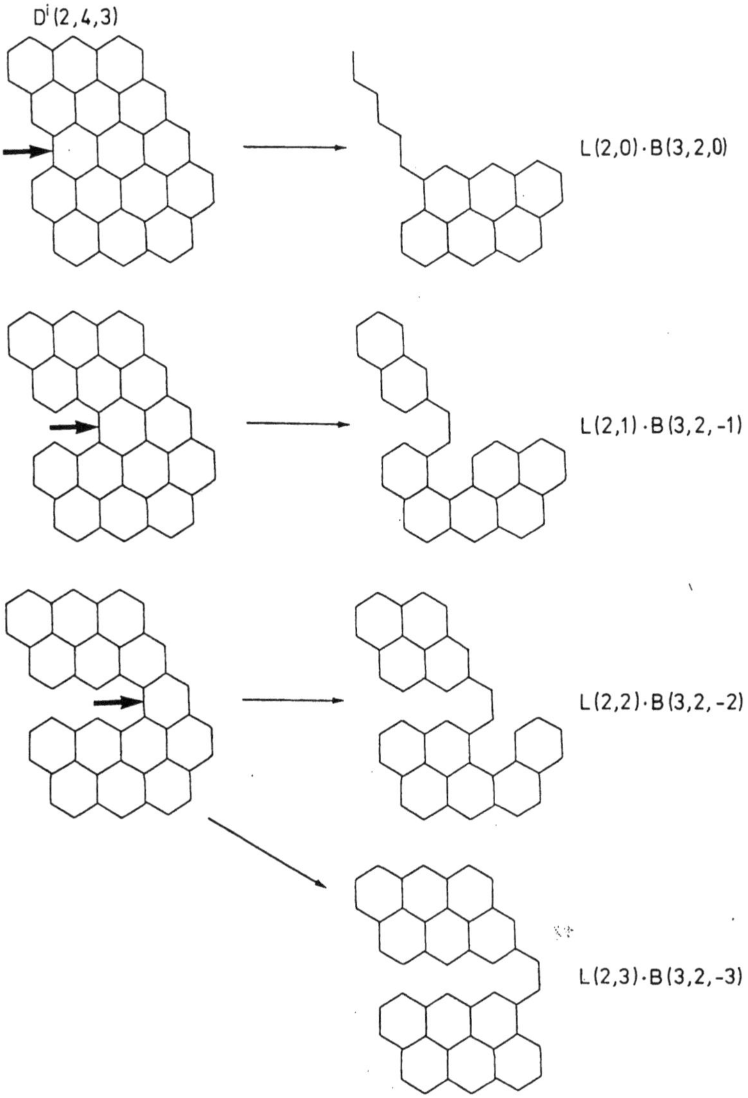

Fig. 4. Example of chopping: the method of fragmentation applied to a pentagon

Based on the chopping a fundamental relation for the *K* numbers of oblate rectangles was deduced, viz. [2, 9]

$$K\{\mathrm{R}^j(m, n)\} = \sum_{i=0}^{n} K\{\mathrm{B}(n, 2p, -i)\}\, K\{\mathrm{B}(n, 2q, -i)\}; \quad p + q = m - 1 \tag{15}$$

This relation is also applicable to $q = 0$ in a degenerate form, viz.

$$K\{R^j(m, n)\} = \sum_{i=0}^{n} K\{B(n, 2m - 2, -i)\} \tag{16}$$

The relations (15) and (16) applied to the 5-tier oblate rectangle ($m = 3$) yield

$$K\{R^j(3, n)\} = \sum_{i=0}^{n} [K\{B(n, 2, -i)\}]^2 \tag{17}$$

and

$$K\{R^j(3, n)\} = \sum_{i=0}^{n} K\{B(n, 4, -i)\} \tag{18}$$

respectively. As an illustration of (18) we observe that the sum of the four K numbers in Fig. 2, viz. 125 + 200 + 200 + 125 = 650, indeed is equal to the K number of the right-hand system in Fig. 1.

We are now prepared for a demonstration of the advanced method of chopping. Starting with Eq. (17) we insert from (12) into only one of the factors and obtain

$$K\{R^j(3, n)\} = \sum_{i=0}^{n} \left[\binom{n+2}{2}(i + 1) - (n + 2)\binom{i+1}{2}\right] K\{B(n, 2, -i)\} \tag{19}$$

By some elementary manipulations of the binomial coefficients one arrives at

$$K\{R^j(3, n)\} = \binom{n+3}{2} \sum_{i=0}^{n} (i + 1) K\{B(n, 2, -i)\} - (n + 2) \sum_{i=0}^{n} \binom{i+2}{2} K\{B(n, 2, -i)\} \tag{20}$$

By virtue of Eqs. (14) and (13), combined with (3) and (11), respectively, one finds that the summations in (20) may be identified with the appropriate K formulas so that

$$K\{R^j(3, n)\} = \binom{n+3}{2} K\{D(2, 3, n)\} - (n + 2) K\{D^i(2, 4, n)\} \tag{21}$$

The rest of the analysis depends on the methods chosen for the deduction of the K formulas on the right-hand side of (21). Different approaches are possible. Here we shall only suggest the stripping (see Sect. 2.2 above), from which one obtains

$$K\{D(2, 3, n)\} = \sum_{i=0}^{n} K\{Z(4, i)\} \tag{22}$$

and

$$K\{D^i(2, 4, n)\} = \sum_{i=0}^{n} K\{M_i(LLAAL)\} \quad (23)$$

Consequently the problem is reduced to finding the K formulas for two classes of multiple chains [1, 2], one of them consisting of zigzag chains [2, 25]. Here we do not give the intermediate results for $K\{Z(4, n)\}$ [2, 6, 25] and $K\{M_n(LLAAL)\}$ [1, 2]. For the classes invoked in (21) it was found [1, 2, 6]

$$K\{D(2, 3, n)\} = \frac{1}{24}(n + 1)(n + 2)^3(n + 3) \quad (24)$$

and

$$K\{D^i(2, 4, n)\} = \frac{1}{240}(n + 1)(n + 2)^2(n + 3)(n + 4)(3n + 5) \quad (25)$$

On inserting the expressions (24) and (25) into (21) the result indeed becomes equivalent to (9).

Members of four classes employed in this paragraph are found in Fig. 5.

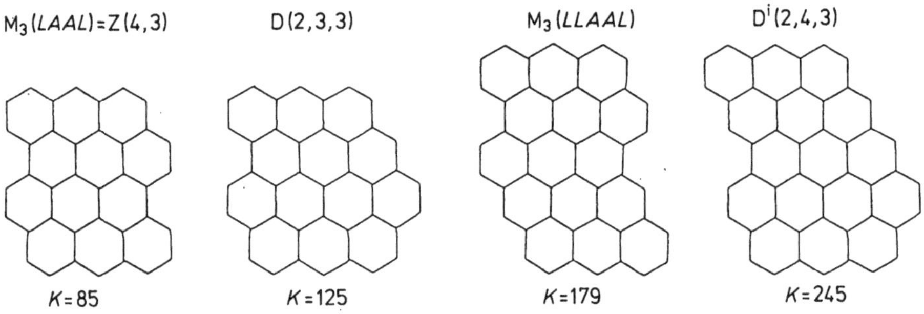

Fig. 5. Four members of certain benzenoid classes with relevance to rectangles

2.4 Fully Computerized Method

The fully computerized method is actually a numerical coefficient fitting for polynomials in general, but was developed in connection with Kekulé structure counts. It was used to reproduce the K formula (10) for $R^j(4, n)$ [9]. Furthermore, it allowed to proceed to the m value one unit larger, yielding [10]

$$K\{R^j(5, n)\} = \frac{1}{362880}(n + 1)(n + 2)^5(n + 3)(31n^6 + 372n^5 + 1942n^4 + 5616n^3 + 9511n^2 + 8988n + 3780) \quad (26)$$

From Eqs. (15) and (12) it is easily verified that $K\{\mathrm{R}^{\mathrm{j}}(m, n)\}$ is a polynominal in n. Furthermore, it was proved [9] that the degree of this polynomial is not greater than $3m - 2$. We write it

$$K\{\mathrm{R}^{\mathrm{j}}(m, n)\} = P_{3m-2}(n) \tag{27}$$

This property is consistent with $K\{\mathrm{R}^{\mathrm{j}}(3, n)\}$ being a polynomial of degree 7 in n; cf. Eq. (9). Let this polynomial be written

$$P_7(n) = A + Bn + C\binom{n}{2} + D\binom{n}{3} + E\binom{n}{4} + F\binom{n}{5} + G\binom{n}{6} + H\binom{n}{7} \tag{28}$$

The eight unknown coefficients of (28) are to be determined by means of eight numerical values of $P_7(n)$. They can be computed by a data program or otherwise. It is practical to include the degenerate case (4). Here we shall use the values: $P_7(0) = 1$, $P_7(1) = 18$, $P_7(2) = 136$, $P_7(3) = 650$, $P_7(4) = 2331$, $P_7(5) = 6860$, $P_7(6) = 17472$ and $P_7(7) = 39852$. The coefficients are found successively on inserting $n = 0, 1, 2, 3, 4, 5, 6$ and 7 into (28). It is expedient to set up the equations in the shape of Pascal's triangle as shown below.

$1 =$	$A;$	$A = 1$
$18 =$	$A + B;$	$B = 17$
$136 =$	$A + 2B + C;$	$C = 101$
$650 =$	$A + 3B + 3C + D;$	$D = 295$
$2331 =$	$A + 4B + 6C + 4D + E;$	$E = 476$
$6860 =$	$A + 5B + 10C + 10D + 5E + F;$	$F = 434$
$17472 =$	$A + 6B + 15C + 20D + 15E + 6F + G;$	$G = 210$
$39852 =$	$A + 7B + 21C + 35D + 35E + 21F + 7G + H;$	$H = 42$

The answer is:

$$K\{\mathrm{R}^{\mathrm{j}}(3, n)\} = P_7(n) = 42\binom{n}{7} + 210\binom{n}{6} + 434\binom{n}{5} + 476\binom{n}{4} + 295\binom{n}{3} + 101\binom{n}{2} + 17n + 1 \quad , \tag{29}$$

This expression is again equivalent to Eq. (9).

In the application of the fully computerized method it is advantageous to know some factors of the polynominal, which is to be derived. From the expressions (5), (9), (10) and (26) it is tempting to guess that $P_{3m-2}(n)$ *for* $m > 1$ has $m + 2$ linear factors so that

$$P_{3m-2}(n) = (n + 1)(n + 2)^m (n + 3)\, Q_{2m-4}(n); \qquad m > 1 \tag{30}$$

where $Q_{2m-4}(n)$ is a polynomial in n of the degree $2m - 4$. During the original applications of the fully computerized method for $R^j(4, n)$ [9] and $R^j(5, n)$ [10] the property (3) was only a conjecture, but proved for the special cases ($m = 4$ and 5). Later a rigorous proof of the general validity ($m = 2, 3, 4, \ldots$) of (30) was given [11], and the fully computerized method in factored form was used to derive $K\{R^j(6, n)\}$. The result [11], after correcting a misprint [2, 14], reads

$$K\{R^j(6, n)\} = \frac{1}{79833600}(n + 1)(n + 2)^6(n + 3)(691n^8 + 11056n^7 + 79788n^6 + 338320n^5 + 921759n^4 + 1654264n^3 + 1915562n^2 + 1315560n + 415800) \quad (31)$$

We shall demonstrate this variant of the fully computerized method for $R^j(3, n)$. In consistence with (30) we have

$$P_7(n) = (n + 1)(n + 2)^3(n + 3)\,Q_2(n) \quad (32)$$

where we set

$$Q_2(n) = A + Bn + C\binom{n}{2} \quad (33)$$

Hence only three numerical values of $P_7(n)$, and consequently of $Q_2(n)$, are required in order to determine the coefficients. We shall use $Q_2(0) = 1/24$, $Q_2(1) = 18/216 = 1/12$ and $Q_2(2) = 136/960 = 17/120$. Now the scheme of computation is:

$$\begin{array}{rlr} 1/24 = & A; & A = 1/24 \\ 1/12 = & A + B; & B = 1/24 \\ 17/120 = & A + 2B + C; & C = 1/60 \end{array}$$

Consequently:

$$K\{R^j(3, n)\} = \frac{1}{120}(n + 1)(n + 2)^3(n + 3)\left[2\binom{n}{2} + 5n + 5\right] \quad (34)$$

Again, as easily is seen, the answer is equivalent to Eq. (9).

2.5 Summation Method

The next milestone, i.e. the derivation of $K\{R^j(7, n)\}$, was achieved by a new method referred to as the summation method. It yielded [12]

$$K\{\mathrm{R}^{j}(7, n)\} = \frac{1}{6227020800}(n+1)(n+2)^7(n+3)(5461n^{10} + 109220n^9 + 1006407n^8 + 5617392n^7 + 21022809n^6 + 55133100n^5 + 102705053n^4 + 134421928n^3 + 118632870n^2 + 64047960n + 16216200) \quad (35)$$

The summation method is based on relation (16) together with [2, 11]

$$K\{\mathrm{B}(n, 2m-2s, -l)\} = (n-l+1)\sum_{i=0}^{l}(i+1)K\{\mathrm{B}(n, 2m-2s-2, -i)\} + (l+1)\sum_{i=l+1}^{n}(n-i+1)K\{\mathrm{B}(n, 2m-2s-2, -i)\}; \quad 1 \leqq s \leqq m-2 \quad (36)$$

Let us at once show the application to $\mathrm{R}^{j}(3, n)$. Then Eq. (18) is to be used together with

$$K\{\mathrm{B}(n, 4, -l)\} = (n-l+1)\sum_{i=0}^{l}(i+1)K\{\mathrm{B}(n, 2, -i)\} + (l+1)\sum_{i=l+1}^{n}(n-i+1)K\{\mathrm{B}(n, 2, -i)\} \quad (37)$$

After inserting from (12) the relation (37) was rendered into the form

$$K\{\mathrm{B}(n, 4, -l)\} = \frac{1}{2}(n+2)(n-l+1)\sum_{i=0}^{l}(i+1)^2 \times [(n+2)-(i+1)] + \frac{1}{2}(n+2) \times (l+1)\sum_{i=l+1}^{n}(i+1)[(n+2)-(i+1)]^2 \quad (38)$$

and consequently:

$$K\{\mathrm{B}(n, 4, -l)\} = \frac{1}{2}(n+2)^2(n-l+1)\sum_{i=0}^{l}(i+1)^2 - \frac{1}{2}(n+2) \times (n-l+1)\sum_{i=0}^{l}(i+1)^3 + \frac{1}{2}(n+2)^3$$

$$\times (l+1) \sum_{i=l+1}^{n} (i+1) - (n+2)^2$$
$$\times (l+1) \sum_{i=l+1}^{n} (i+1)^2 + \frac{1}{2}(n+2)$$
$$\times (l+1) \sum_{i=l+1}^{n} (i+1)^3 \tag{39}$$

After some elementary manipulations of the summations it was achieved

$$K\{B(n, 4, -l)\} = \frac{1}{2}(n+2)^3 (l+1) \sum_{i=0}^{n} (i+1) - \frac{1}{2}(n+2)^3$$
$$\times (l+1) \sum_{i=0}^{l} (i+1) - (n+2)^2 (l+1) \sum_{i=0}^{n} (i+1)^2$$
$$+ \frac{1}{2}(n+2)^2 (n+l+3) \sum_{i=0}^{l} (i+1)^2$$
$$+ \frac{1}{2}(n+2)(l+1) \sum_{i=0}^{n} (i+1)^3$$
$$- \frac{1}{2}(n+2)^2 \sum_{i=0}^{l} (i+1)^3 \tag{40}$$

The summations of (40) were expanded in terms of known formulas and the result simplified into:

$$K\{B(n, 4, -l)\} = \frac{1}{24}(n+2)^3 (n^2+4n+5)(l+1) - \frac{1}{24}(n+2)^2$$
$$\times (l+1)^2 - \frac{1}{12}(n+2)^3 (l+1)^3$$
$$+ \frac{1}{24}(n+2)^2 (l+1)^4 \tag{41}$$

The form is now ready for executing another set of summations in accord with Eq. (18). One has

$$K\{R^j(3, n)\} = \frac{1}{24}(n+2)^3 (n^2+4n+5) \sum_{i=0}^{n} (i+1)$$
$$- \frac{1}{24}(n+2)^2 \sum_{i=0}^{n} (i+1)^2 - \frac{1}{12}(n+2)^3 \sum_{i=0}^{n} (i+1)^3$$
$$+ \frac{1}{24}(n+2)^2 \sum_{i=0}^{n} (i+1)^4 \tag{42}$$

Again the known formulas for the summations may be inserted and the result simplified to the form (9).

2.6 The John-Sachs Theorem

The famous John-Sachs theorem [26] gives the Kekulé structure count of a benzenoid in terms of an $n_p \times n_p$ determinant, where n_p is the number of peaks [27], equal to the number of valleys [27]. As pointed out by Gutman and Cyvin [28] the elements of this John-Sachs determinant may be identified with K numbers of certain benzenoids, occasionally degenerated to an acyclic chain (polyene), or zero. An application to the oblate rectangles gave the result [2, 13]:

$$K\{R^j(m,n)\} = \begin{vmatrix} \binom{n+2}{2} & \binom{n+3}{4} & \binom{n+4}{6} & \cdots & \binom{n+m}{2m-2} & \binom{n+m}{2m-1} \\ (n+2) & \binom{n+3}{3} & \binom{n+4}{5} & \cdots & \binom{n+m}{2m-3} & \binom{n+m}{2m-2} \\ 0 & (n+2) & \binom{n+3}{3} & \cdots & \binom{n+m-1}{2m-5} & \binom{n+m-1}{2m-4} \\ \cdot & \cdot & \cdot & \cdots & \cdot & \cdot \\ 0 & 0 & 0 & \cdots & (n+2) & \binom{n+2}{2} \end{vmatrix} \quad (43)$$

This is an $m \times m$ determinant. The first row and last column are special and therefore separated by broken lines.

An efficient technique was developed for the expansion of the determinant. In essence, it is given by [14]

$$K\{R^j(m,n)\} = (-1)^{m-1}(n+2)^{m-1}\binom{n+m}{n-m+1} + \sum_{i=0}^{m-2}(-1)^i \times (n+2)^i\binom{n+i+2}{n-i} K\{B(n, 2m-2i-2, 0)\} \quad (44)$$

where the K formulas for the pertinent auxiliary classes are found according to the recurrence relation

$$K\{B(n, 2m-2, 0)\} = (-1)^m (n+2)^{m-2}\binom{n+m}{n-m+2} + \sum_{i=0}^{m-3}(-1)^i(n+2)^i\binom{n+i+3}{n-i} \times K\{B(n, 2m-2i-4, 0)\} \quad (45)$$

Eqs. (44) and (45) are especially convenient for practical applications because of the possible occurrence of vanishing binomial coefficients.

It was achieved to expand the determinant (43) for $m = 8$ with the aid of the relations (44) and (45). The result is the following K formula for the 15-tier oblate rectangle [14]:

$$K\{R^j(8, n)\} = \frac{1}{10461394944000}(n+1)(n+2)^8(n+3)$$
$$\times(929569n^{12} + 22309656n^{11} + 250158485n^{10}$$
$$+ 1731086820n^9 + 8229767127n^8 + 28315930608n^7$$
$$+ 72322500575n^6 + 138258580980n^5$$
$$+ 196559445604n^4 + 203012336736n^3$$
$$+ 144957849840n^2 + 64500408000n$$
$$+ 13621608000) \tag{46}$$

In our standard example, the 5-tier oblate rectangle, the John-Sachs determinant is of the order 3×3. The number of peaks (or valleys) is three in a special orientation of the benzenoid, as is shown in Fig. 6. For an arbitrary n (and $m = 3$) the John-Sachs determinant reads

$$K\{R^j(3, n)\} = \begin{vmatrix} \binom{n+2}{2} & \binom{n+3}{4} & \binom{n+3}{5} \\ (n+2) & \binom{n+3}{3} & \binom{n+3}{4} \\ 0 & (n+2) & \binom{n+2}{2} \end{vmatrix} \tag{47}$$

It is of course easy to execute the expansion of (47) directly, but we shall use the relations (44) and (45) for the sake of illustration.

From (44) with $m = 3$ one obtains

$$K\{R^j(3, n)\} = (n+2)^2\binom{n+3}{5} + \binom{n+2}{2}K\{B(n, 4, 0)\}$$
$$- (n+2)\binom{n+3}{4}K\{B(n, 2, 0)\} \tag{48}$$

The last K number, viz. $K\{B(n, 2, 0)\}$ pertains to the $L(2, n)$ parallelogram, for which one has

$$K\{L(2, n)\} = \binom{n+2}{2} = K\{B(n, 2, 0)\} \tag{49}$$

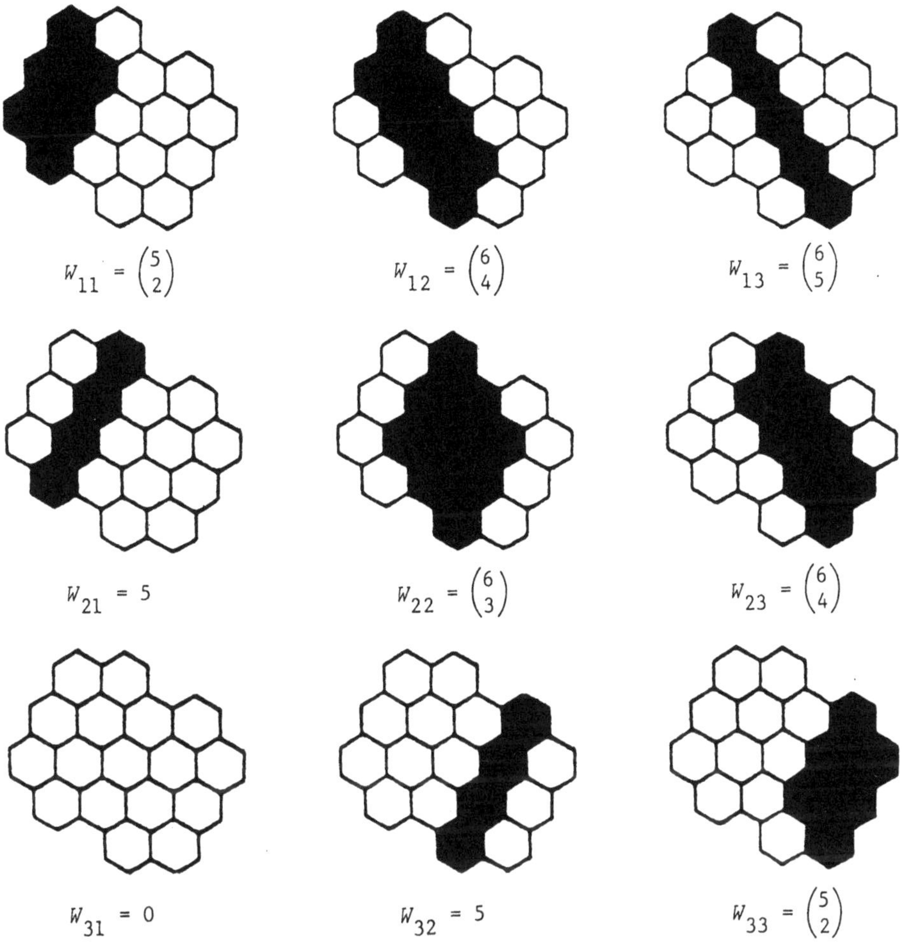

Fig. 6. Illustration of the John-Sachs method for $R^j(3, 3)$. The K numbers of the black benzenoids are the pertinent elements (W_{ij}) of the John-Sachs determinant

see also Eq. (11). The quantity (49) may be interpreted as the initial condition in (45). It gives

$$K\{B(n, 4, 0)\} = -(n+2)\binom{n+3}{4} + \binom{n+3}{3} K\{L(2, n)\} \tag{50}$$

This member of an auxiliary class, viz. B(n, 4, 0), is depicted in Fig. 2 for $n = 3$. It is also seen to be identical with D(2, 3, n); cf. Fig. 5. Indeed, on inserting from (49) into (50) the latter equation turns out to be equivalent with (24). Finally, with the aid of (49) and (50) Eqn. (48) becomes a function of n and equivalent to (9).

2.7 Transfer-Matrix Method

The transfer-matrix method is a powerful tool for studying Kekulé structures and their numbers [29–33]. In this approach one studies the manner in which a Kekulé structure may be propagated from one position at one side of a monomer (or unit) cell to the other side of the cell. For the present oblate rectangles $R^j(m, n)$ one might choose the cells to be the regions lying between the broken lines in Fig. 7. Next the different possible local characters (of the Kekulé structures) at the boundary of a cell are to be specified. In fact, one can see that exactly one π-bond of a Kekulé structure cuts through any of the cell boundaries we have chosen or, in other words, each broken line cuts exactly one double bond. This is a manifestation of a type of "long-range order" that we have much discussed elsewhere. Thus the local structure at a boundary may be labelled by an integer locating the bond on the boundary that is made double. See for example Fig. 7. We then specify the "local state" (or structure) at each boundary by a column vector (of length $n + 1$) with zeroes in all positions except for a one in the location corresponding to the double bond. Then the local states in Fig. 7 are $\{1, 0, 0, 0\}$, $\{0, 0, 0, 1\}$, $\{0, 1, 0, 0\}$, or symbolically: $|1)$, $|4)$, $|2)$.

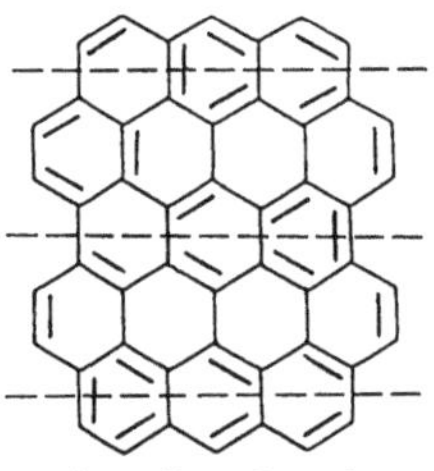

Fig. 7. The $R^j(3, 3)$ rectangle divided into two unit cell regions, plus two end regions. A particular Kekulé structure is indicated. The double-bond locations on the three boundaries of the two cells are identified by: 1, 4, 2

The next step is to determine the number of ways it is permissible to propagate from one local state to another at an adjacent boundary. We shall give a detailed derivation of the transfer matrix for an oblate rectangle $R^j(m, n)$ in general since the transfer-matrix method has not been applied to this particular case before. Consider the boundaries on two sides of a cell as in Fig. 8(a), with local states $|p)$ and $|q)$ at the bottom and top. The other bonds on the boundaries are necessarily single, and are deleted in Fig. 8(b), where also we note that there is only one permissible placement for double bonds in the region between p and q (indicated by dotted lines). Deletion of these intervening bonds that are determined leaves us with the structure of Fig. 8(c), which is seen to consist of two polyacenic regions. For $p \leqq q$ one has p ways to place double bonds on the left-hand region and $n + 2 - q$ ways to place them on the right-hand region. Thence, the number of ways to propagate from $|p)$ to $|q)$ is

$$(q| \mathbf{T} |p) = p(n + 2 - q); \qquad p \leqq q \tag{51}$$

If $p \geqq q$, one simply interchanges p and q on the left- (or right-) hand side of the above equation;

$$(q| \mathbf{T} |p) = (n + 2 - p)\, q; \qquad p \geqq q \tag{52}$$

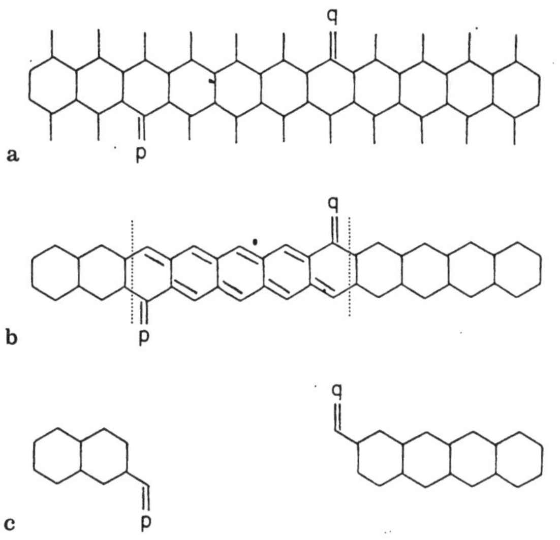

Fig. 8. Constructions to determine (via the development in the text) the number of ways to propagate from a local state $|p)$ to an adjacent one $|q)$

The **T** matrix with q,p-th element $(q|\,\mathbf{T}\,|p)$ is our transfer matrix, which counts manners of propagation between possible pairs of adjacent local states. According to (51) and (52) the transfer matrix (**T**) for oblate rectangles has the appearance:

$$\mathbf{T} = \begin{bmatrix} n+1 & n & n-1 & \dots & 1 \\ & 2n & 2(n-1) & \dots & 2 \\ & & 3(n-1) & \dots & 3 \\ & & & \dots & \cdot \\ \text{(symmetric)} & & & & n+1 \end{bmatrix} \tag{53}$$

Powers of the transfer matrix account for propagation between local states that are more distant. That is, $(q|\,\mathbf{T}^l\,|p)$ gives the number of ways of propagating from $|p)$ across l cells to $|q)$. To count Kekulé structures on $R^j(m, n)$ we note that there are $m - 1$ cells and that any one of the local states may occur at the boundaries of the initial and final cells. Thus the total Kekulé structure count is

$$K\{R^j(m, n)\} = \sum_{i=1}^{n+1} \sum_{j=1}^{n+1} (j|\,\mathbf{T}^{m-1}\,|i) \tag{54}$$

The transfer matrix formula can often be further manipulated to yield other formulas of interest.

Let us first consider, as a numerical example, the $n = 3$ case. From Eq. (53) one has

$$\mathbf{T} = \begin{bmatrix} 4 & 3 & 2 & 1 \\ 3 & 6 & 4 & 2 \\ 2 & 4 & 6 & 3 \\ 1 & 2 & 3 & 4 \end{bmatrix} \tag{55}$$

and for the second and third powers:

$$\mathbf{T}^2 = \begin{bmatrix} 30 & 40 & 35 & 20 \\ 40 & 65 & 60 & 35 \\ 35 & 60 & 65 & 40 \\ 20 & 35 & 40 & 30 \end{bmatrix} \tag{56}$$

and

$$\mathbf{T}^3 = \begin{bmatrix} 330 & 510 & 490 & 295 \\ 510 & 820 & 805 & 490 \\ 490 & 805 & 820 & 510 \\ 295 & 490 & 510 & 330 \end{bmatrix} \tag{57}$$

respectively. Then upon summation over the elements of $\mathbf{T}$, $\mathbf{T}^2$ and $\mathbf{T}^3$ one obtains the Kekulé structure counts for the $m = 2$, 3 and 4 cases;

$$K\{\mathrm{R}^{\mathrm{j}}(2, 3)\} = 10 + 15 + 15 + 10 = 50 \tag{58}$$

$$K\{\mathrm{R}^{\mathrm{j}}(3, 3)\} = 125 + 200 + 200 + 125 = 650 \tag{59}$$

$$K\{\mathrm{R}^{\mathrm{j}}(4, 3)\} = 1625 + 2625 + 2625 + 1625 = 8500 \tag{60}$$

Here the intermediate steps simply indicate row sums.

In the above examples, Eqs. (56) and (59) pertain to $m = 3$ (and $n = 3$). In the following we shall treat the general case of $m = 3$ (arbitrary n) and once more arrive at the K formula for $\mathrm{R}^{\mathrm{j}}(3, n)$, the standard example used for demonstration of the methods in preceding paragraphs.

First we derive a general expression for an element of $\mathbf{T}^2$. With the aid of Eqs. (51) and (52) it is found:

$$\begin{aligned}(p|\,\mathbf{T}^2\,|q) = (q|\,\mathbf{T}^2\,|p) &= \sum_{i=1}^{p-1} (n + 2 - p)\, i(n + 2 - q)\, i \\ &+ \sum_{i=p}^{q-1} p(n + 2 - i)\,(n + 2 - q)\, i \\ &+ \sum_{i=q}^{n+1} p(n + 2 - i)\, q(n + 2 - i); \qquad p \leqq q\end{aligned} \tag{61}$$

A tedious, but elementary expansion of these summations yields the expression

$$\begin{aligned}(p|\,\mathbf{T}^2\,|q) = (q|\,\mathbf{T}^2\,|p) &= \frac{1}{6}(n + 2)\, p\{2(n + 2)^2 q - (n + 2 - q) \\ &\times (p^2 - 1) - [3(n + 2) - q]\, q^2\}; \\ &p \leqq q\end{aligned} \tag{62}$$

The next step is to execute the summation of the elements according to (54), viz.

$$K\{R^j(m, n)\} = \sum_{i=1}^{n+1} (i|\,T^2\,|i) + 2 \sum_{i=1}^{n} \sum_{j=i+1}^{n+1} (j|\,\mathbf{T}^2\,|i) \tag{63}$$

The first one of these summations is

$$\sum_{i=1}^{n+1} (i|\,\mathbf{T}^2\,|i) = \frac{1}{3}(n+2)\sum_{i=1}^{n+1} i^4 - \frac{2}{3}(n+2)^2 \sum_{i=1}^{n+1} i^3 + \frac{1}{6}(n+2) \times (2n^2+8n+7)\sum_{i=1}^{n+1} i^2 + \frac{1}{6}(n+2)^2 \sum_{i=1}^{n+1} i \tag{64}$$

where the summations were worked out and the result simplified to:

$$\sum_{i=1}^{n+1} (i|\,\mathbf{T}^2\,|i) = \frac{1}{180}(n+1)(n+2)^2(n+3)(2n^2+8n+15) \tag{65}$$

The second (double) summation in (63) was executed in two steps. First we have the summation over j, viz.

$$\sum_{j=i+1}^{n+1} (j|\,\mathbf{T}^2\,|i) = \frac{1}{6}(n+2)\,i \sum_{j=i+1}^{n+1} j^3 - \frac{1}{2}(n+2)^2\, i \sum_{j=i+1}^{n+1} j^2 + \frac{1}{6}(n+2)(2n^2+8n+7+i^2)\,i \sum_{j=i+1}^{n+1} j - \frac{1}{6}(n+2)^2(i^3-i)\sum_{j=i+1}^{n+1} 1 \tag{66}$$

where the summations again were worked out, and the result simplified to the following form, which is amenable for the subsequent summation over i.

$$\sum_{j=i+1}^{n+1} (j|\,\mathbf{T}^2\,|i) = -\frac{1}{8}(n+2)\,i^5 + \frac{1}{6}(n+2)(2n+3)\,i^4 - \frac{1}{24}(n+2)(6n^2+16n+7)\,i^3 - \frac{1}{12}(n+2)(2n^2+9n+9)\,i^2 + \frac{1}{24}(n+1)(n+2)^2(n^2+5n+8)\,i \tag{67}$$

After the summation over i and simplification of the result we arrive at the expression:

$$\sum_{i=1}^{n} \sum_{j=i+1}^{n+1} (j|\mathbf{T}^2|i) = \frac{1}{720} n(n+1)(n+2)^2(n+3) \times (3n^2 + 14n + 23) \tag{68}$$

Finally, by inserting from (65) and (68) into (63) the known formula for $K\{\mathrm{R}^j(3, n)\}$ is achieved; see Eq. (9).

2.8 Conclusion

We have demonstrated a number of methods for deducing the K formula of $\mathrm{R}^j(3, n)$ as a steadily recurring example. Most of the methods have been developed during the intensive studies of the Kekulé structure count (K) for oblate rectangles, $\mathrm{R}^j(m, n)$, with fixed values of m. All of the computations of $K\{\mathrm{R}^j(3, n)\}$ demonstrated in the preceding sections evidently require a degree of patient work to develop. However, each of the methods, but the transfer-matrix method, was applied to derive a K formula for the class of oblate rectangles with an unprecedentedly large m value. The use of the transfer-matrix method to treat oblate rectangles with fixed values of n is among the topics for the next section.

3 Oblate Rectangles with Fixed Values of *n*

3.1 Linearly Coupled Recurrence Relations

The problem of finding $K\{\mathrm{R}^j(m, n)\}$ for some of the lowest (fixed) values of n was first attacked by Gutman [15], who employed a method of linearly coupled recurrence relations. He emphasized the necessity to invoke auxiliary benzenoid classes related to the oblate rectangles.

In the systematic treatment of Chen et al. [18] the same problem was treated by means of the auxiliary classes $\mathrm{B}(n, 2m-2, -l)$ exclusively; they are defined in Sect. 1.1. Here we shall briefly demonstrate the application of this method to the case of $n = 3$. From Eq. (16) it is obtained

$$K\{\mathrm{R}^j(m, 3)\} = 2K\{\mathrm{B}(3, 2m-2, 0)\} + 2K\{\mathrm{B}(3, 2m-2, -1)\} \tag{69}$$

From (17) one obtains similarly (for $p = 1$ and $p = 2$)

$$\begin{aligned} K\{\mathrm{R}^j(m, 3)\} &= 20K\{\mathrm{B}(3, 2m-4, 0)\} + 30K\{\mathrm{B}(3, 2m-4, -1)\} \\ &= 250K\{\mathrm{B}(3, 2m-6, 0\} + 400K\{\mathrm{B}(3, 2m-6, -1)\} \end{aligned} \tag{70}$$

Hence

$$K\{\mathrm{R^j}(m+1,3)\} = 20K\{\mathrm{B}(3,2m-2,0)\} + 30K\{\mathrm{B}(3,2m-4,-1)\} \tag{71}$$

$$K\{\mathrm{R^j}(m+2,3)\} = 250K\{\mathrm{B}(3,2m-6,0)\} + 400K\{\mathrm{B}(3,2m-6,-1)\} \tag{72}$$

On eliminating the two quantities on the right-hand sides of (69), (71) and (72), one attains at a relationship between the left-hand side quantities, which virtually represents a recurrence relation. We write it in the form:

$$K\{\mathrm{R^j}(m+1,3)\} = 15K\{\mathrm{R^j}(m,3)\} - 25K\{\mathrm{R^j}(m-1,3)\}; \quad m \geqq 2 \tag{73}$$

3.2 Explicit Formulas

From a recurrence relation like (73) and a sufficient number of initial conditions an explicit formula for the quantity in question may be obtained by the standard method of difference equations. In the present case $K\{\mathrm{R^j}(1,3)\} = 4$ and $K\{\mathrm{R^j}(2,3)\} = 50$ may serve as the initial conditions. The resulting explicit formula for $K\{\mathrm{R^j}(m,3)\}$ reads

$$K\{\mathrm{R^j}(m,3)\} = \frac{2}{5}\left[(5+2\sqrt{5})\left(\frac{15+5\sqrt{5}}{2}\right)^{m-1} + (5-2\sqrt{5})\left(\frac{15-5\sqrt{5}}{2}\right)^{m-1}\right] \tag{74}$$

For reasons to be apparent later this equation is given in a form slightly different from the earlier expressions [9, 16].

3.3 General Formulations

The number of terms in the recurrence relation for $K\{\mathrm{R^j}(m,n)\}$ increases in general with increasing values of n. It was found that it has exactly $\frac{1}{2}(n+4)$ terms when n is even and $\frac{1}{2}(n+3)$ when n is odd. In general we write:

$$K\{\mathrm{R^j}(m+1,n)\} = \sum_{j=0}^{n'} c_j K\{\mathrm{R^j}(m-j,n)\}; \qquad n' = \left[\frac{n}{2}\right] \tag{75}$$

The bracket symbol means that $[x]$ is the largest integer smaller than or equal to x. The quantities c_j are constants; in the example of Eq. (73): $c_0 = 15$, $c_1 = -25$.

In the book of Cyvin and Gutman [2] explicit expressions for the coefficients c_j are quoted without proof as unpublished results of Chen and Cyvin:

$$c_j = \frac{(-1)^j (n+2)^{j+2}}{4(j+1)} \begin{bmatrix} \frac{n}{2} + j + 1 \\ 2j+1 \end{bmatrix}; \qquad n = 0, 2, 4, \ldots \tag{76a}$$

$$c_j = (-1)^j (n+2)^{j+1} \begin{bmatrix} \frac{n+1}{2} + j + 1 \\ 2j+2 \end{bmatrix}; \qquad n = 1, 3, 5, \ldots \tag{76b}$$

The characteristic equation to be solved in order to find the explicit formula for $K\{R^j(m, n)\}$ has the degree $\frac{1}{2}(n+2)$ when n is even and $\frac{1}{2}(n+1)$ when n is odd. Accordingly, the solution of (74) was found from a quadratic equation. It is clear that the possibilities to produce exact explicit formulas along this line are limited; the solutions for $n = 4$ and $n = 5$ have been accomplished [17, 18] by means of the relevant cubic equations, but the answer for larger n values have not been found previously.

In this connection the transfer-matrix method (Sect. 2.7) shows its superiority to the other methods, which have been treated here. By an elaborate expansion of the eigenvectors and eigenvalues of **T** it was arrived at a general formula for $K\{R^j(m, n)\}$, which is suitable for producing the explicit formulas of $K\{R^j(m, n)\}$ with fixed values of n. It reads

$$K\{R^j(m, n)\} = \frac{2}{n+2} \sum_{p=1}^{n+1} \left[\cot \frac{\pi p}{2(n+2)}\right]^2 \lambda_p^{m-1} \tag{77}$$

where only odd values of p give nonvanishing terms, and λ_p, the eigenvalues, are given by

$$\lambda_p = \frac{n+2}{4}\left[\sin \frac{\pi p}{2(n+2)}\right]^{-2} \tag{78}$$

The application to $n = 3$ gives

$$K\{R^j(m, 3)\} = \frac{2}{5}\left[\left(\cot \frac{\pi}{10}\right)^2 \lambda_1^{m-1} + \left(\cot \frac{3\pi}{10}\right)^2 \lambda_3^{m-1}\right] \tag{79}$$

where

$$\lambda_1 = \frac{5}{4}\left(\sin \frac{\pi}{10}\right)^{-2}, \qquad \lambda_3 = \frac{5}{4}\left(\sin \frac{3\pi}{10}\right)^{-2} \tag{80}$$

The equivalence with Eq. (74) in the form it was chosen is easily established by realizing the fact that

$$\left(\cot\frac{\pi}{10}\right)^2 = 5 + 2\sqrt{5} \tag{81}$$

etc.

In consistence with (75) the characteristic equation for an arbitrary n reads

$$\lambda^{n'+1} - \sum_{j=0}^{n'} c_j \lambda^{n'-j} = 0\,; \qquad n' = \left[\frac{n}{2}\right] \tag{82}$$

with the c_j coefficients from (76). The general solution of this equation is given by (78).

3.4 Discussion

Among the results of Kekulé structure counts for oblate rectangles the most general formulas which have been found, are given by Eqs. (43) and (77). We shall not call these formulas explicit because of the m-dependent order of the determinant in the former case (43) and the n-dependent summation in the latter (77). Equation (54) is also general, but not explicit in the strict sense. The determinant formula (43) is amenable for producing explicit formulas of $K\{R^j(m, n)\}$ with fixed values of m. On the other hand, the summation formula (77) gives straightforwardly explicit formulas of $K\{R^j(m, n)\}$ with fixed values of n. Also from Eq. (54) the K formulas for $R^j(m, n)$ with fixed values of m can be deduced, but it requires far more labor than the application of (43).

A "perfectly" explicit formula for $K\{R^j(m, n)\}$, where both m and n are arbitrary, is not known and seems not likely ever to be found. In the last section we shall show an example of what we mean by a "perfectly" explicit formula.

4 A Perfectly Explicit Formula: Perforated Rectangles

Consider a "perforated" oblate rectangle as depicted in Fig. 9. Like a rectangle it is a system with two parameters (m, n). A perforated rectangle, $\varrho(m, n)$, belongs to the multiple coronoids; it has $m - 1$ corona holes.

A study of the Kekulé structure counts for perforated rectangles gave the result:

$$K\{\varrho(m, n)\} = \frac{1}{4r}\left[(r + 1)(2n + 2r)^m + (r - 1)(2n - 2r)^m\right] \tag{83}$$

where

$$r = \sqrt{n^2 - 2n + 2} \tag{84}$$

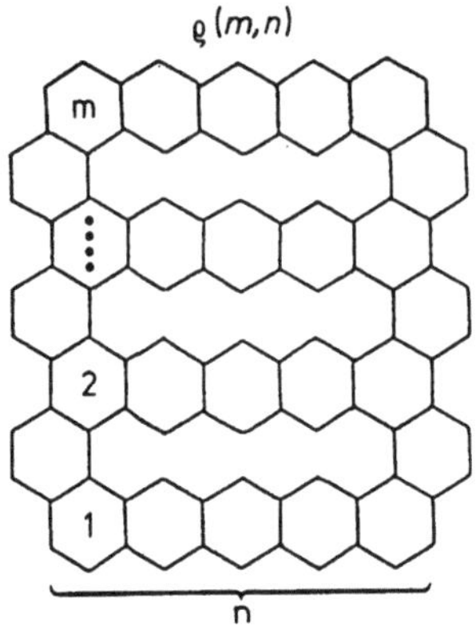

Fig. 9. Definition of the class of perforated (oblate) rectangles.

We characterize this formula as being "perfectly" explicit because it gives straightforwardly the explicit formulas of K for $\varrho(m, n)$ with fixed values of m as well as fixed values of n. For example, one finds for $m = 3$ (n arbitrary)

$$K\{\varrho(3, n)\} = 8(2n^3 - n^2 + 2n + 1) \tag{85}$$

and for $n = 3$ (m arbitrary)

$$K\{\varrho(m, 3)\} = \frac{1}{4\sqrt{5}}[(\sqrt{5} + 1)(6 + 2\sqrt{5})^m + (\sqrt{5} - 1)(6 - 2\sqrt{5})^m] \tag{86}$$

5 Acknowledgements

Financial support to BNC from The Norwegian Research Council for Science and the Humanities is gratefully acknowledged. DJK acknowledges the support from The Welch Foundation of Houston, Texas.

6 References

1. Cyvin SJ, Cyvin BN, Gutman I (1985) Z. Naturforsch. 40a: 1253
2. Cyvin SJ, Gutman I (1988) Kekulé structures in benzenoid hydrocarbons, Springer, Berlin Heidelberg New York
3. Brunvoll J, Cyvin BN, Cyvin SJ, Gutman I (1988) Match 23: 209
4. Yen TF (1971) Theor. Chim. Acta 20: 399
5. Gordon M, Davison WHT (1952) J. Chem. Phys. 20: 428
6. Cyvin SJ (1986) Monatsh. Chem. 117: 33
7. Jiang YS (1980) Scient. Sinica 23: 847
8. Ohkami N, Hosoya H (1983) Theor. Chim. Acta 64: 153
9. Cyvin SJ, Cyvin BN, Bergan JL (1986) Match 19: 189
10. Cyvin SJ (1986) Match 19: 213
11. Chen RS (1986) Match 21: 259
12. Chen RS (1986) Match 21: 277

13. Cyvin SJ, Cyvin BN, Brunvoll J, Chen RS, Su LX (1987) Match 22: 141
14. Cyvin SJ, Cyvin BN, Chen RS (1987) Match 22: 151
15. Gutman I (1985) Match 17: 3
16. Chen RS (1986) J. Xinjiang Univ. 3(2): 13
17. Su LX (1986) Match 20: 229
18. Chen RS, Cyvin SJ, Cyvin BN (1987) Match 22: 111
19. Zhang FJ, Chen RS (1986) J. Xinjiang Univ. 3(3): 10
20. Cyvin SJ, Chen RS, Cyvin BN (1987) Match 22: 129
21. Chen RS, Cyvin SJ (1988) Match 23: 179
22. Chen RS (1987) In: King RB, Rouvray DH (eds) Graph theory and topology in chemistry, Elsevier, Amsterdam, p 552
23. Randić M (1976) J. Chem. Soc. Faraday Trans. 2 72: 232
24. Cyvin SJ (1988) Monatsh. Chem. 119: 41
25. Gutman I, Cyvin SJ (1987) Monatsh. Chem. 118: 541
26. John P, Sachs H (1985) In: Bodendiek R, Schumacher H, Walter G (eds) Graphen in Forschung und Unterricht, Verlag Barbara Franzbecker, Bad Salzdetfurth, p 85
27. Sachs H (1984) Combinatorica 4: 89
28. Gutman I, Cyvin SJ (1987) Chem. Phys. Letters 136: 137
29. Klein DJ, Hite GE, Seitz WA, Schmalz TG (1986) Theor. Chim. Acta 69: 409
30. Klein DJ, Hite GE, Schmalz TG (1986) J. Comput. Chem. 7: 443
31. Klein DJ, Živković TP, Trinajstić N (1987) J. Math. Chem. 1: 309
32. Schmalz TG, Seitz WA, Klein DJ, Hite GE (1988) J. Am. Chem. Soc. 110: 1113
33. Hite GE, Živković TP, Klein DJ (1988) Theor. Chim. Acta 74: 349

Clar's Aromatic Sextet and Sextet Polynomial

Haruo Hosoya

Department of Chemistry, Ochanomizu University, Bunkyo-ku, Tokyo 112, Japan

Table of Contents

Several illustrations of the Clar's aromatic sextet theory for the electronic properties of benzenoid hydrocarbons are demonstrated. It is shown how various techniques and concepts of the graph theory are useful for realizing and formulating not only this purely empirical theory but also the mathematical beauty of the structural formula of aromatic hydrocarbons. It is proposed that the sextet polynomial $B_G(x)$ be defined in terms of the resonant sextet number p(G, k). For a "thin" polyhex graph, $B_G(x)$ is shown to be equal to the number of the Kekulé structures K(G), while for "fat" polyhex the concept of the super-sextet needs to be introduced. Proper and improper sextets, Clar transformation, and sextet rotation are defined so that the relevant graph-theoretical manipulations can be transformed into algebra. By using the sextet polynomial and other graph-theoretical concepts thus defined, novel mathematical relations among several resonance-theoretical quantities which have been proposed by other researchers were found. Correlation between Clar's aromatic sextet and the benzene character proposed by Polansky and also the partial electron density map is pointed out.

1 Clar's Aromatic Sextet

Hückel molecular orbital calculation clearly shows the difference in the π-electronic energy of anthracene (Ia) and phenanthrene (IIa) in accordance with their thermodynamic stability.

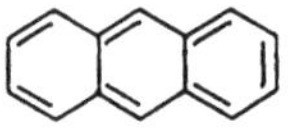

Ia

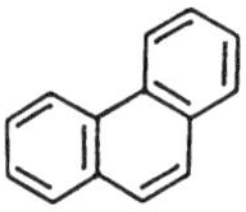

IIa

The π-electronic energy is obtained by adding up the solutions of the characteristic, or secular, polynomial of the graph representing the carbon atom skeleton of the molecule. Namely, for I and II we have

$$P_{\mathrm{I}}(x) = x^{16} - 16x^{12} + 98x^{10} - 296x^{8} + 473x^{6} - 392x^{4} + 148x^{2} - 16$$

$$E_{\pi} = 19.3137$$

$$P_{\mathrm{II}}(x) = x^{14} - 16x^{12} + 98x^{10} - 297x^{8} + 479x^{6} - 407x^{4} + 166x^{2} - 25$$

$$E_{\pi} = 19.4483$$

The absolute value of the last term of the characteristic polynomial is known to be the square of the number of the Kekulé structures, $K(G)$ [1].

Nowadays it is very difficult to pinpoint in the classical literatures in organic chemistry the credit of attributing the relative stability of unsaturated hydrocarbon molecules to $K(G)$ [2, 3]. On the other hand, long before these quantum-chemical theories were introduced Robinson proposed using a circle inside each benzene ring of an aromatic hydrocarbon molecule to represent the six mobile electrons and also the derived aromatic stability [4]. However, his symbol does not reflect any difference in the stability between I and II as,

Ib

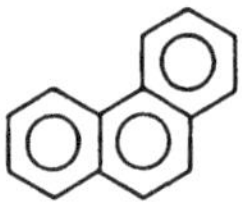

IIb

It was Clar [5] who invented a novel usage of the circles and arrows on the Kekulé structure of aromatic hydrocarbons to represent the stability of the molecule and local aromatic characters based on his own numerous experimental results and empirical rules [6], but not on sophisticated quantum-chemical treatment. According to his scheme the ground state of I and II can well be represented by the following Clar formulas, Ic and IIc,

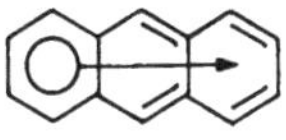

Ic

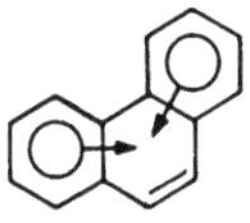

IIc

where a circle represents exactly the six π-electrons moving in a certain hexagon and no two neighboring hexagons are allowed to share π-electrons simultaneously to form their own sextets. Thus only one circle can be drawn for I, while for II two circles can coexist or can be resonant with each other. The arrow indicates that the aromatic sextet can move along the hexagons spanned by that arrow suggesting the "dilution" of the aromatic character over the region spanned by that arrow. That an alternative structure IId has very small contribution to the ground state of

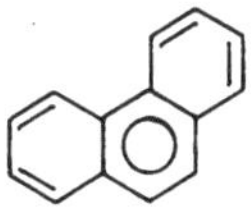

IId

phenanthrene can be inferred from ready addition of bromine to the "fixed" double bond in the central hexagon as depicted in IIc. If we call the maximum number of resonant sextets mnrs, the so-called Clar formula should have mnrs sextets.

The Clar formula, the $K(G)$ number, mnrs, and properties of various aromatic hydrocarbon molecules are given in Table 1. As deduced from the comparison of I and II and also of III and IV, in every member of linear polyacenes the mnrs remains to be unity regardless of the size of the molecule, while in zigzag polyacenes every kink contributes to increase the mnrs of the molecule. These different behaviors are in parallel with the relative stabilities of these two classes of benzenoid hydrocarbons [7].

Table 1. Typical benzenoid hydrocarbons and their resonance-theoretical and chemical properties.

No.	Clar Formula	$K(G)$	mnrs[a]	color	m.p. (°C)	H_2SO_4[b]
III	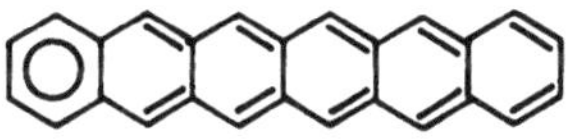	7	1	dark green	400 (decomp.)	○
IV	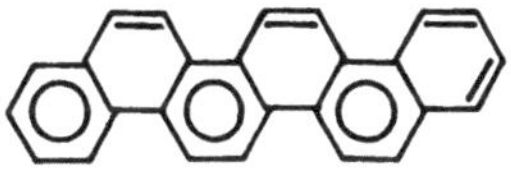	21	3	colorless	364	○
V		20	3	pale yellow	439	×

Table 1. (continued)

VI	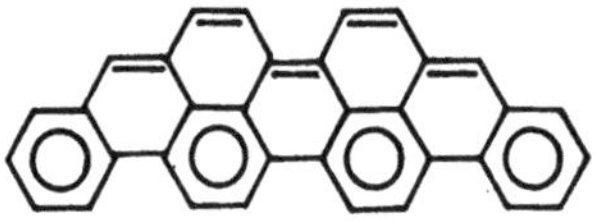	41	4	bright red	478	○
VII		50	4	orange	473	○
VIII	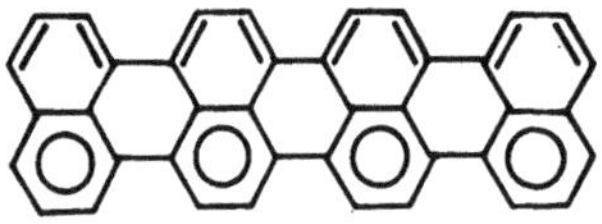	81	4	green	>570	○
IX	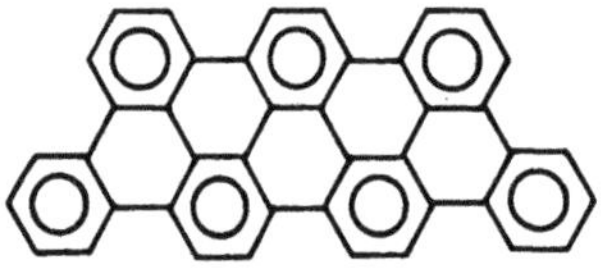	227	7	orange yellow	>540	×
X	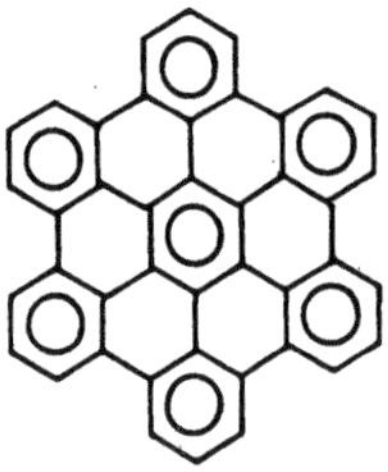	250	7	yellow	>700	×
XI	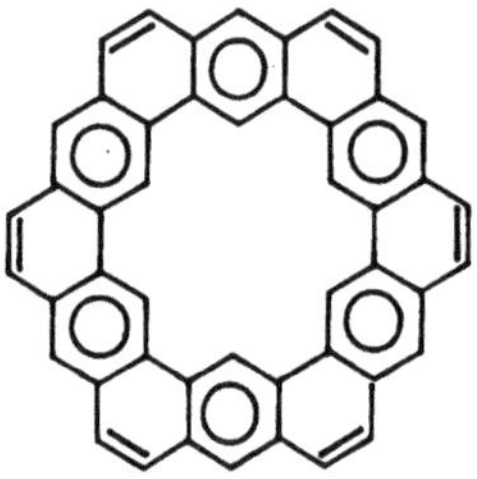	200	6	greenish yellow	>620	×

a Maximum number of resonance sextets.
b Solubility in concentrated sulfuric acid.

Further, branching of hexagonal units gives additional increase in the mnrs. Triphenylene, XIIa, has the largest mnrs

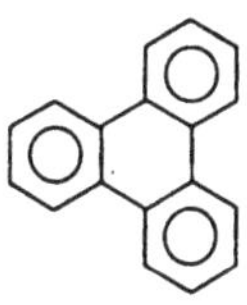

XIIa

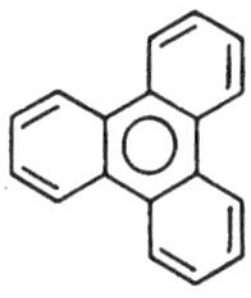

XIIb

(=3) and is the most stable among the isomeric tetrahexes, i.e., benzenoid hydrocarbons with four hexagons. In the central hexagon of XIIa there is drawn no aromatic sextet nor any fixed double bond. In other words triphenylene gains its stability by the conjugation of aromatic and "vacant" sextets. Although practically no aromatic character is observed in the central vacant hexagon, the structure XIIb seems to play an important role to make the terminal aromatic sextets in XIIa conjugate with each other (vide infra), just as the relation of IId with IIc. One can expand the network of XIIa by alternately adding the aromatic and vacant sextets and leaving no fixed double bond as IX and X in Table 1. Clar called such stable benzenoid hydrocarbon molecules fully benzenoid hydrocarbons [5].

Remarkably high stability of X, hexabenzocoronene, prevents us from measuring its melting point by a conventional apparatus. The glass capillary for sealing the sample melts before the crystal of X undergoes any change on heating. Its high stability can be predicted from its "fully aromatic" structure. The so-called kekulene, XI, is not a fully benzenoid hydrocarbon, but is known to be rather stable. The electronic spectrum of XI is similar to that of II but not of I [8]. Thus XI may be deemed as cyclic phenanthrene but not cyclic anthracene.

Many researchers tried to explain the secret of the Clar's aromatic sextet theory, or hypothesis from quantum-chemical points of view. However, those trials have been failing until the graph and combinatorial theories came to be applied to this challenging problem [9, 10]. In the following discussion it will be shown how various techniques and concepts of the graph theory are useful for realizing and formulating not only the fantastic theory of Clar but also the mathematical beauty of the structural formula of aromatic hydrocarbons.

2 Definition of Sextet Polynomial [9, 10]

We will mainly be concerned with the benzenoid and coronoid systems, which are meant to be the carbon atom skeletons of aromatic hydrocarbon molecules irrespctive of their real existence. In some cases branches may be attached to the benzenoid and coronoid skeletons, representing the extension of the π-electronic network toward the outer region of the polycyclic system. In any case the system is alternant. The thermodynamic stability of the electronic ground state of these series of molecules can be predicted from the following graph-theoretical discussion. Although Clar found many empirical relations between the wavelengths of the electronic spectra of an aromatic hydrocarbon and its Clar formula, extension of the graph-theoretical discussion toward the properties of the excited states has not been developed so far [5].

For the majority of the benzenoid and coronoid systems with an even number ($n = 2k$) of carbon atoms one can draw one or more Kekulé structures spanning all the carbon atoms with a set of k disjoint double bond. Kekulé structure is synonymous either with perfect matching, or 1-factor. In this chapter each of such patterns, k_i, is called a Kekulé pattern. Let the number of the Kekulé patterns, or the maximum matching number, of a graph G be denoted as $|k_i|$, or $K(G)$. A Kekuléan is a graph with non-zero $K(G)$. When two isomeric Kekuléan aromatic hydrocarbons are compared, generally the one with more Kekulé patterns is more stable than the other, as in the cases of I and II and of III and IV.

A graph with an odd number of points is non-Kekuléan by definition. No benzenoid hydrocarbon molecule or radical corresponding to non-Kekuléan graph has ever been synthesized. The phenalene skeleton, XIIIa, is the smallest non-Kekuléan benzenoid. However, it is regrettably true that even phenalenyl radical is

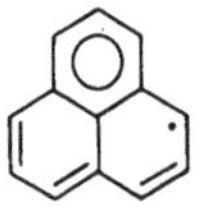

XIII a

XIII b

sometimes represented by an obviously misleading structure as XIIIb. There are two groups of non-Kekuléans in benzenoid and coronoid systems with even *n*. That the so-called Clar's hydrocarbon, XIVa, is non-Kekuléan is a direct outcome of the fact

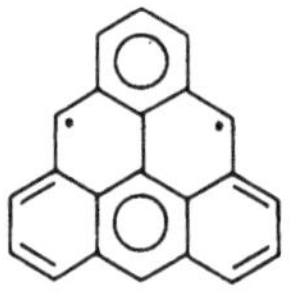

XIV a

XIV b

that the numbers of the starred (=12) and unstarred (=10) atoms are different. On the other hand, graph XV does not have any

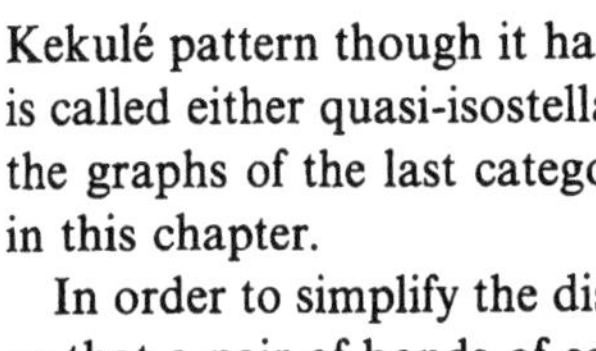

XV

Kekulé pattern though it has the same numbers of starred and unstarred atoms and is called either quasi-isostellar [11] or concealed non-Kekuléan graph [12]. Although the graphs of the last category are interesting targets, we will no longer treat them in this chapter.

In order to simplify the discussion, a benzenoid or coronoid system is to be drawn so that a pair of bonds of each hexagon lie in parallel with the vertical line. Let the sets of the circularly arranged three double bonds as shown below in a given Kekulé pattern be called, respectively, proper and improper sextets.

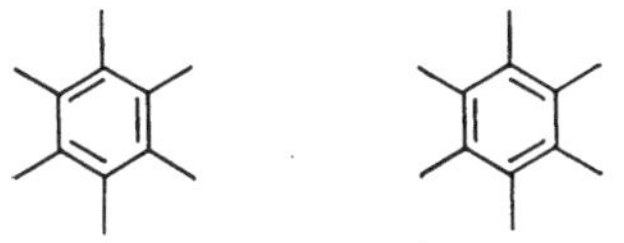

For a given benzenoid or coronoid system one can draw the set of all the $K(G)$ Kekulé patterns as exemplified in Fig. 1 for benzanthracene, XVI. The pair of patterns k_1 and k_2 differ only in the arrangement of proper and improper sextets in a certain hexagon. According to Clar one may draw a circle representing an aromatic sextet in that hexagon as below. The sextet pattern

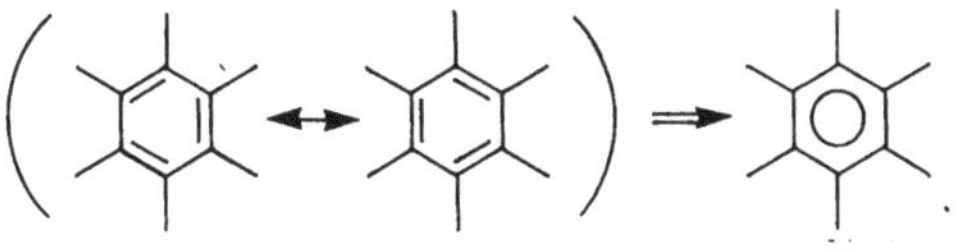

is defined as the pattern obtained by suppressing the remaining double bonds from this Kekulé pattern with an aromatic sextet. The pair of Kekulé patterns (k_1, k_2) gives the sextet pattern s_4. Similarly one can find other sets of k_i's to give various sextet patterns, such as, $(k_1, k_3) \rightarrow s_6$, $(k_1, k_6) \rightarrow s_3$, $(k_2, k_4) \rightarrow s_6$, $(k_2, k_5) \rightarrow s_5$, etc. One may also draw a set of more than two aromatic sextets on disjoint hexagons from the combination of the Kekulé patterns as (k_1, k_3, k_6, k_7) to give s_1. The two aromatic sextets can be said to be resonant with each other. However, no two aromatic sextets on the hexagons with a common bond can be resonant. For any benzenoid or coronoid system a zero-sextet pattern is defined as the sextet pattern with no aromatic sextet. Thus one can prepare the set of all the possible sextet patterns as in Fig. 1 for benzanthracene. Note that the following pattern is not a sextet pattern, since its parent pattern does not belong to the family of the Kekulé patterns.

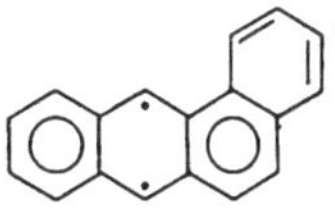

As see from Fig. 1 the numbers of the Kekulé and sextet patterns are the same. In a later discussion we will show that for "thin" benzenoid and coronoid systems this is always the case.

For a given polyhex graph one can draw a set of sextet patterns with various numbers of resonant sextets including the zero-sextet pattern. Let the number of the sextet patterns of G with k resonant sextets be denoted as $r(G, k)$. The total number of the sextet patterns $|s_i|$ is

$$|s_i| = \sum_{k=0}^{m} r(G, k), \tag{1}$$

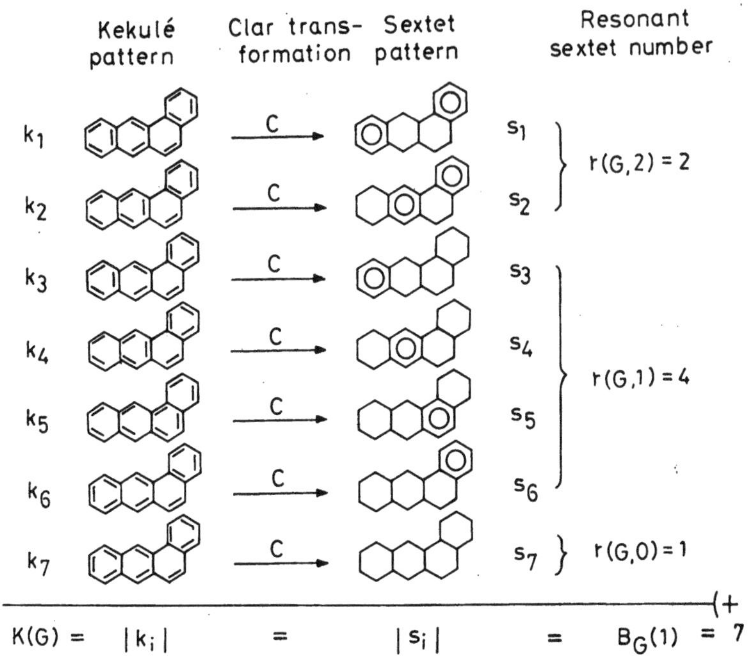

Fig. 1 One-to-one correspondence between the Kekulé and sextet patterns of benzanthracene to give its sextet polynomial

where m is the largest number of k. The sextet polynomial $B_G(x)$ for a benzenoid and coronoid system G is defined as

$$B_G(x) = \sum_{k=0}^{m} r(G, K)\, x^k\,. \tag{2}$$

It is obvious from the above definition that for any benzenoid and coronoid system there is one and only one zero-sextet pattern, i.e.,

$$r(G, 0) = 1\,. \tag{3}$$

In Table 2 are given the sextet polynomials for the lower members of aromatic hydrocarbons together with their $K(G)$ numbers. The sextet polynomials for larger benzenoid graphs are extensively tabulated and discussed [13, 14]. As already mentioned, for a "thin" benzenoid or coronoid system there is exactly a one-to-one relation between the Kekulé patterns and sextet patterns. In other words the following equality is obeyed,

$$|k_i| = |s_i|\,. \tag{4}$$

This can be stated as a theorem.

Table 2. Sextet polynomials of lower members of benzenoid hydrocarbons.

Benzenoid	$B_G(x)$	$B_G(1)$	$B'_G(1)$	m[a]
	$1 + x$	2	1	1
	$1 + 2x$	3	2	1
	$1 + 3x$	4	3	1
	$1 + 3x + x^2$	5	5	2
	$1 + 4x$	5	4	1
	$1 + 4x + 2x^2$	7	8	2
 	$1 + 4x + 3x^2$	8	10	2
	$1 + 4x + 3x^2 + x^3$	9	13	3
	$1 + 4x + x^2$	6	6	2
	$1 + 5x$	6	5	1
	$1 + 5x + 3x^2$	9	11	2
	$1 + 5x + 4x^2$	10	13	2
 	$1 + 5x + 5x^2$	11	15	2
 	$1 + 5x + 5x^2 + x^3$	12	18	3
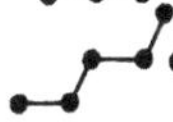  				
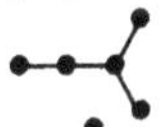	$1 + 5x + 6x^2 + x^3$	13	20	3
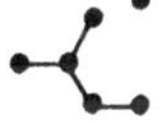	$1 + 5x + 5x^2 + 2x^3$	13	21	3
	$1 + 5x + 6x^2 + 2x^3$	14	23	3
	$1 + 5x + 3x^2$	9	11	2
	$1 + 4x + 4x^2$	9	12	2
	$1 + 5x + 4x^2 + x^3$	11	16	3
	$1 + 7x + 9x^2 + 2x^3$	19	31	3
	$1 + 8x + 9x^2 + 2x^3$ [b]	20	32	3

a Maximum number of resonant sextets.
b Super-sextet is added.

Theorem 1
For a thin graph

$$K(\mathrm{G}) = B_{\mathrm{G}}(1)\,. \tag{5}$$

For an unbranched catacondensed benzenoid graph G, or a catahex, Gutman showed that several different types of graphs, such as "caterpillar graph" and "Clar graph" can be defined so as to yield the same counting polynomial as the sextet polynomial of G [15]. This problem is extensively analyzed by El-Basil et al. [16–18]. It is very difficult to give a rigorous mathematical proof for a general case here. However, we can observe interesting mathematical relation between the two sets of patterns.

3 Clar Transformation and Sextet Rotation [10]

Define the Clar transformation (C) as a simultaneous substitution of all the proper sextets by circles in a given Kekulé pattern k_i followed by the suppression of the remaining double bonds into single bonds, as exemplified for graph XVI given in Fig. 1.

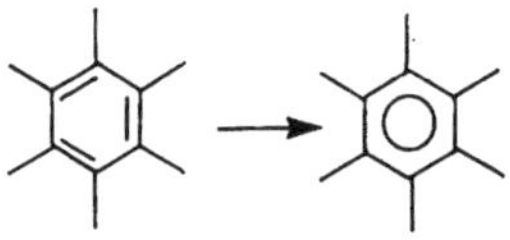

(6)

It can symbolically be written as

$$C(k_i) = s_i . \quad (7)$$

Define the sextet rotation (R) as a simultaneous rotation of all the proper sextets in a given Kekulé pattern k_i into the improper sextets to give another Kekulé pattern k_j,

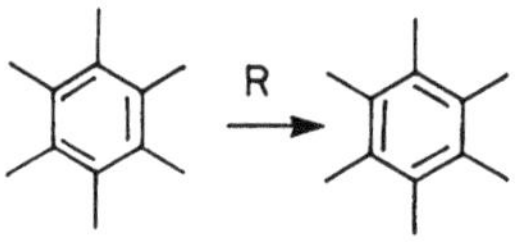

(8)

or symbolically as

$$R(k_i) = k_j . \quad (9)$$

For example, we get $R(k_1) = k_7$ for XVI in Fig. 1. Note that for such k_i with no proper sextet, e.g. k_7 in Fig. 1, one cannot operate the sextet rotation. In this case let us put it down as

$$R(k_i) = \varnothing ,$$

and call such k_i the root Kekulé pattern.

Similarly the counter-sextet rotation ($\tilde{R}$) is defined as follows:

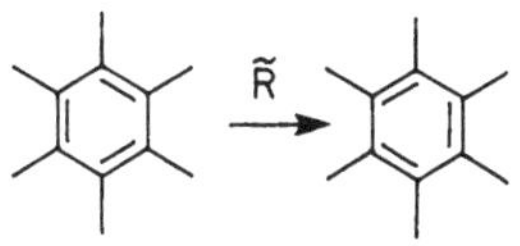

(10)

Note that the operation R and $\tilde{R}$ are not the inverses each other. As evident from Fig. 1 the one-to-one correspondence between the sets of k_i and s_i is observed through the Clar transformation. Namely, the relation (5) is shown.

Although for thin graphs the proof is straightforward, for fat graphs we have to define the super-sextet, such as Va, which

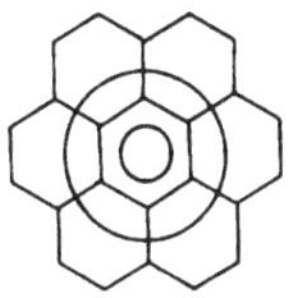

Va

corresponds to the pattern 3 in Fig. 2 for coronene. As seen in Table 2 the number of the sextet patterns without supersextet is smaller than $K(\mathrm{G})$ by one. Ohkami gave a complete proof for Eq. (5) by defining the super-sextet properly [19], while He and He also succeeded in proving this relation from a different approach [20]. Here we will just point out that the key of these proofs lies in the following Lemma:

Lemma

For each benzenoid or coronoid system, there exists one and only one root Kekulé pattern.

Although the variable x in the sextet polynomial $B_{\mathrm{G}}(x)$ does not mean anything other than what holds the power k and the coefficient $r(\mathrm{G}, k)$, $B_{\mathrm{G}}(x)$ can be formally differentiated with respect to x as

$$B'_{\mathrm{G}}(x) = \frac{d}{dx} B_{\mathrm{G}}(x) . \tag{11}$$

Then it is easy to get the second Theorem:

Theorem 2

$$B'_{\mathrm{G}}(1) = \sum_{k=1}^{m} k\, r(\mathrm{G}, k) = \sum_{i}^{\text{hexagon in G}} K(\mathrm{G} \ominus \mathrm{r}_i) . \tag{12}$$

where $K(\mathrm{G} \ominus \mathrm{r}_i)$ represents the number of the Kekulé patterns for the subgraph of G obtained by deleting ring r_i from G together with all the lines incident to r_i.

This theorem states that the summation of $K(\mathrm{G} \ominus \mathrm{r}_i)$ over all the entries $\{\mathrm{s}_i\}$ of the sextet pattern gives the total number of the aromatic sextets in $\{\mathrm{s}_i\}$ and can be obtained as $B'_{\mathrm{G}}(1)$. The proof is easy and not given here.

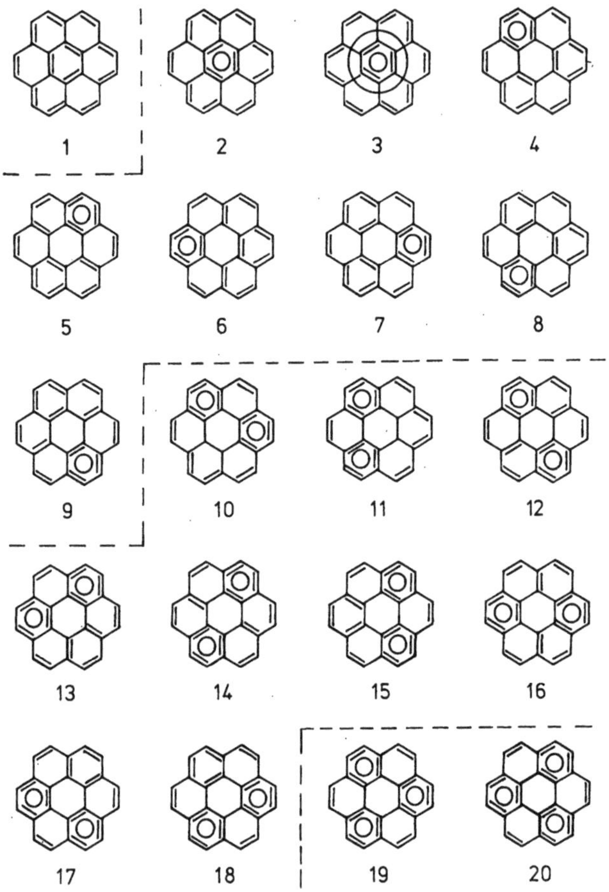

Fig. 2. Twenty Kekulé and sextet patterns of coronene to give its sextet polynomial, $B_G(x) = 1 + 8x + 9x^2 + 2x^3$. The pattern 3 is the super-sextet pattern. See Table 2 and Fig. 4

The number B'_G (1) can be partitioned into the component hexagons in G by assigning the value $K(G \ominus r_i)$ for hexagon i as exemplified below for XVI:

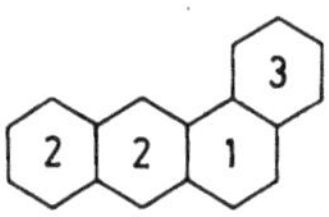

XVI a

Randić [21] and Aihara [22] independently proposed an idea of the index of local aromaticity, ILA, and overall index of aromaticity, OIA, based on the counting of the Kekulé patterns. However, these concepts were found to be closely related to the sextet polynomial and its derivative as [9]

$$\mathrm{ILA_i} = 2K(\mathrm{G} \ominus \mathrm{R}_i)/K(\mathrm{G}) \tag{13}$$

$$\mathrm{OIA} = 2\sum_i \{K(\mathrm{G} \ominus \mathrm{R}_i)/K(\mathrm{G})\} = 2B'_{\mathrm{G}}(1)/B_{\mathrm{G}}(1) \tag{14}$$

While the Clar pattern clearly distinguishes between the two kinds of benzene rings in an aromatic hydrocarbon network with and without aromatic character, the diagram as XVIa or the index as ILA give finer differences in the aromatic character among the component hexagons. This line of reasoning is also taken into consideration by Herndon and Hosoya in their parametrized valence bond calculation which will be explained later [23]. Further, it will also be shown that by plotting the partial π-electron density map we can clearly observe the interesting features of the local π-electronic structure in the ground state of an aromatic hydrocarbon molecule, such as the aromatic character and bond fixation, as predicted from the valence bond structures [24].

Before going into these discussions let us again focus our attention on the mathematical structure of the Kekulé and sextet patterns of benzenoid hydrocarbon graphs. The third theorem reads as follows:

Theorem 3

The sextet rotation to the set of the Kekulé patterns $\{k_i\}$ gives a directed tree graph with a root, or the root Kekulé pattern, representing the hierarchical structure of $\{k_i\}$, where each point corresponds to a Kekulé pattern.

In Figs. 3 and 4 are given the rooted directed trees derived by joining all the entries of the Kekulé patterns of XVI and V with the sextet and counter-sextet rotations. One can realize how all the Kekulé and sextet patterns are related to each other. Theorem 3 can be proved by showing that there is one and only one root Kekulé pattern, a Kekulé pattern without any proper sextet, (see Lemma) and no cyclic relation among the Kekulé patterns with respect to the sextet rotation. This graph-theoretical discussion does not necessarily mean that the root Kekulé pattern is the most chemically-important in the family of the Kekulé patterns, but that all the patterns are mathematically related with each other.

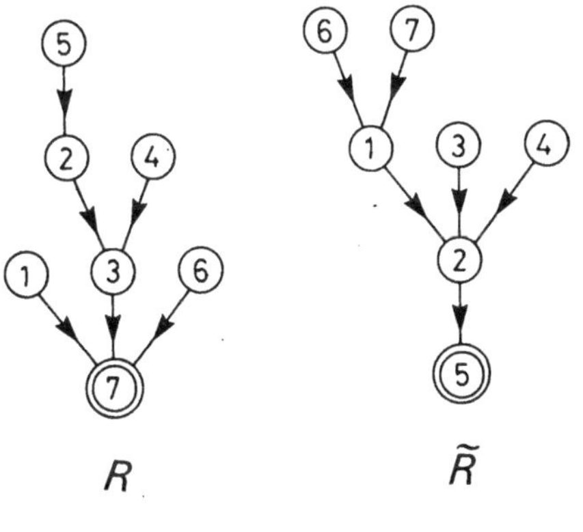

Fig. 3. Directed rooted trees of the Kekulé patterns of benzanthracene derived by ($\tilde{R}$) performing the sextet and counter-sextet rotations. The double (R) circle represents the root Kekulé pattern

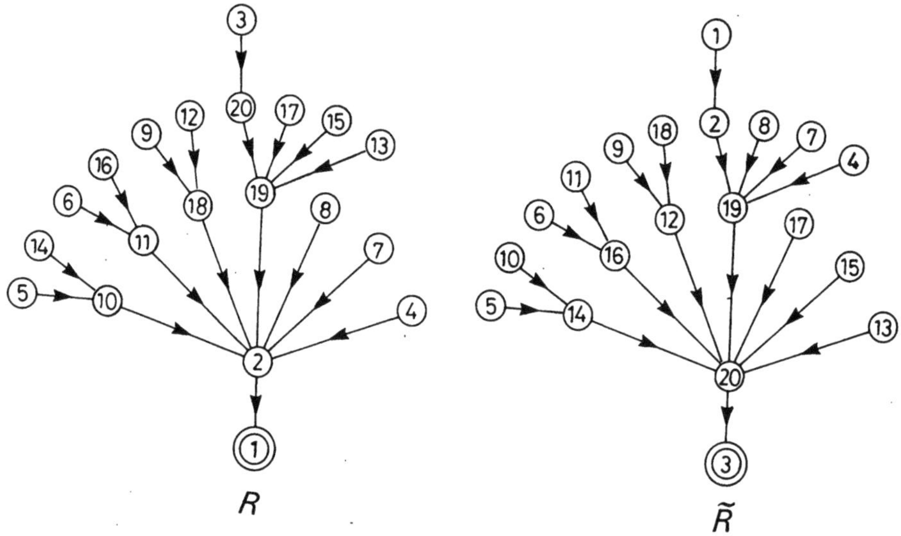

Fig. 4. Directed rooted trees of the Kekulé patterns of coronene. Note their isomorphic structures and the similarity in their local structures

Although the graph-theoretical discussion of the aromatic sextet is not complete without formulating explicitly the definition of the super-sextet, we will skip over to the discussion of more chemical relevance [25].

4 Realization of The Clar Formula

In quantum-chemical terminology the ground state of a benzenoid hydrocarbon is expressed in terms of the linear combination of the wavefunctions corresponding to the set of the Kekulé patterns. The π-electronic energy of each Kekulé pattern can be approximated as the sum of the contributions from those of the aromatic sextets, $a\,(<0)$, and isolated double bonds, $b\,(<0)$. For example, in the case of triphenylene, the energies of XIIa and XIIb are, respectively, expressed as $3a$ and $a + 6b$. That the two sextet patterns are connected by the sextet rotation of the central sextet can be interpreted as the resonance between the two structures with an exchange integral k. Then the energy of the ground state of the compound is to be obtained as the lowest root of the following determinant,

$$\begin{vmatrix} 3a - x & k \\ k & a + 6b - x \end{vmatrix} = 0$$

This is the central idea of the theory proposed by Herndon and Hosoya to simulate the resonance energy obtained from the SCFMO calculation by parametrizing the values of a, b, and k [23]. They took the reference values from the PPP type calculation by Dewar and de Llano [26] for the set of ten typical benzenoid hydrocarbons, and they could simulate the resonance energies of nearly thirty benzenoid hydrocarbons with rather high correlation coefficient. According to their results the ground state of XII can be expressed as the mixture of 95% of XIIa and 5% of XIIb, and for coronene, V, the contribution of the Kekulé pattern with one aromatic sextet in the central hexagon is as low as two per cent.

Polansky and Derflinger proposed a useful concept of "benzene character" which is the projection of the ocupied π-MO's in a given hexagon L of a polycyclic benzenoid hydrocarbon onto the three occupied MO's of a benzene molecule located on that position [27]. This quantity is shown to be expressed as a linear combination of the Coulson bond order for the component six bonds and for the three para-bonds in L. Their original definition of the benzene character for L can be transformed into the "normalized" benzene character as [28]

$$r_L = \left(2 \sum_{r>s}^{\text{ortho in L}} p_{rs} - \sum_{r>s}^{\text{para in L}} p_{rs}\right) \Big/ 3 - 2 \tag{15}$$

For example, the values of $1000 r_L$ calculated with Hückel MO for XVI and XII are as follows:

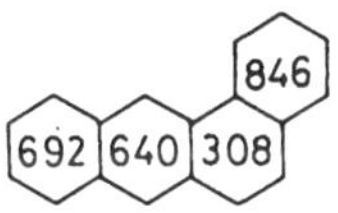

XVI b

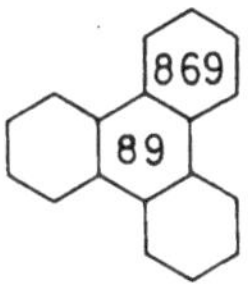

XII c

which is to be compared, respectively, with XVIa and the values quoted in the above discussion. Extensive analysis has been done for a number of benzenoid hydrocarbons [28, 29]. One can decompose the MO benzene character (15) into the sum of the orbital contributions of the occupied orbitals. It was then found that the local aromaticity of benzenoid hydrocarbons is largely determined by the combined contribution of the highest three occupied MO's, whereas thermal reactivities are known to be largely determined by HOMO [30–33].

Based on this conclusion one can introduce the concept of the partial electron density and draw its contour map in the plane just above and below, say one Bohr radius, the molecular plane [24, 29, 34]. Again for XVI and XII the results of ϱ_3 are given in Fig. 5, which is the contribution of the highest three occupied Hückel MO's. Note that due to the pairing theorem proved by Coulson and Rushbrooke [35] the π-electron densities on all the component carbon $2p\pi$ orbitals are the same and the contour map of the conventional electron density cannot differentiate any of the local aromaticity of alternant hydrocarbon molecules.

In Fig. 6 we will show the results of ϱ_3 for X, XI, and another hexabenzocoronene, XVII [29]. Although ϱ_3 is not an

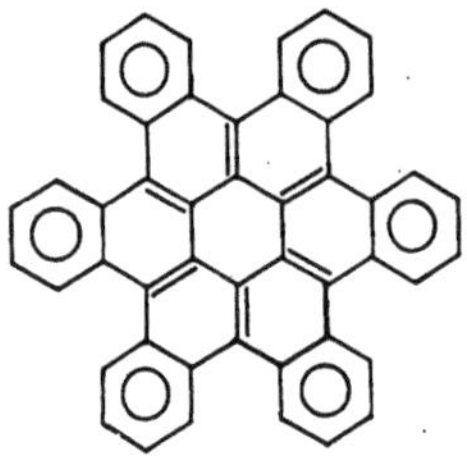

XVII

observable quantity, in these maps we can now realize what Eric Clar perceived amidst of his vast amount of chaotic experimental results for benzenoid hydrocarbons.

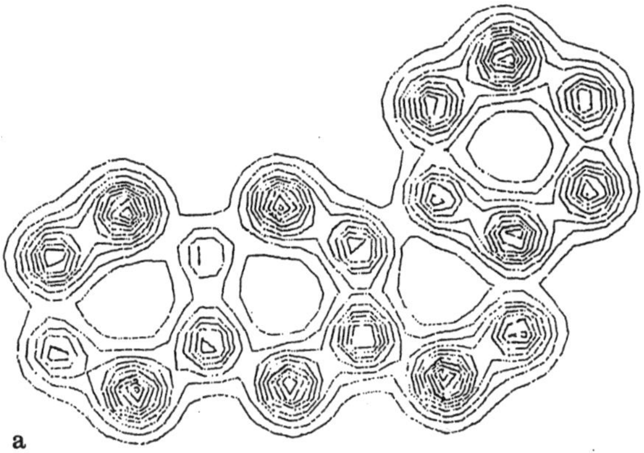

a

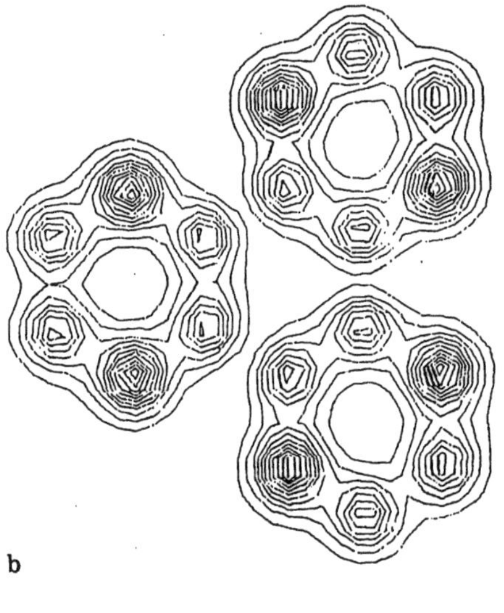

b

Fig. 5. The contour maps of the partial π-electron density, ϱ_3, of (**a**) benzanthracene and (**b**) triphenylene obtained from the highest three occupied orbitals. The values are taken on the plane one Bohr radius above and parallel to the molecular plane

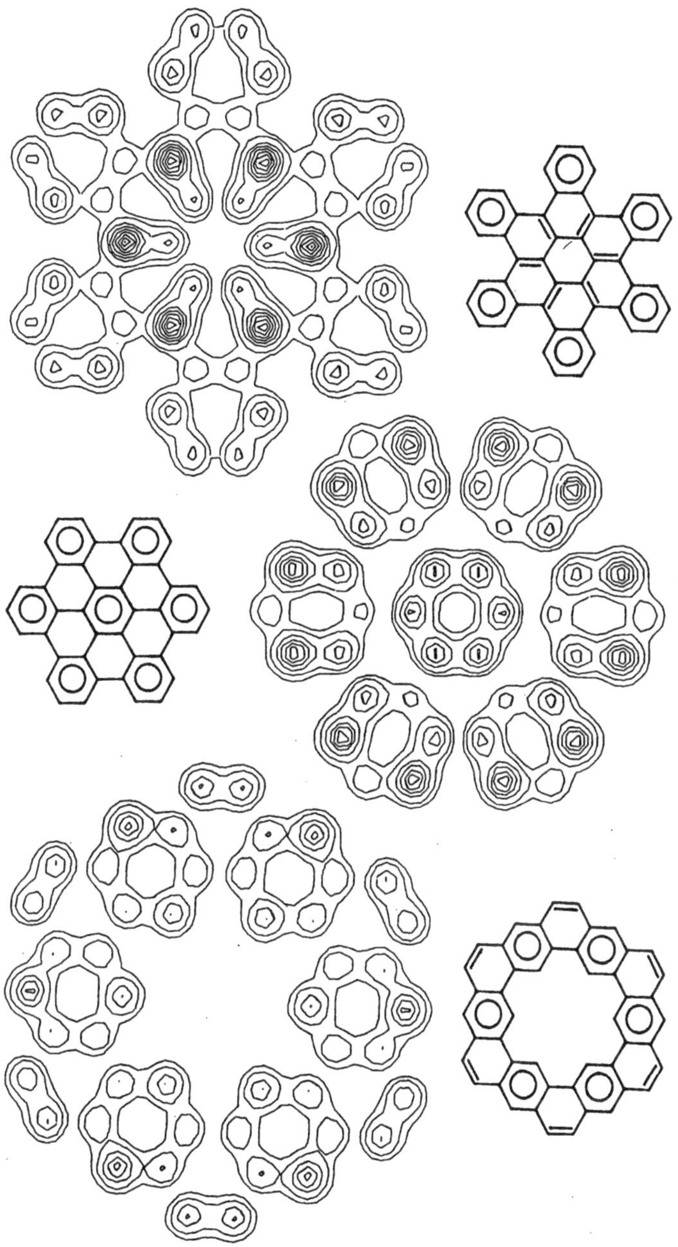

Fig. 6. The contour maps of ϱ_3 of the two hexabenzocoronenes and kekulene showing the characteristic features of their Clar patterns

5 References

1. Longuet-Higgins HC (1950) J. Chem. Phys. 18: 265
2. Pauling L (1948) The nature of the chemical bond, 2nd ed, Cornell Univ. Press, Ithaca, New York
3. Wheland GW (1953) The theory of resonance and its application to organic chemistry, Wiley, New York
4. Robinson R (1966) Aromaticity, Intern. Symposium, Sheffield 1966, p 47
5. Clar E (1972) The aromatic sextet, Wiley, London
6. Clar E (1964) Polycyclic hydrocarbons, Academic, London
7. Yamaguchi T, Suzuki M, Hosoya H (1975) Natl. Sci. Rept. Ochanomizu Univ. 26: 39
8. Diederich F, Staab HA (1978) Angew. Chem. Int. Ed. 17: 372
9. Hosoya H, Yamaguchi T: Tetrahedron Lett. 1975: 4659
10. Ohkami N, Motoyama A, Yamaguchi T, Hosoya H, Gutman I (1980) Tetrahedron 37: 1113
11. Hosoya H (1986) Croat. Chem. Acta 59: 583
12. Cyvin SJ, Gutman I (1987) J. Mol. Struct. (Theochem) 150: 157
13. Ohkami N, Hosoya H (1983) Theor. Chim. Acta 64: 153
14. Ohkami N, Hosoya H (1984) Natl. Sci. Rept. Ochanomizu Univ. 35: 71
15. Gutman I (1977) Theor. Chim. Acta 45: 309
16. Gutman I, El-Basil S (1983) Z. Naturforsch. 39a: 276
17. El-Basil S (1987) J. Math. Chem. 1: 153
18. El-Basil S, Randić M (1987) J. Math. Chem. 1: 281
19. Ohkami N (1985) Thesis, Ochanomizu Univ.
20. He WJ, He WC (1986) Theoret. Chim. Acta 70: 43
21. Randić M (1975) Tetrahedron 31: 1477
22. Aihara J (1976) Bull. Chem. Soc. Jpn. 49: 1429
23. Herndon WC, Hosoya H (1984) Tetrahedron 40: 3987
24. Hosoya H, Shobu M, Takano K, Fujii Y (1983) Pure and Appl. Chem. 55: 269
25. Zhang FJ and Chen RS (1986) Match 19: 179
26. Dewar MJS, de Llano C (1969) J. Am. Chem. Soc. 91: 789
27. Polansky O, Derflinger G (1967) Int. J. Quant. Chem. 1: 379
28. Aida M, Hosoya H (1980) Tetrahedron 36: 1317
29. Fujii Y (1983) Master's Thesis, Ochanomizu Univ.
30. Fukui K, Yonezawa T, Shingu H (1952) J. Chem. Phys. 20: 722
31. Fukui K (1971) Acc. Chem. Res. 4: 57
32. Woodward RB, Hoffmann R (1965) J. Am. Chem. Soc. 87: 395
33. Hoffmann R, Woodward RB (1968) Acc. Chem. Soc. 1: 11
34. Shobu M, Hosoya H (1981) Nat. Sci. Rept. Ochanomizu Univ. 32: 55
35. Coulson CA, Rushbrooke GS (1940) Proc. Cambridge Phil. Soc. 36: 193

Caterpillar (Gutman) Trees in Chemical Graph Theory

Sherif El-Basil

Faculty of Pharmacy, Kasr El-Aini Street, Cairo, Egypt 11562

Table of Contents

Gutman trees (also known as caterpillar trees and benzenoid trees) are demonstrated to be elegant storage devices of information on graph-theoretical properties of many mathematical objects including benzenoid graphs, rook boards, king polyomino graphs, Clar graphs, and Young diagrams. The notion of a "restricted" *equivalence relation* is considered to relate these graphs. The uses of caterpillars in various areas of physico-chemical interests such as modeling of interactions, computational chemistry and ordering of graphs are discussed. Namely, caterpillar trees are used to model wreath product groups, Clar structures and nonadjacency relations in graphs of chemical interest. Further, they are used to study several properties of benzenoid hydrocarbons including UV absorption spectrum, molecular susceptibility, anisotropy, heat of atomization as well as Diels-Alder addition rate constant. Several novel relations correlating the connectivity of a caterpillar tree with several combinatorial properties of benzenoid hydrocarbons, such as, Kekulé counts, conjugated circuits, Sachs graphs, and self-avoiding paths, are presented.

Topics in Current Chemistry, Vol. 153

1 Introduction

There seems to be something "magic" about the number 6. In 540 B.C. the Society of Pythagoras held that earth had been produced from the regular *hexahedron*. In most religions it is written that God created the Universe in *six* days. Kekulé's famous dream (a snake swallowing its own tail) implies a *6-membered* ring. The *hexagon*, being a fundamental structural unit in nature, enters into the constitution of epidermal plant cells, honey comb, graphite and many others, but most importantly of benzenoid hydrocarbons (also known as benzenoid systems). Indeed, the peculiar features usually associated with "aromaticity" find some answers by considering the connectivity of a regular *hexagon*. Here, we consider a graph-theoretical treatment of a particular class of benzenoid hydrocarbons which may be represented by certain tree graphs known as *Caterpillar trees* or *Gutman trees*. We start by a definition of such objects: When all the end points of a graph G are delected another graph, G′, results which is called the *derivative* [1] of G. In particular when G is a tree graph T and G′ is a path, L, then G is called a *caterpillar tree* [2]. Thus such a tree results when a set containing m_1 (monovalent) vertices is added to vertex v_1 of a path, a set of m_2 vertices is added to v_2 of the same path, ..., and finally at set containing m_n vertices is added to the *n*th vertex of the path. Figure 1 illustrates the definition of a caterpillar tree. A convenient notation of such a tree which is generated on L_n, a path containing *n* vertices, is $T_n(m_1, m_2, \ldots, m_n)$. Naturally $T_n(0, 0, \ldots, 0)$ is simply L_n. Gutman [3] introduced the uses of these trees in chemistry nearly a decade ago and the present author suggested the name "*Gutman Trees*", in the chemical literature [4–6]. Also, because these trees can be made to generate resonance relations among the rings (hexagons) of unbranched [7] benzenoid hydrocarbons the name "*Benzenoid Trees*" has been suggested [8]. In this review we shall consider the chemical applications of caterpillar trees in several active areas, namely, modeling of interactions, computational applications and general ordering of graphs.

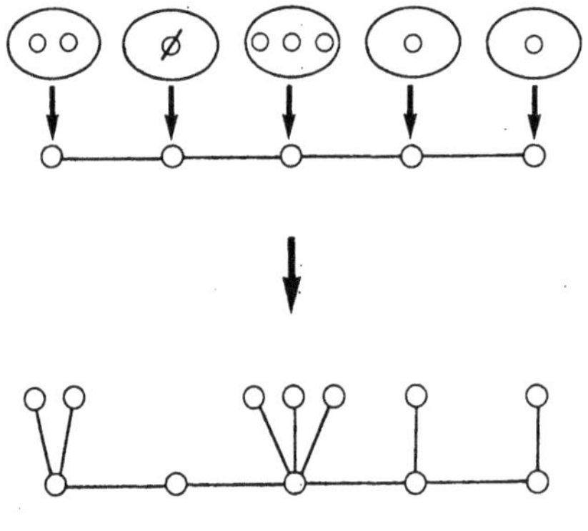

Fig. 1. Illustration of the definition of the Gutman tree (caterpillar), $T_5(2, 0, 3, 1, 1)$, the derivative of which is L_5

2 Modeling of Interactions

2.1 Wreath Product Groups [9]

Certain structures of groups of chemical interest which may be difficult to see, can be easily envisaged by using caterpillar trees as models. We use as an example the *composition* [9] of two permutation groups S_i and S_j (involving the permutation of i and j objects respectively) which is denoted by $S_i[S_j]$ (and is read: S_i around S_j). This group is known as the "*wreath product*" or the "Gruppenkranz". As a simple illustration we show how a caterpillar tree can be used to model the elements of $S_2[S_3]$. Then we have two sets, viz.,

$$A = \{a, b\} \tag{1}$$

$$X = \{1, 2, 3\} \tag{2}$$

We let α be an operation in S_2 and β an operation in S_3, then the permutations in $S_2[S_3]$ are defined by

$$(\alpha;\ \beta_1\beta_2)\,(a, k) = (\alpha a,\ \beta_1 k) \tag{3}$$

$$(\alpha;\ \beta_1\beta_2)\,(b, k) = (\alpha b,\ \beta_2 k) \tag{4}$$

where k is either 1, 2 or 3, i.e., $k \in X$. The operations β_1 and β_2 are not necessarily distinct. There are two elements in S_2 and six elements in S_3 and thus $2 \times 6^2 = 72$ elements in $S_2[S_3]$. Six such elements are shown below:

1) ((a) (b); (1) (2) (3), (1) (2) (3))
2) ((a) (b); (1) (2) (3), (12) (3))
3) ((a) (b); (1) (2) (3), (13) (2))
4) ((a) (b); (1) (2) (3), (23) (1))
5) ((a) (b); (1) (2) (3), (123))
6) ((a) (b); (1) (2) (3), (132))

The cycle structure contribution of the above elements is generated by operation on the cross product $A \times X$. As an illustration we take the operation of the fourth element, namely,

((a) (b); (1) (2) (3), (23) (1)) (a1) = (a1)
((a) (b); (1) (2) (3), (23) (1)) (a2) = (a2)
((a) (b); (1) (2) (3), (23) (1)) (a3) = (a3)
((a) (b); (1) (2) (3), (23) (1)) (b1) = (b1)
((a) (b); (1) (2) (3), (23) (1)) (b2) = (b3)
((a) (b); (1) (2) (3); (23) (1)) (b3) = (b2)

The above formal description can be pictorially simplified by using the Gutman tree $T_2(3, 3)$ as shown below.

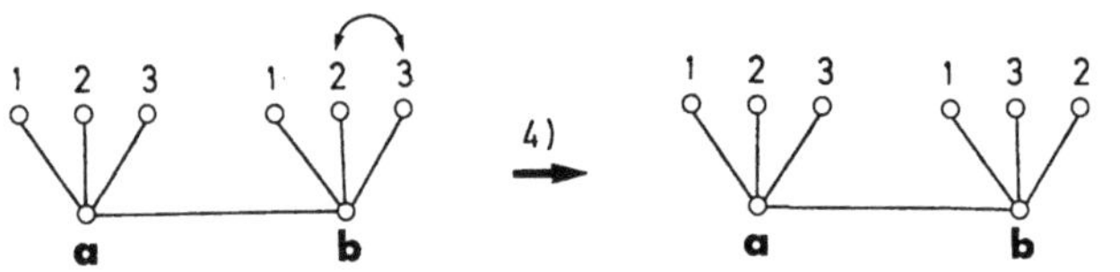

The operation results in the permutation of 2 and 3 leaving other objects unchanged and therefore this element generates four one-cycles and a one two-cycle, i.e., $s_1^4 s_2$. In general for $S_i[S_j]$ a tree of the form $T_i(j, j)$ can be used. Extended wreath products introduced by Balasubramanian [10] can also be modeled using appropriate Gutman trees [11].

2.2 Clar Structures

In the theory known nowadays as the Clar theory of the aromatic sextet [12] a benzenoid system is represented by a "Clar structure" which is obtained by drawing circles in some of the hexagons of the corresponding benzenoid graph. These circles represent the "aromatic sextets" in the hydrocarbon. We consider here *only Clar structures containing the maximum number of circles* which are some times referred to as proper (or correct) Clar formulas. The rules for constructing such Clar structures are as follows [13].

(a) It is not allowed to draw circles in adjacent hexagons.
(b) Circles can be drawn in hexagons if the rest of the conjugated system has at least one Kekulé structure.

This type of Clar structure can be generated for a limited number of classes of catacondensed benzenoid hydrocarbons by defining an operation which resembles *coloring* [14] of the vertices of certain caterpillars but allows more flexibility. Namely, we define a "Clar coloring" [15] of the vertices of a caterpillar by the following steps:

(a) Vertices are colored in exactly two ways: *black* (arbitrarily), corresponding to a hexagon (in the relevant hydrocarbon) which contains an aromatic sextet of π-electrons (i.e., a circle in Clar notation), *white* vertices, then, correspond to hexagons containing no circles.
(b) No two black vertices may be adjacent.

The resulting colorings are then in one-to-one correspondence with Clar structures. Figure 2 shows the five Clar structures of a zigzag polyacene, fulminene, and the Clar colorings of the corresponding Gutman tree, a path. If $\zeta(L_n)$ is the number of Clar colorings of path L_n, then it can be shown that [15]

$$\zeta(L_n) + \zeta(L_{n+1}) = \zeta(L_{n+3}) \qquad (5)$$
$$n = 1, 2, 3, \ldots$$

Fig. 3 depicts all five Clar structures of a branched hydrocarbon and the five Clar colorings of the corresponding caterpillar, in this case, a "comb tree". It has been shown recently [16] that $\zeta(T_n(1, 1, \ldots, 1))$, the number of Clar colorings of a comb tree recurs in the following manner:

$$\zeta(T_n(1, 1, \ldots, 1)) = \zeta(T_{n-1}(1, 1, \ldots, 1)) + \zeta(T_{n-2}(1, 1, \ldots, 1)) \tag{6}$$

where, interestingly [17]

$$\zeta(T_n(1, 1, \ldots, 1)) = F_{n+1}, \tag{7}$$

F_j being a jth Fibonacci number [18] ($F_0 = F_1 = 1$, $F_2 = 2$, $F_3 = 3, \ldots, F_n = F_{n-1} + F_{n-2}$).

It is worthy of observation that for both classes of benzenoid hydrocarbons represented in Figs. 2, 3 the corresponding trees (whose number of colorings = the number of Clar structures) coincide with the inner duals [19] of the benzenoid graphs.

Relation (7) holds for any other regular or irregular caterpillars [17], e.g. it holds for $T_3(2, 2, 2)$ as well as $T_3(1, 3, 2)$. The described type of coloring when applied to a general Gutman tree $T_n(a, b, c, \ldots)$ where $a, b, c, \ldots,$ are all ≥ 1 generate interactions and distribution of objects in boards of chemical-physical interest. Figure 4 depicts the five colorings of $T_3(2, 2, 2)$ and the corresponding distribution of objects in king, dimer, domino and polyomino patterns [20, 21] as well as the set of all matchings in the path L_4. One observes the number of any of the interactions shown in Fig. 4 is F_4, cf. Eq. (7). In analogy of Clar coloring of Gutman trees one can define two rules of placing kings (circles) in a king polyomino in the following way:

(a′) No two kings are allowed to occupy adjacent cells.
(b′) Every empty cell is adjacent to at least one occupied cell.

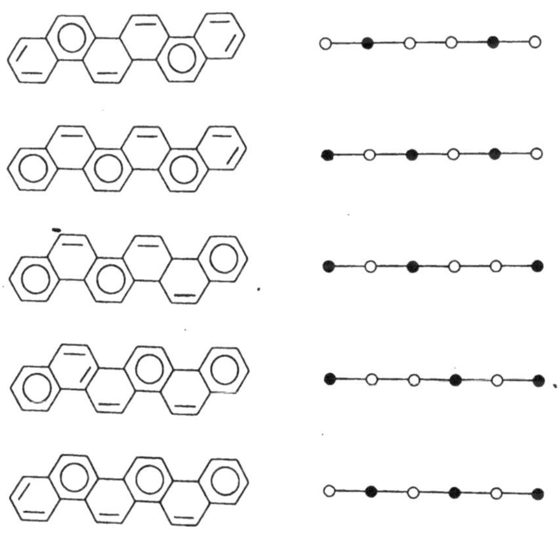

Fig. 2. The five Clar structures of fulminene and the corresponding five colorings of the corresponding caterpillar

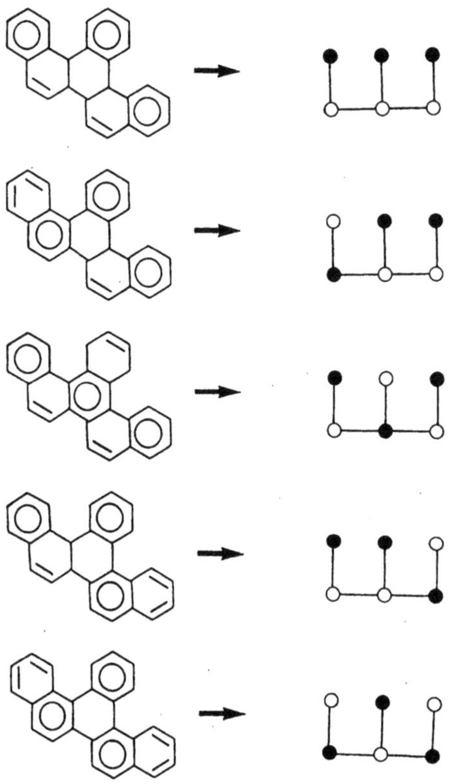

Fig. 3. All five Clar structures of a branched benzenoid hydrocarbon and the corresponding five colored Gutman trees. The indicated coloring generates F_4, the fourth Fibonacci number

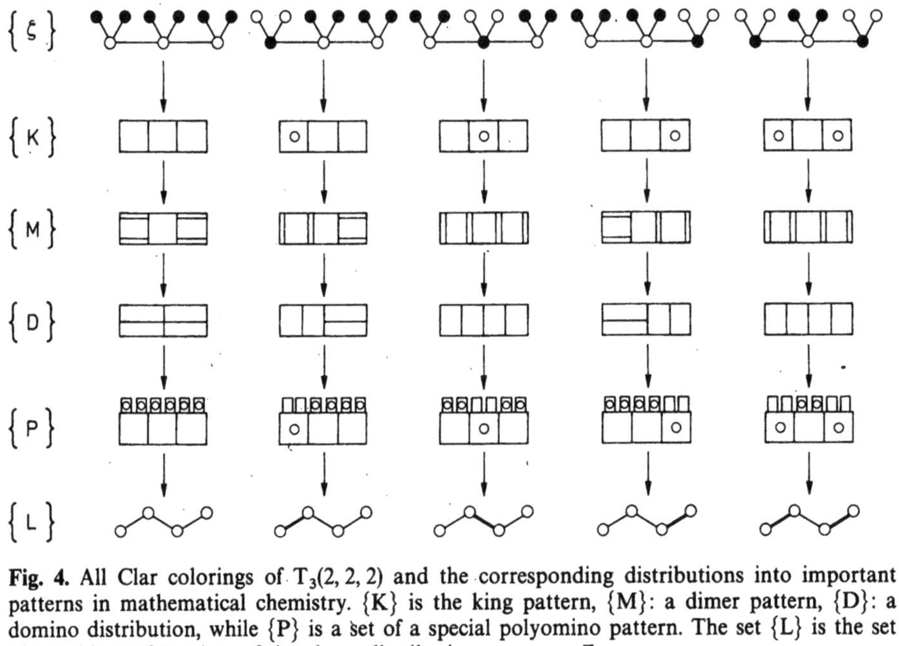

Fig. 4. All Clar colorings of $T_3(2,2,2)$ and the corresponding distributions into important patterns in mathematical chemistry. {K} is the king pattern, {M}: a dimer pattern, {D}: a domino distribution, while {P} is a set of a special polyomino pattern. The set {L} is the set of matchings of L_4. Any of the above distributions generate F_4

2.3 Nonadjacency Relations

The revival of interest in the applications of graph theory to chemistry and physics led to the use of already known techniques in mathematics to problems of chemical and physical interests. Two texts appeared on these developments [22, 23] in addition to an early book edited by Balaban [24].

One common method which proved successful in analyzing the connectivity relations in molecules (or molecular graphs in general) is the definition of nonadjacency relations between selected sets of graph invariants. The latter may be the *edges* [25] of a tree, the *hexagons* [26] of a benzenoid hydrocarbon, the vertices [27] of a graph or the *cells* [20] of a board. Chart 1 portrays some of the more important nonadjacency relations already known in chemical graph theory. The function $\Theta(G, k)$ in Chart 1 is a general nonadjacent function which generates a sequence of integers $\Theta(G, 0)$, $\Theta(G, 1)$, $\Theta(G, 2)$, ..., $\Theta(G, m)$ where m is the maximum value of k and $\Theta(G, 0)$ is defined to be unity. One of the earlier explicit forms of $\Theta(G, k)$ is $p(T, k)$; the number of k-matchings in a tree T which was introduced by Hosoya [25] in the early seventies to study the physical properties of alkanes. Stated in different words $p(T, k)$ counts the number of selections of k nonadjacent (i.e. nonincident) edges in T. Thus one normally speaks of $p(T, 1)$ as the number of 1-matchings, $p(T, 2)$, the number of 2-matchings and so on.

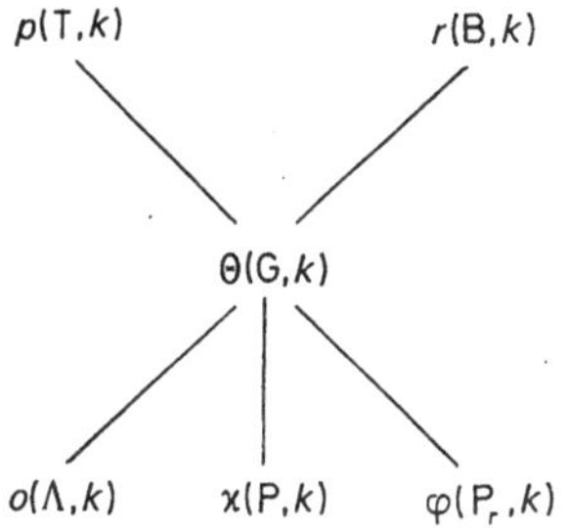

Chart. 1. Nonadjacency relations. $\Theta(G, k)$ is a general relation which counts the number of k nonadjacent objects in G. $p(T, k)$ = the number of k-matchings in T, $o(\Lambda, k)$ is the number of selections of k nonadjacent vertices in Λ, $r(B, k)$ is the kth resonant number of B, $\varphi(P_r, k)$ = the number of distributions of objects on k nonadjacent rook-board cells, and $\varkappa(P, k)$ is the number of ways in which one can distribute k nonadjacent objects on a king polyomino graph

If the number of vertices in T (or any general G) is even then $p(T, m)$ is known as the number of perfect matchings, or in mathematics as the number of "*one-factors*" or "*linear factors*". When T is substituted by a benzenoid graph, B, $p(B, m)$ is the number of Kekulé structures. In an analogous fashion the function $r(B, k)$ known as kth resonant numbers, were introduced by Hosoya and Yamaguchi [26] to study benzenoid hydrocarbons. Thus $r(B, k)$ is the number of selections of k *resonant* hexagons in B. In his study of the topological properties of benzenoid systems, Gutman [27] defined the combinatorial function $o(\Lambda, k)$ to count the number of selections of k nonadjacent vertices in special types of graphs which he called Clar graphs, Λ (see below). Other analogous quantities can be inferred from Chart 1.

The general function can be used to generate polynomials of the form

$$F(G;x) = \delta \sum_{k}^{m} \Theta(G,k)\, x^{f(k,n)} \tag{8}$$

where δ is either 1 or $(-1)^k$, $f(k, n)$ is either $n - 2k$ or k, and n is the number of vertices in G. A summary of the $F(G; x)$ functions may be found elsewhere [4, 8].

Gutman trees as graph-generators (Modeling Θ(G, k)). Given a particular Gutman tree with a given $p(T, o), \ldots, p(T, m)$ sequence, one can generate non-isomorphic graphs possessing identical nonadjacency sequences, e.g., we take Gutman tree $T = T_4(2, 1, 0, 1)$. It is easy to check that the sequence 1, 7, 13, 6 corresponds respectively to $p(T, 0)$, $p(T, 1)$, $p(T, 2)$, $p(T, 3)$. Now we wish to use T to generate a benzenoid system for which 1, 7, 13, 6 correspond respectively to r(B, 0), r(B, 1), r(B, 2) r(B, 3). Such hydrocarbon exists and *will never be branched* [3], but unbranched catacondensed systems. To construct the desired B, we map the edges of T into the hexagons of B such that two edges are *incident* in T only if the corresponding two hexagons in B are *nonresonant.* Similarly one can construct a Clar graph [27], Λ, by mapping the edges of T into the vertices of Λ such that two vertices in Λ are *adjacent* only if the corresponding two edges are incident. In a king polyomino [20, 21] two cells are *defined* to be adjacent if they share at least one common vertex while in a rook board [28], P_r, two cells are adjacent if they share the same row or the same column in the board. Figure 5 shows five important graph-theoretical objects: a Gutman tree, a benzenoid hydrocarbon, a Clar graph, a king polyomino and a rook board, all sharing the nonadjacency sequence 1, 7, 13, 6. Here T is used to generate other graphs by "modeling" the nonadjacency relations in these graphs. The name "*equinumerical*" graphs [29] has been suggested to describe the graphs of Fig. 5. In Fig. 6 one-to-one correspondence between all possible 3-matchings, i.e., $p(T, 3)$, of $T_4(2, 1, 0, 1)$ and other corresponding interactions in other equinumerical graphs is illustrated.

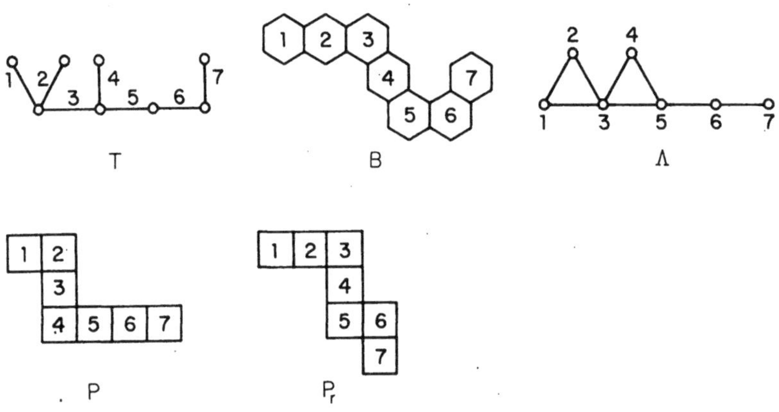

Fig. 5. A Gutman tree, T, an unbranched benzenoid system, B, a Clar graph, Λ, a king polyominography, P, and a rook board, P_r

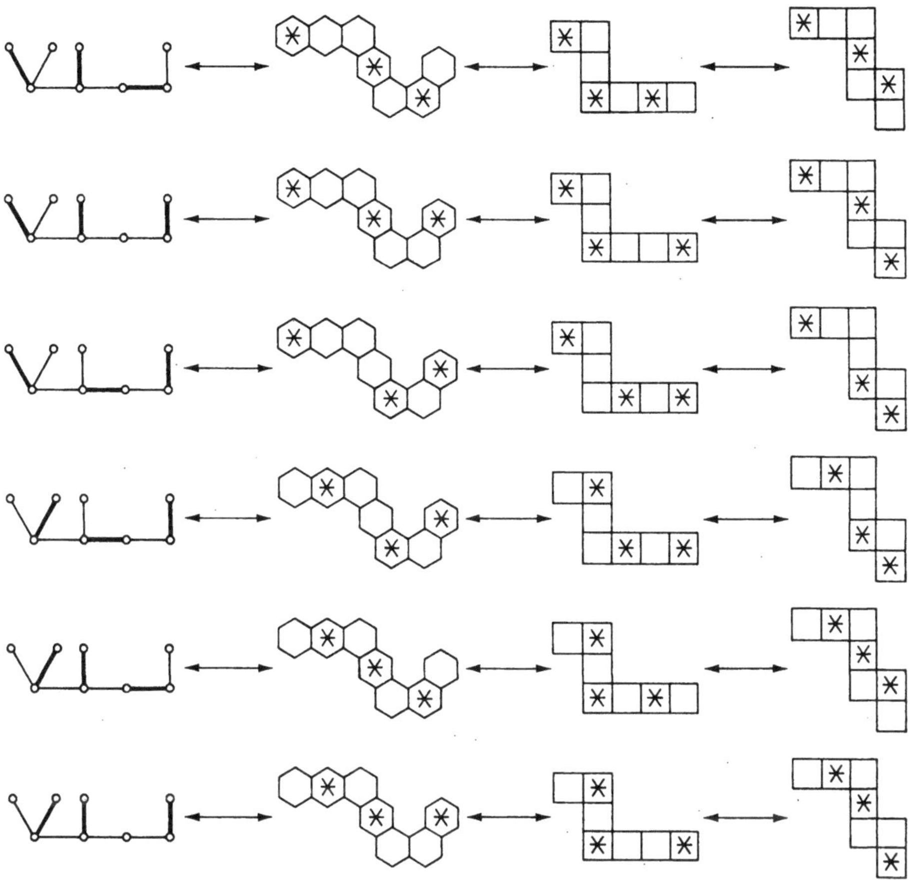

Fig. 6. All 3-matchings in $T_4(2, 1, 0, 1)$ and the corresponding object-distribution in other "equinumerical" graphs, viz., an unbranched benzenoid hydrocarbon, a king polyomino and a rook board

2.4 On a "Restricted" Equivalence Relation

It is almost trivial to see that an equivalence relation exists between the elements of such sets of graphs as the one shown in Fig. 5, i.e., between what is called here equinumerical graphs. Let this relation be R. Then R is evidently *reflexive* because every graph is related to itself. Also R is *symmetric* because for every Gutman tree T we can generate a Clar graph Λ. In fact T and Λ are related as

$$L(T) = \Lambda \tag{9a}$$

or

$$L^{-1}(\Lambda) = T \tag{9b}$$

Namely Λ is the *line graph* [30] of T. Other less transparent relations can be envisaged for other pairs of the set. For example let $\hat{O}$ be an operator which maps the edges of T into hexagons of B so that two incident edges in T will always correspond to two nonresonant hexagons in B. Formally

$$\hat{O}\mathrm{T} = \mathrm{B} \tag{10a}$$

or

$$\hat{O}^{-1}(\mathrm{B}) = \mathrm{T} \tag{10b}$$

For our 5-element set there will be $\binom{5}{2} = 10$ such relations and their inverses, i.e. a total of 20 relations can be written in forms similar to (9), (10). Further, R is also *transitive* because one can define subsets of equinumerical graphs such as $\{\mathrm{B}, \mathrm{T}\}$, $\{\mathrm{T}, \Lambda\}$ and $\{\mathrm{B}, \Lambda\}$ whence R is an equivalence relation on $\{\mathrm{T}, \Lambda, \mathrm{B}, \mathrm{P}_r, \mathrm{P}\}$.

Three obvious restrictions exist on the structure of the elements of the above set, viz.:

(a) Λ is *not* one of the nine forbidden line graphs [30].
(b) Because of the definition of adjacent cells in a king polyomino graph, P, no such board exists with more than 4 adjacent cells. Hence Gutman trees with more than 4 simultaneously incident edges are excluded from the above set, and thus no benzenoid hydrocarbon with more than 4 linearly annelated hexagons belongs to the above set. Naturally if P is eliminated this restriction is removed.
(c) B is an unbranched catacondensed hydrocarbon; branched and pericondensed systems are not included in the above relation.

The above notion of equivalence suggests another name for equinumerical graphs, namely *equivalent graphs*. One then speaks of an unbranched catacondensed benzenoid hydrocarbon, B_i, and its *equivalent* Gutman tree, T_i. More precisely one speaks of a set $\{\mathrm{B}_i\}$ of all isoarithmic [31] unbranched benzenoids and their equivalent tree. The function that maps B_i into T_i is then an *onto (surjective)* function. Analogously one considers a B_i and its equivalent Clar graph, rook board and king polyomino. Of course the latter may not exist, e.g., T_4 (4, 1, 2, 1) does not have an equivalent king polyomino graph.

3 Correlation of Physical and Chemical Properties of Unbranched Benzenoid Hydrocarbons

Several physical properties of benzenoid hydrocarbons can be correlated with a function of the molecular connectivity of their equivalent Gutman trees. Namely, we use the Randić connectivity index [32], χ, viz.,

$$\chi(\mathrm{T}) \equiv \chi = \sum_{\mathrm{v}_i, \mathrm{v}_j \in \mathrm{T}} (d_i d_j)^{-\frac{1}{2}}, \qquad i \neq j \tag{11}$$

where d_i is the degree of vertex i and the summation is taken over all vertices v_i, in T. The following physical properties are correlated.

(1) The *UV absorption spectra* [33]: Figure 7 portrays three families of benzenoid hydrocarbons which are identified by their LA sequences [3, 34] (of Gutman). In all cases correlation coefficients are >0.9. More details can be found elsewhere [35].

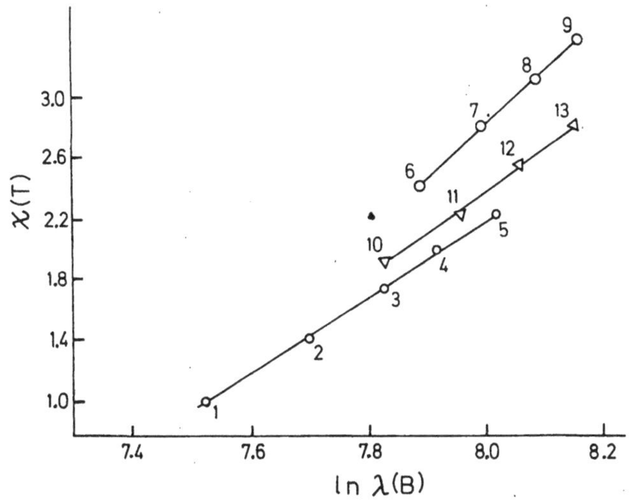

Fig. 7. Correlation of the UV spectra (in ln units), ln λ(B), with connectivity index, χ(T), (Eq. 11) for three homologous series of hydrocarbons: 1 = benzene; 5 = pentacene; 6 = chrysene, 9 = $L_1A_2(L_2)_3A_2L_1$; 10 = phenanthrene; 13 = $L_1(L_2)_3A_2L_1$. Circles correspond to beta absorption bands while triangles correspond to para bands

(2) Figure 8 displays a correlation of the *molecular succeptibility* [36] of the benzenoid hydrocarbon, x_m(B) and *anisotropy* [36], ΔK, for linear benzenoid systems, i.e., benzene, naphthalene, anthracene, ..., hexacene, and χ of Eq. (11) of the corresponding equivalent Gutman trees.
(3) Figure 9 shows a correlation between the *heats of atomizations* [37] of two families of unbranched hydrocarbons and the Randić connectivity index of their equinumerical (equivalent) caterpillars.
(4) *A Chemical Property*: Figure 10 considers two plots correlating rate constants of *Diels-Alder addition* [38] of maleic anhydride of two series of hydrocarbons with the connectivity indices of their corresponding trees.

4 Computational Applications

4.1 Kekulé Counts

The number of Kekulé structures of benzenoid hydrocarbons, K, sometimes referred to as Kekulé counts is a parameter of great interest to the mathematical chemist. A recent book by Cyvin and Gutman [34] is very informative of the size of this problem. Quite recently publications on these ounts *even for unbranched catacondensed systems*

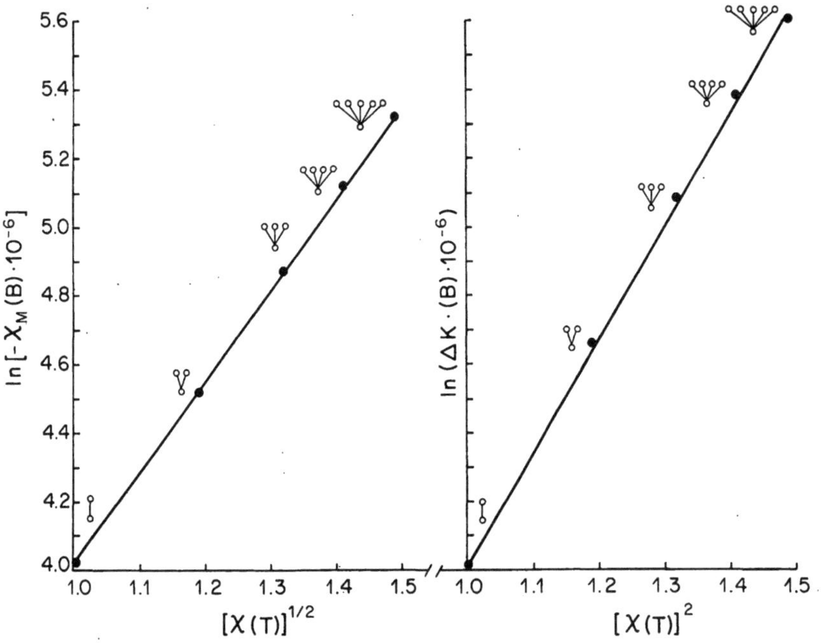

Fig. 8. Correlation of the natural logarithms of molecular succeptibility, x_m(B) and anisotropy, ΔK, for the linear acenes with the corresponding connectivity indices of the equivalent Gutman trees, χ(T), Eq. 11, drawn to the left of the correlation lines

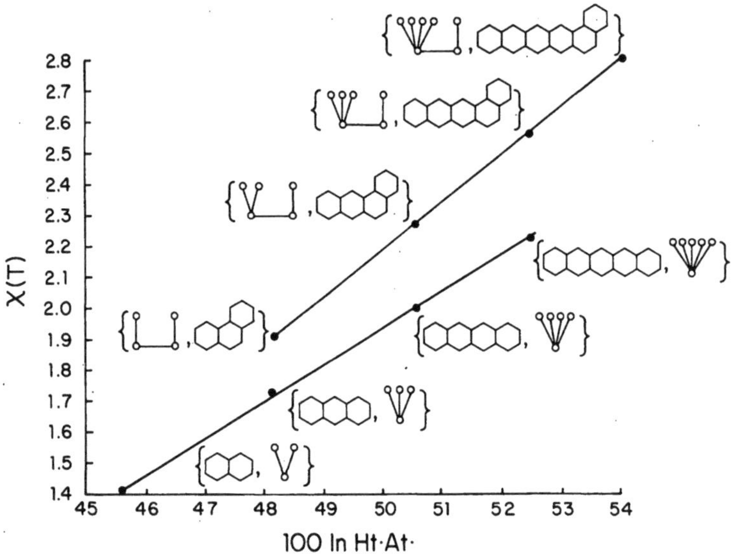

Fig. 9. Correlation of the heats of atomizations in ln units, ln Ht. At., for two families of unbranched benzenoids and χ(T)'s, Eq. 11, of their equivalent caterpillars

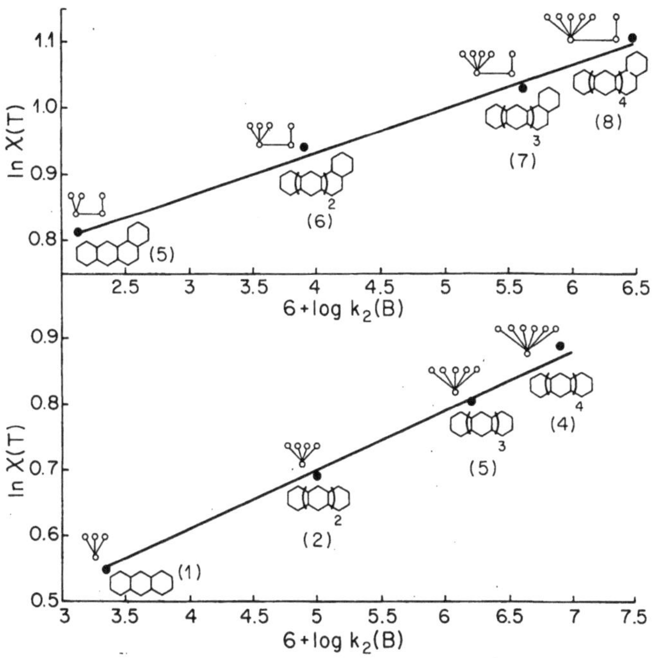

Fig. 10. Plots correlating rate constants of maleic anhydride addition, log k_2, of two series of hydrocarbons versus connectivity indices, χ(T)'s, Eq. (11), of their equivalent Gutman trees

were still appearing [39]. Here we outline some unpublished results by this author [40] along these lines. We define a connectivity index, x', which is much simpler than Randić's χ (11), namely

$$x' = \sum d_i d_j \tag{12}$$

where d_i is the degree of vertex i in T, and the summation is taken over all vertices in T. Simple linear correlations can be obtained between x' of a caterpillar tree and K of the corresponding equivalent unbranched benzenoid hydrocarbons. As illustrations we consider a few families of such hydrocarbons represented by their LA sequences [34] (D is maximum vertex degree in the relevant Gutman tree).

(i) The family $L_1A_2L_1$, $L_1L_2A_2L_1$, $L_1L_2L_2A_2L_1$, ...

$$K = 2(x' - D - 2)^{\frac{1}{2}} + 1 \tag{13}$$

(ii) The family $L_1A_2A_2L_1$, $L_1L_2A_2A_2L_1$, $L_1L_2L_2A_2A_2L_1$, ...

$$K = 3(x' - D - 6)^{\frac{1}{2}} + 2 \tag{14}$$

(iii) The family $L_1A_2A_2A_2L_1$, $L_1L_2A_2A_2A_2L_1$, $L_1L_2L_2A_2A_2A_2L_1$, ...

$$K = 5(x' - D - 10)^{\frac{1}{2}} + 3 \quad (15)$$

4.2 Graph-Theoretical Properties

Several graph-theoretical properties of unbranched benzenoid hydrocarbons have been correlated with the Randić molecular connectivity indices of the equivalent Gutman trees [3], $\chi(T)$. As illustrations we consider the following families of hydrocarbons (In all cases the correlation coefficient = 1).

(i) The set $L_1L_2A_2L_1$, $L_1L_2L_2A_2L_1$, ..., $L_1L_2L_2L_2L_2L_2L_2L_2A_2L_1$

$$\ln \sum S^{ac}(B) = 2.549 + 1.004(\chi(T))^{2.3} \quad (16)$$

where $\sum S^{ac}(B)$ is the sum of all acyclic Sachs graphs in the benzenoid system B [41].

(ii) The set $L_1L_2A_2L_2L_1$, $L_1L_2L_2A_2L_2L_1$, ..., $L_1L_2L_2L_2L_2L_2L_2A_2L_2L_1$

$$\ln \sum R(B) = -9.461 + 10.049(\chi(T))^{0.30} \quad (17)$$

where $\sum R(B)$ is the total number of conjugated circuits [42] in B.

(iii) The set $L_1L_2A_2A_2L_1$, $L_1L_2L_2A_2A_2L_1$, ..., $L_1L_2L_2L_2L_2L_2L_2A_2A_2L_1$

$$\ln \sum p(B) = 4.374 + 0.484(\chi(T))^2 \quad (18)$$

where $\sum p(B)$ is the total number of self-avoiding paths [43] in B.

5 General Ordering of Graphs

Nearly a decade ago Gutman and Randić [44] rigorously considered the problem of branching in acyclic graphs. They used conditions of Muirhead [45] as criteria of ordering a set of trees (caterpillars and non-caterpillars) where they employed a non-ascending sequence composed of the vertex-degrees of a given tree. Namely, for two trees whose sequences are $(d_1, d_2, \ldots, d_n)$ and $(d'_1, d'_2, \ldots, d'_n)$ and, if the partial sums of one tree, T, are always greater than or equal to (but not smaller than) those of T′, one says that T majorizes T′; i.e. T precedes T′ in the hierarchy. If this condition is violated, T and T′ are said to be non-comparable, and they lead to a bifurcation site in the ordering diagram. Such sites appear in hierarchical diagrams of trees containing eight vertices or more. Further, Gutman and Randić [44] observed that their ordering of trees can be made to overlap ordering schemes of Young diagrams by Ruch and Schönhofer [46] by the following steps. (a) Suppress the information on terminal vertices (i.e. those of degree 1), (b) reduce the degree of each of the remaining vertices by one. For example, $T_4(2, 0, 2, 1)$ has the non-ascending degree sequence (4, 3, 2, 2, 1, 1, 1, 1, 1) and thus corresponds to the Young diagram Y(3, 2, 1, 1), i.e. to a Young diagram possessing four rows such that the first (top) row has 3 boxes, the second row 2 boxes, the third row one box, and finally the last (bottom) row one box. The transformation of a tree into a Young diagram is an onto (surjective) function as can be shown from the following diagram.

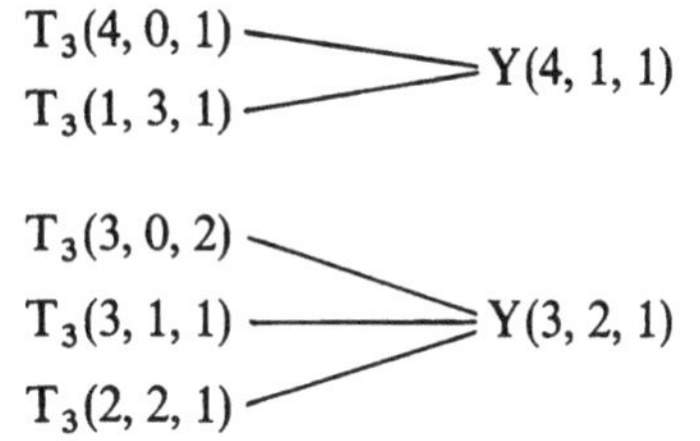

We can make use of the overlapping of the two ordering schems to order a set of *equivalent* objects to the set of Gutman trees. (For example a set of unbranched benzenoid hydrocarbons, Clar graphs, rook boards, or king polyomino graphs, if defined). Figure 11 illustrates how this is used to order the set of all Clar graphs containing seven vertices.

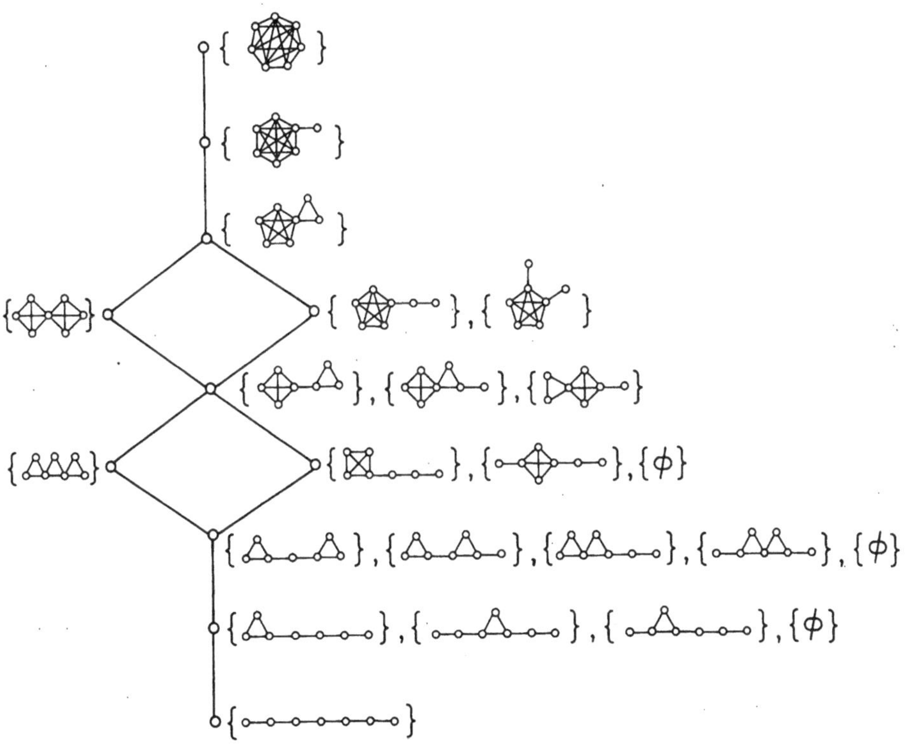

Fig. 11. Ordering of all Clar graphs containing 7 vertices according to Ruch's greater relations defined for a set of Young diagrams containing six squares. The ordering makes use of the equivalence relation defined between a Young diagram, a caterpillar (Gutman) tree and Clar graphs. The $\{\varnothing\}$'s correspond to places on the diagram which correspond to non-caterpillar trees and thus do not have an equivalent Clar graph

6 References and Notes

1. Harary F, Schwenk AJ (1973) Discrete Math. 6: 359
2. Harary F, Schwenk AJ (1973) Discrete Math. 6: 361
3. Gutman I (1977) Theoret. Chim. Acta 45: 309
4. El-Basil S (1986) J. Chem. Soc., Faraday Trans. II 82: 299
5. El-Basil S (1984) Theoret. Chim. Acta 65: 199
6. El-Basil S (1984) Theoret. Chim. Acta 65: 191
7. A benzenoid hydrocarbon in which no hexagon is adjacent to more than two hexagons is called unbranched (or nonbranched). cf. Ref. [3]
8. El-Basil S (1987) J. Math. Chem. 1: 153
9. Harary F (1969) Graph theory, Addison-Wesley, Reading, p 164
10. Balasubramanian K (1982) Studies in physical and theoretical chemistry 23: 149
11. El-Basil S: Unpublished Results
12. Clar E (1972) The aromatic sextet, Wiley, London
13. Gutman I (1982) Bull. Soc. Chim. Beograd 47: 465
14. Harary F, Palmer EM (1973) Graphical Enumeration, Academic, New York, p 16
15. El-Basil S (1988) Discrete Appl. Math. 19: 145
16. El-Basil S (1988) Bull. Chem. Soc. Japan, 61: 533
17. Eq. (7) holds for any tree of the form $T_n(a, b, c, d, \ldots)$ where $a, b, c, d, \ldots, \geqq 1$. Further $a, b, c, d, \ldots$ do not have to be equal
18. Cohen DIA (1978) Basic Techniques of Combinatorial Theory, Wiley, New York, p 63
19. Trinajstić N (1983) Chemical graph theory, CRC Press, Boca Raton, FL vol I, p 24
20. El-Basil S (1987) J. Comput. Chem. 8: 956
21. Motoyama A, Hosoya H (1985) J. Math. Phys. 18: 1485
22. Gutman I, Polansky OE (1986) Mathematical concepts in organic chemistry, Springer Berlin Heidelberg New York
23. Trinajstić N (1983) Chemical graph theory, CRC Press, Boca Raton FL vols I, II
24. Balaban AT (1976) Chemical applications of graph theory, Academic, London
25. Hosoya H (1971) Bull. Chem. Soc. Japan 44: 2332
26. Hosoya H, Yamaguchi T: Tetrahedron Lett. 1975: 4659
27. Gutman I (1982) Z. Naturforsch. 37a: 69; Gutman I, El-Basil S (1984) Z. Naturforsch. 39a: 276
28. Godsil CG, Gutman I (1981) Croat. Chem. Acta. 54: 53
29. Randić M (1988) Private communication
30. Harary F (1969) Graph theory, Addison-Wesley, Reading, chapt. 8
31. Isoarithmicity is discussed in: Cyvin SJ, Gutman I (1988) Kekulé Structures in Benzenoid Hydrocarbons, Springer, Berlin Heidelberg New York, p 29; (Lecture Notes in Chemistry 46) also: Balaban AT, Tomescu I (1983) Match 14: 155; Croat. Chem. Acta. 57: 391 The name isoconjugate is used sometimes to describe hydrocarbons which are isoarithmic and contain identical numbers of conjugated circuits: Randić M (1977) Tetrahedron 33: 1905
32. Randić M (1975) J. Am. Chem. Soc. 97: 6609
33. Clar E (1972) The aromatic sextet, Wiley, London; Rouvray DH, El-Basil S (1988) J. Mol. Struct. (Theochem) 165: 9
34. Cyvin SJ, Gutman I (1988) Kekulé Structures in Benzenoid Hydrocarbons, Springer, Berlin Heidelberg New York, pp 16–17 (Lecture Notes in Chemistry 46)
35. El-Basil S: Unpublished results (available on request)
36. Clar E (1964) Polycyclic hydrocarbons, Academic, London, vol 1, p 114
37. Dewar MJS, Hashmall JA, Trinajstić N (1970) J. Am. Chem. Soc. 92: 5555
38. Balaban AT (1982) Pure and Appl. Chem. 54: 1077
39. Živković TP, Trinajstić N (1987) Chem. Phys. Lett. 136: 141; Klein DJ, Schmalz TG, El-Basil S, Randić M, Trinajstić N (in press) J. Mol. Struct. (Theochem.)
40. El-Basil S (Unpublished results)
41. Hosoya H, Ohkami N (1983) J. Comput. Chem. 4: 585

42. Randić M (1977) J. Am. Chem. Soc. 99: 444
43. Randić M, Brissey GM, Spencer RB, Wilkins CL (1979) Computers, and Chemistry 3: 5
44. Gutman I, Randić M (1977) Chem. Phys. Lett. 47: 15
45. Muirhead RF (1901) Proc. Edinburgh Math. Soc. 19: 36; (1903) 21: 144; (1906) 24: 45; also: Randić M (1978) Chem. Phys. Lett. 55: 547
46. Ruch E, Schönhofer A (1970) Theoret. Chim. Acta. 19: 225

Author Index Volumes 151–153

Author Index Vols. 26–50 see Vol. 50
Author Index Vols. 50–100 see Vol. 100
Author Index Vols. 101–150 see Vol. 150

The volume numbers are printed in italics

Bei Fragen zur Produktsicherheit wenden Sie sich bitte an:
If you have any questions regarding product safety,
please contact:

Walter de Gruyter GmbH
Genthiner Straße 13
10785 Berlin
productsafety@degruyterbrill.com